Lecture Notes in Bioinformatics   11925

Subseries of Lecture Notes in Computer Science

More information about this series at http://www.springer.com/series/5381

Maria Raposo · Paulo Ribeiro · Susana Sério ·
Antonino Staiano · Angelo Ciaramella (Eds.)

Computational Intelligence Methods for Bioinformatics and Biostatistics

15th International Meeting, CIBB 2018
Caparica, Portugal, September 6–8, 2018
Revised Selected Papers

 Springer

Editors
Maria Raposo 📵
Universidade Nova de Lisboa
Caparica, Portugal

Paulo Ribeiro 📵
Universidade Nova de Lisboa
Caparica, Portugal

Susana Sério 📵
University of Lisbon
Caparica, Portugal

Antonino Staiano 📵
Università degli Studi di Napoli Parthenope
Naples, Italy

Angelo Ciaramella 📵
Università degli Studi di Napoli Parthenope
Naples, Italy

ISSN 0302-9743 ISSN 1611-3349 (electronic)
Lecture Notes in Bioinformatics
ISBN 978-3-030-34584-6 ISBN 978-3-030-34585-3 (eBook)
https://doi.org/10.1007/978-3-030-34585-3

LNCS Sublibrary: SL8 – Bioinformatics

This Springer imprint is published by the registered company Springer Nature Switzerland AG
The registered company address is: Gewerbestrasse 11, 6330 Cham, Switzerland

Preface

This book contains the revised selected papers of CIBB 2018, the 15th International Conference on Computational Intelligence Methods for Bioinformatics and Biostatistics. This international conference provided a multi-disciplinary forum for researchers interested in the application of computational intelligence, in a broad sense, opened problems in bioinformatics, biostatistics, systems and synthetic biology, and medical informatics, and presented cutting edge methodologies capable of accelerating life science discoveries. Following its tradition and roots, this year's meeting brought together researchers with different backgrounds, including mathematics, computer science, materials science, medicine, and biology, from the international scientific community interested in advancements and future perspectives in bioinformatics and biostatistics. Under this compliance, current trends and future opportunities at the edge of computer and life sciences, the application of computational intelligence to a system and synthetic biology and the consequent impact on innovative medicine were presented. Theoretical and experimental biologists also presented novel challenges and fostered multidisciplinary collaboration aiming to blend theory and practice, where the founding theories of the techniques used for modeling and analyzing biological systems are investigated and used for practical applications and the supporting technologies. The conference program only included oral presentations, among which cutting-edge plenary keynote lectures were given by prominent speakers such as Alberto Paccanaro (University of London, UK), Alexandra Carvalho (Universidade de Lisboa, Portugal), Benoit Liquet (University of Pau and Pays de l'Adour, France), Fernando L. Ferreira (Universidade Nova de Lisboa, Portugal) and Veronica Vinciotti (Brunel University London, UK).

The conference program was organized in papers on a main conference track with heterogeneous open problems at the forefront of current research and in eight further special sessions on specific themes: computational methods for neuroimaging analysis, machine learning in health informatics and biological systems, soft computing methods for characterizing diseases from omics data, engineering bio-interfaces and rudimentary cells as a way to develop synthetic biology, modeling and simulation methods for system biology and system medicine, fast and efficient solutions for computational intelligence methods in bioinformatics, systems, and computational biology, networking biostatistics and bioinformatics, machine explanation – interpretation of machine learning models for medicine and bioinformatics.

The conference was held at the Faculdade de Ciências e Tecnologia, Universidade NOVA de Lisboa, Caparica, Portugal, during September 6–8, 2018, (https://eventos.fct.unl.pt/cibb2018/) and was made possible by the efforts of the Organizing, Program, and Steering Committees and by the support of sponsors and participants. The organization attributed four fees grants to PhD students. These grants were conferred by taking into account the average punctuation given by the reviewers (Program Committee).

Researchers from Europe, Asia, the USA, Brazil, and Africa attended the conference. Overall, 56 contributions were submitted for consideration by CIBB 2018 and after a round of reviews each paper had received an average of 2.5 reviews from the Program Committee and additional reviewers. From these contributions, 51 were invited for an oral presentation. After feedback and a discussion during the conference, selected papers were invited for further submissions. After a second round of reviews with an average of 2.4 reviews for each paper, the accepted papers were collected into this volume being divided by four major sections: Computational Intelligence Methods for Bioinformatics and Biostatistics, Modeling and Simulation Methods in System Biology, Computational Models in Health Informatics and Medicine, and Engineering Bio-Interfaces and Rudimentary Cells as a Way to Develop Synthetic Biology.

With the continued support from the scientific community, the next edition of CIBB will be held in Bergamo, Italy.

October 2019

Maria Raposo
Paulo Ribeiro
Susana Sério
Antonino Staiano
Angelo Ciaramella

The original version of the book was revised: The affiliations of the volume editors Antonino Staiano and Angelo Ciaramella have been corrected. The correction to the book is available at https://doi.org/10.1007/978-3-030-34585-3_31

Organization

CIBB 2018 was organized by the Departamento de Física, Faculdade de Ciências e Tecnologia, Universidade NOVA de Lisboa, Portugal.

General Chairs

Maria Raposo — Universidade NOVA de Lisboa, Portugal
Paulo Lisboa — Liverpool John Moores University, UK
Giorgio Valentini — Università degli Studi di Milano, Italy

Biostatistics Technical Chair

Chiara Brombin — Università Vita-Salute San Raffaele, Italy

Bioinformatics Technical Chair

Angelo Ciaramella — Università degli Studi di Napoli Parthenope, Italy

Organizing Committee

Maria Raposo — Universidade NOVA de Lisboa, Portugal
Paulo A. Ribeiro — Universidade NOVA de Lisboa, Portugal
Susana Sério — Universidade NOVA de Lisboa, Portugal
Andrea Bracciali — University of Stirling, UK
Filipa Pires — Universidade NOVA de Lisboa, Portugal
Gonçalo Magalhães-Mota — Universidade NOVA de Lisboa, Portugal
Telma Marques — Universidade NOVA de Lisboa, Portugal
Sara Pereira — Universidade NOVA de Lisboa, Portugal
Jeniffer Farias dos Santos — Universidade de São Paulo, Brazil
Paulo Zagalo — Universidade NOVA de Lisboa, Portugal
Thais P. Pivetta — Universidade NOVA de Lisboa, Portugal
João Pereira da Silva — Universidade NOVA de Lisboa, Portugal
Ana Cruz (Secretariat) — Universidade NOVA de Lisboa, Portugal

Special Session Chairs

Andrea Bracciali — University of Stirling, UK
Antonino Staiano — Università degli Studi di Napoli Parthenope, Italy

Steering Committee

Pierre Baldi	University of California, Irvine, USA
Elia Biganzoli	University of Milan, Italy
Clelia Di Serio	University Vita-Salute San Raffaele, Italy
Alexandru Floares	Oncological Institute Cluj-Napoca, Romania
Jon Garibaldi	University of Nottingham, UK
Nikola Kasabov	Auckland University of Technology, New Zealand
Francesco Masulli	University of Genova, Italy, and Temple University, USA
Leif Peterson	TMHRI, USA
Roberto Tagliaferri	University of Salerno, Italy

Program Committee

Claudio Angione	Teesside University, UK
Sansanee Auephanwiriyakul	Chiang Mai University, Thailand
Gilles Bernot	University of Nice Sophia Antipolis, France
Andrea Bracciali	University of Stirling, UK
Stefan Canzar	Saarland University, The Netherlands
Giulio Caravagna	University of Edinburgh, UK
Davide Chicco	Princess Margaret Cancer Centre, Canada
Angelo Ciaramella	University of Naples Parthenope, Italy
Luisa Cutillo	University of Sheffield, UK
Angelo Facchiano	CNR, Istituto di Scienze dell'Alimentazione, Italy
Enrico Formenti	Nice Sophia Antipolis University, France
Christoph M. Friedrich	University of Applied Science and Arts Dortmund, Germany
Yair Goldberg	University of Haifa, Israel
Marco Grzegorczyk	Groningen University, The Netherlands
Giosue' Lo Bosco	University of Palermo, Italy
Hassan Mahmoud	University of Genova, Italy
Anna Marabotti	University of Salerno, Italy
Elena Marchiori	Radboud University, The Netherlands
Tobias Marschall	Saarland University, Max Planck Institute for Informatics, Germany
Mauri Giancarlo	University of Milano-Bicocca, Italy
Bud Mishra	Courant Institute, NYU, and School of Medicine, Mt. Sinai, USA
Marianna Pensky	University of Central Florida, USA
Nadia Pisanti	Università di Pisa, Erable Team, Inria, Italy
Vilda Purutçuoğlu	Middle East Technical University, Turkey
Paolo Romano	IRCCS University Hospital San Martino IST, Italy
Simona Ester Rombo	University of Palermo, Italy
Stefano Rovetta	University of Genoa, Italy
Antonino Staiano	University of Napoli Parthenope, Italy

Francesco Stingo	MD Anderson, USA
Paolo Tieri	Consiglio Nazionale delle Ricerche, Italy
Alfonso Urso	ICAR-CNR, Italy
Filippo Utro	IBM T.J. Watson Research Center, USA
Alfredo Vellido	Universitat Politecnica de Catalunya, Spain
Pawel P. Labaj	Boku University Vienna, Austria
Maria Raposo	Universidade NOVA de Lisboa, Portugal
Paulo A. Ribeiro	Universidade NOVA de Lisboa, Portugal
Susana Sério	Universidade NOVA de Lisboa, Portugal
Marco Frasca	Università degli Studi di Milano, Italy
Armando Blanco	Universidad de Granada, Spain
Matteo Re	Università degli Studi di Milano, Italy

Special Session Organizers

Tiago Azevedo	University of Cambridge, UK
Giovanna Maria Dimitri	University of Cambridge, UK
Pietro Liò	University of Cambridge, UK
Angela Serra	University of Salerno, Italy
Simeon Spasov	University of Cambridge, UK
Davide Chicco	Princess Margaret Cancer Centre, Canada
Marco Masseroli	Politecnico di Milano, Milan, Italy
Annalisa Barla	Università di Genova, Genoa, Italy
Anne-Christin Hauschild	Krembil Research Institute, Canada
Angelo Ciaramella	Università di Napoli Parthenope, Italy
Giosuè Lo Bosco	Università di Palermo, Italy
Riccardo Rizzo	ICAR-CNR, Italy
Antonino Staiano	Università di Napoli Parthenope, Italy
Maria Raposo	Universidade NOVA de Lisboa, Portugal
Quirina Ferreira	Universidade de Lisboa, Portugal
Paulo A. Ribeiro	Universidade NOVA de Lisboa, Portugal
Susana Sério	Universidade NOVA de Lisboa, Portugal
Chiara Damiani	University of Milano-Bicocca, Italy
Marco S. Nobile	University of Milano-Bicocca, Italy
Riccardo Colombo	University of Milano-Bicocca, Italy
Giancarlo Mauri	University of Milano-Bicocca, Italy
Alex Graudenzi	University of Milano-Bicocca, Italy
Marzia Di Filippo	University of Milano-Bicocca, Italy
Dario Pescini	University of Milano-Bicocca, Italy
Stefano Beretta	University of Milano-Bicocca, Italy
Paolo Cazzaniga	University of Bergamo, Italy
Ivan Merelli	Institute for Biomedical Technologies, National Research Council, Italy
Clelia Di Serio	Vita-Salute San Raffaele University, Italy
Cugnata Federica	Vita-Salute San Raffaele University, Italy
Ian H. Jarman	Liverpool John Moores University, UK

Alfredo Vellido   Universitat Politècnica de Catalunya, Spain
José D. Martìn-Guerrero   University of Valencia, Spain
Davide Bacciu   Università of Pisa, Italy

Keynote Abstracts

A Unified Regularized Group PLS Algorithm Scalable to Big Data. Application on Genomics Data

Benoit Liquet

University of Pau and Pays de l'Adour, France

Partial Least Squares (PLS) methods have been heavily exploited to analyse the association between two blocs of data. These powerful approaches can be applied to data sets where the number of variables is greater than the number of observations and in presence of high collinearity between variables. Different sparse versions of PLS have been developed to integrate multiple data sets while simultaneously selecting the contributing variables. Sparse modelling is a key factor in obtaining better estimators and identifying associations between multiple data sets. The cornerstone of the sparsity version of PLS methods is the link between the SVD of a matrix (constructed from deflated versions of the original matrices of data) and least squares minimisation in linear regression. We present here an accurate description of the most popular PLS methods, alongside their mathematical proofs. A unified algorithm is proposed to perform all four types of PLS including their regularised versions. Our methods enable us to identify important relationships between genomic expression and cytokine data from an HIV vaccination trial. We also proposed a new methodology by accounting for both grouping of genetic markers (e.g. genesets) and temporal effects. Finally, various approaches to decrease the computation time are offered, and we show how the whole procedure can be scalable to big data sets.

Prof. Benoit Liquet, is Professor at the University of Pau et Pays de l'Adour, (France) and member of LMAP. Research Output: Award 2009, 2016 Scientific Excellence Award from the University of Bordeaux II; More than 50 articles in international peer-reviewed journals, 3 book chapters and two books on R software (one in French and one in English), of which the Chinese and the Indonesian version is now available. The French version Le logiciel R: maitriser le langage: effectuer des analyses statistiques has been nominated in 2011 for the Roberval prize which recognises the best books for the category "university education". This book is currently a best seller in mathematics applied to informatics (Amazon.fr; number one in November 2013). Two Books on Dynamical Biostatistical models (one in French and one in English).

Sparse Graphical Models in Genomics: An Application to Censored qPCR Data

Veronica Vinciotti

Department of Mathematics, Brunel University London, Uxbridge UB8 3PH, London, UK

Regularized inference of networks using graphical modelling approaches has seen many applications in biology, most notably in the recovery of regulatory networks from high-dimensional gene expression data. Various extensions to the standard graphical lasso approach have been proposed, such as dynamic and hierarchical graphical models. In this talk, I will focus on a latest extension to censored graphical models in order to deal with censored data such as qPCR data. We propose a computationally efficient EM-like algorithm for the estimation of the conditional independence graph and thus the recovery of the underlying regulatory network.

Veronica Vinciotti is professor in the Department of Mathematics, Brunel University London, London, UK. The main focus of her research is on statistical network science, with particular emphasis on the development of statistical models for the detection of protein binding sites from ChIP-seq data and for the recovery of regulatory networks from gene expression data (microarray, RNA-seq, qPCR). Active member of the Royal Statistical Society and is currently in the management team of the European Cooperation for Statistics of Network Data Science (COSTNET–COST Action CA15109).

Answering Questions in Biology and Medicine by Making Inferences on Networks

Alberto Paccanaro

University of London, UK

An important idea that has emerged recently is that a cell can be viewed as a complex network of interrelating proteins, nucleic acids and other bio-molecules. At the same time, data generated by large-scale experiments often have a natural representation as networks such as protein-protein interaction networks, genetic interaction networks, co-expression networks. From a computational point of view, a central objective for systems biology and medicine is therefore the development of methods for making inferences and discovering structure in biological networks possibly using data which are also in the form of networks. In this talk, I'll present novel computational methods for solving biological problems which can all be phrased in terms of inference and structure discovery in large scale networks. These methods are based and extend recent developments in the areas of machine learning (particularly semi-supervised learning and matrix factorization), graph theory and network science. I will show how these computational techniques can provide effective solutions for: (1) quantifying similarity between heritable diseases at molecular level using exclusively disease phenotype information; (2) disease gene prediction; (3) drug side-effect prediction.

Alberto Paccanaro is professor at the University of London, UK. His research interests are in applying and developing machine learning and pattern recognition techniques for solving problems in molecular biology and medicine. His recent work has focused on the development of methods for analysis and inference in large scale biological networks.

Model Selection for Temporal Biomedical Data

Alexandra Carvalho

Universidade de Lisboa, Portugal

Human health care is changing rapidly, pressing the development of machine learning techniques for automatic diagnoses and prognosis, as well as personalized therapies for individual patients. The emerging availability of temporal data, namely via electronic medical records, is triggering this line of research. One of the main problems is to model the dynamic process underlying the data evolution. We detail how to learn efficiently Markovian data, when the dependencies can be expressed as a dynamic Bayesian network. We follow a score-based approach, and guarantee that the learned model is optimal according to several model selection criteria. Finally, we address the problem of early classification, which is essential in time-sensitive applications, such as personalized therapies.

Alexandra Carvalho is assistant professor at the Universidade de Lisboa and is interested on algorithms to machine learning, with applications in bioinformatics and biomedicine and she participated in more than 15 national and European projects in the area of Bioinformatics and Machine Learning. Currently, she coordinates a project in personalized medicine bringing together four institutions to apply machine learning techniques to predict treatment outcome of rheumatic and musculoskeletal diseases.

Ethics and Our Moral in Research, Let's Think About It!

Fernando Luís Ferreira

Universidade NOVA de Lisboa, Portugal

As researchers, it is our will is to pursue knowledge, to contribute to society and to open new roads for the Future. Ethics is a theme always present in our minds but probably remains outside the central concerns of researchers while main subjects are developed. Sometimes we come across a formal consent or an ethics statement seen mostly as a bureaucratic task. However, lately with the so called exponential technologies, we find ourselves dueling with a variety of controversial questions resulting from the different branches of artificial intelligence as those applied to self-driving cars' decisions the exposure of privacy and Decision support systems in medicine, etc. Some are arguing that risks become clear and, one of this days, we may face a singularity and, eventually, becoming too late to stop. This short talk aims to rise some questions about present ethical issues aiming at promoting the intervention and discussion among participants at this Conference.

Fernando Luis Ferreira is Senior Researcher at UNINOVA, Portugal, and is interested in Engineering with Health and Human Sciences. He has been involved in research projects in the Aerospace domain FP6-AEROSPACE- 502917, FP7-234344 CRESCENDO and health applications including knowledge management and clinical research on cancer and Brain studies FP6-508803 BIOPATTERN NOE and AAL-2016 CARELINK for Dementia sufferers and their community. With two books published on computer Science and more than 40 scientific publications, he is teaching for more than 15 years and he keeps is research activities focused on eHealth, Futuristic Engineering, Knowledge Management and Cognitive Sciences.

Contents

Modeling and Simulation Methods in System Biology

Computational Models in Health Informatics and Medicine

**Engineering Bio-Interfaces and Rudimentary Cells as a Way
to Develop Synthetic Biology**

Computational Intelligence Methods for Bioinformatics and Biostatistics

Compressive Sensing and Hierarchical Clustering for Microarray Data with Missing Values

Angelo Ciaramella(ID), Davide Nardone(ID), and Antonino Staiano(✉)(ID)

Department of Science and Technology, University of Naples "Parthenope",
Isola C4, Centro Direzionale, 80143 Naples, Italy
{angelo.ciaramella,antonino.staiano}@uniparthenope.it,
davide.nardone@studenti.uniparthenope.it

Abstract. Commonly, in gene expression microarray measurements multiple missing expression values are generated, and the proper handling of missing values is a critical task. To address the issue, in this paper a novel methodology, based on compressive sensing mechanism, is proposed in order to analyze gene expression data on the basis of topological characteristics of gene expression time series. The approach conceives, when data are recovered, their processing through a non-linear PCA for dimensional reduction and a Hierarchical Clustering Algorithm for agglomeration and visualization. Experiments have been performed on the yeast *Saccharomyces cerevisiae* dataset by considering different percentages of information loss. The approach highlights robust performance when high percentage of loss of information occurs and when few sampling data are available.

Keywords: Microarray gene expression · Missing data · Compressive Sensing · Hierarchical clustering · Saccharomyces Cerevisiae sequences

1 Scientific Background

DNA microarray technology permits the simultaneous measurement of expression profiles of thousands of genes under different experimental conditions [1]. For this reason, microarray data provides an alternative to the identification of disease genes [2–6], for instance tumor sub-types at the molecular level [7]. Nonetheless, microarray technologies are very complex and, in the measurement phase, missing values in gene expression profiles are very frequent [1,7]. Concretely, irregular use of microarray chips, insufficient resolution and contamination of microarray surface, just to mention a few cases, would lead to gene expression profiles with missing values. Moreover, several used algorithms for gene selection, clustering and functional annotation [8], directly remove these instances or genes causing a substantial loss of information.

Different approaches proposed in literature handle missing values through imputation [7]. In this paper, it is introduced a novel methodology for analyzing

© Springer Nature Switzerland AG 2020
M. Raposo et al. (Eds.): CIBB 2018, LNBI 11925, pp. 3–10, 2020.
https://doi.org/10.1007/978-3-030-34585-3_1

gene expression data with missing values, on the basis of topological characteristics of time series. The estimation of multiple missing expression values is accomplished by using a Compressive Sensing (CS) mechanism. Furthermore, the recovered data are elaborated by using a non-linear PCA for dimensional reduction and a hierarchical clustering algorithm for agglomeration and visualization. The evaluation of the proposed approach has been performed through experiments on the yeast *Saccharomyces cerevisiae* dataset [9], by considering different percentages of information loss.

The paper is organized as follows. In Sect. 2, the adopted methodologies are introduced, and in Sect. 3 some experimental results are shown and discussed. Finally, in Sect. 4 conclusions and future directions close the paper.

2   Materials and Methods

Before going into the details of the methods involved in the gene expression analysis proposed, it could be beneficial to take a closer look at all the phases involved in that process. In particular, the proposed approach is composed by the following steps: (a) missing gene expression values estimation by CS; (b) gene expression dimensionality reduction by nonlinear PCA; (c) a pre-clustering procedure through a competitive learning technique in order to get a coarse grouping of genes. This step is necessary as initialization for the agglomerative clustering of the next step, in order to get better results; (d) a fine-tuned hierarchical agglomerative clustering based on both Fisher and Negentropy information. Let's take a closer look at the mentioned steps.

– **Compressive Sensing**
  The aim of CS, or Compressed Sensing, is the recovering of signals from far fewer measurements [10–12] using two basic principles: *sparsity* and *incoherence*. A vector $\mathbf{f} \in \mathbf{R}^n$ is expanded in an orthonormal basis $\mathbf{\Psi} = [\psi_1, \dots, \psi_n]$:

$$\mathbf{f} = \sum_{i=1}^{n} x_i \psi_i = \mathbf{\Psi}\mathbf{x}, \tag{1}$$

  where $\mathbf{x} = [x_1, \dots, x_n]^T$ is the representation of $\mathbf{f}$ w.r.t. the basis $\mathbf{\Psi}$. When most of the components of $\mathbf{x}$ are zero, then $\mathbf{x}$ is referred to as a sparse representation of $\mathbf{f}$, and $\mathbf{\Psi}$ is a sparsifying basis.
  On the other hand, denoting by $\mathbf{\Phi}_s$ the $m \times n$ sensing matrix with the vectors $\phi_1^*, \dots, \phi_m^*$ as rows (a^* is the complex conjugate transpose of a), the process of recovering $\mathbf{f} \in \mathbf{R}^n$ from

$$\mathbf{y} = \mathbf{\Phi}_s \mathbf{f} \in \mathbf{R}^m \tag{2}$$

  is ill-posed, in general, when $m < n$, since there are infinitely many candidate signals $\widetilde{\mathbf{f}}$ for which $\mathbf{\Phi}_s \widetilde{\mathbf{f}} = \mathbf{y}$.

  Consider now the pair $(\mathbf{\Phi}, \mathbf{\Psi})$ of orthobases of $\mathbf{R}^n$. The first basis $\mathbf{\Phi}$ is used for sensing the object $\mathbf{f}$ and the second is used to represent $\mathbf{f}$.

One of the main properties is the coherence $\mu(\mathbf{\Phi}, \mathbf{\Psi})$ that measures the largest correlation between any two elements of $\mathbf{\Phi}$ and $\mathbf{\Psi}$.

We suppose to observe a subset of the n coefficients of $\mathbf{f}$ and collect data as in Eq. 2. The source signal is recovered by solving an ℓ_1-norm constrained minimization problem.

A raw signal can be regarded as a vector $\mathbf{f}$ that can be represented as a linear combination of certain basis functions, as in Eq. 1:

$$\mathbf{f} = \mathbf{\Psi}\mathbf{x}. \tag{3}$$

In this work, the matrix $\mathbf{\Psi}$ is a DCT matrix.

- **Non Linear Principal Component Analysis**
  Nonlinear Principal Component Analysis (NLPCA) can be considered a nonlinear generalization of standard principal component analysis (PCA) [13]. It generalizes the principal components, describing the subspace in the original data space, in a curved space. NLPCA can be obtained by using a neural network with an auto-associative architecture also known as autoencoder or a replicator network. In particular, the auto-associative neural network is a multi-layer perceptron where the outputs of the network are required to be identical to the input. However, in the middle of the network there is a layer that works as a bottleneck in which a reduction of the dimension of the data is enforced. This bottleneck-layer provides the desired component values.

- **Competitive learning**
  A preliminary coarse (high number of clusters) grouping of data is computed in the pre-clustering phase, where a number of fixed clusters are determined with a Competitive Neural Network (NN) approach through a *Winner Takes All* (WTA) strategy [14].

- **Agglomerative Hierachical Clustering**
  The raw clusters given by WTA in the previous step, are taken by a hierarchical clustering, namely Negentropy Based Hierarchical Clustering (NEC) [14], where couples of clusters are considered for merging and the merging decision is taken on the basis of both Fisher and Negentropy information. After that the NLPCA dimension reduction and pre-clustering are adopted, the next step is based on both Fisher's and Negentropy information to agglomerate the clusters found by the WTA. In particular, the Fisher's linear discriminant is a classification method that projects high-dimensional data onto a line and performs classification in this one-dimensional space. The projection maximizes the distance between the means of the two classes while minimizing the variance within each class. For two classes we have

$$J_F(\mathbf{w}) = \frac{\mathbf{w}^\mathrm{T}\mathbf{S}_\mathrm{B}\mathbf{w}}{\mathbf{w}^\mathrm{T}\mathbf{S}_\mathrm{W}\mathbf{w}} \tag{4}$$

where $\mathbf{S}_\mathrm{B}$ is the between-class covariance matrix and $\mathbf{S}_\mathrm{W}$ is the total within-class covariance matrix. From Eq. 4, differentiating with respect to $\mathbf{w}$, it can be observed that $J_F(\mathbf{w})$ is maximized when

$$\mathbf{w} \propto \mathbf{S}_\mathrm{W}^{-1}(\mathbf{m}_2 - \mathbf{m}_1). \tag{5}$$

On the other hand, according to the definition of Negentropy, J_N is given by

$$J_N(\mathbf{x}) = H(\mathbf{x}_{Gauss}) - H(\mathbf{x}), \tag{6}$$

where $\mathbf{x}_{Gauss}$ is a Gaussian random vector of the same covariance matrix as $\mathbf{x}$ and $H(.)$ is the differential entropy. Negentropy can also be interpreted as a measure of non-Gaussianity [14]. In this paper we adopted an approximation of Negentropy that gives a compromise between the properties of the two classic non-Gaussianity measures given by kurtosis and skewness.

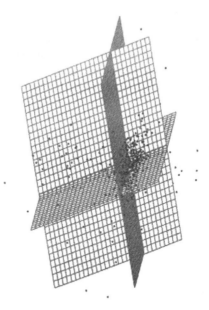

Fig. 1. NLPCA 3D projection of the Saccharomyces cerevisiae dataset.

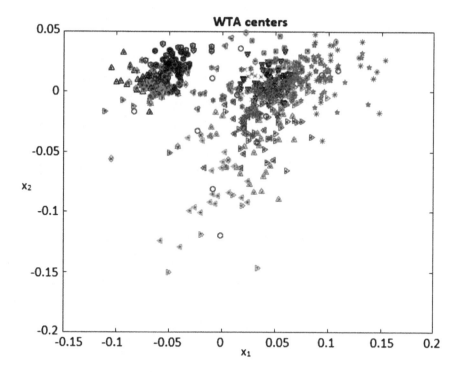

Fig. 2. WTA clustering on the 2D projected data with no loss.

3   Experimental Results

The proposed methodology has been evaluated on the Saccharomyces cerevisiae dataset [9]. DNA microarrays are used to study temporal gene expression of almost all genes in Saccharomyces cerevisiae during the metabolic shift from fermentation to respiration. Expression levels were measured at 7 time points during the diauxic shift. Before the analysis, the dataset underwent a pre-processing phase by removing the genes that are not expressed or do not change, with small variance over time, with very low absolute expression values and with low entropy profiles, thus leading to 614 genes. The number of WTA centers used for all the experiments is 20. The nonlinear NLPCA projection of the dataset with no loss and the clustering obtained after the projection of the data on two dimensions, followed by WTA strategy are shown (see Figs. 1 and 2). Several research works in literature have compared the performance against the information loss on this data set [7]. Indeed, successively, we simulate the loss of one value for each sequence in the dataset, corresponding to 14.3% of information loss on the whole dataset.

An example of the recovered sequence, with one lost value, after applying the CS approach along with the WTA clustering of the genes obtained after the projection of the data on two dimensions is shown (see Fig. 3).

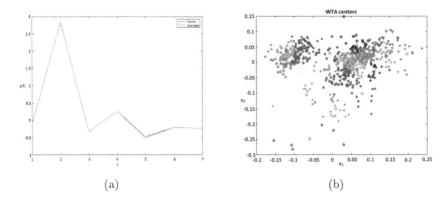

(a)   (b)

Fig. 3. Experiment performed with 14.3% of information loss (one lost value). (a) Comparison between the original gene sequence and the estimated one. (b) WTA clustering on the 2D projected.

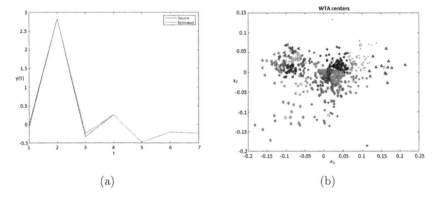

(a)   (b)

Fig. 4. Experiment performed with 28.57% of information loss (two lost values). (a) Comparison between the original gene sequence and the estimated one. (b) WTA clustering on the 2D projected data.

Next, the loss of two values for each sequence in the dataset, obtaining a 28.57% of information loss, was simulated. An example of the recovered sequence, with the loss of two values, after applying the CS approach along with the WTA clustering of the genes obtained after the projection of the data on two dimensions is shown (see Fig. 4). It is worth stressing that the obtained dendrograms are very close to that obtained on the dataset without loss.

In particular, a quantitative metric introduced in [15] (PhyloCore algorithm[1]) has been used to compare the obtained agglomerations. We achieved a score of 1 between the dendogram with lossless information and the one with 14.3% of information loss, and a score of 0.88 with the one with 28.57% of information loss. An example of comparison, in the case of 14.3% of information loss, is visualized

---

[1] http://www.mas.ncl.ac.uk/~ntmwn/compare2trees/.

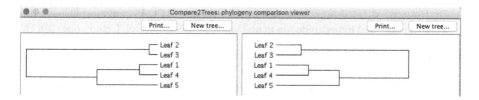

Fig. 5. Comparison between the dendrograms without information loss and with 28.57% of information loss.

where 5 clusters for the WTA approach are adopted (see Fig. 5). Moreover, we performed a cross-correlation analysis between the original sequences and those obtained by adopting the CS approach. We achieved an average correlation coefficient of 0.9866, by considering the original sequences and those estimated by CS with 14.3% of information loss, and 0.8813 with 28.57% of information loss.

4   Conclusions

In this work we introduced a novel methodology for analyzing gene expression data based on topological characteristics of time series, proposing to solve the problem of information loss in data, through a Compressive Sensing mechanism. Furthermore, the recovered data have been processed by using a non-linear PCA for dimensional reduction and a Hierarchical Clustering for agglomeration and visualization. The experiments conducted on the Yeast *Saccharomyces cerevisiae* dataset, by considering different percentages of information loss, namelyL: no loss, 14.3% and 28.57% of information loss, proved that the proposed approach can be considered robust to information loss when few data are available.

Acknowledgments. The research was developed when Davide Nardone was a M.Sc. student in Applied Computer Science at University of Naples Parthenope.

This work was partially funded by the University of Naples Parthenope (*Sostegno alla ricerca individuale per il triennio 2016–2018* project, and supported by Gruppo Nazionale per il Calcolo Scientifico (GNCS-INdAM)).

References

1. Draghici, S., Khatri, P., Eklund, A.C., Szallasi, Z.: Reliability and reproducibility issues in DNA microarray measurements. Trends Genet. **22**(2), 101–109 (2006)
2. Camastra, F., Di Taranto, M.D., Staiano, A., Statistical and computational methods for genetic diseases: an overview. Comput. Math. Methods Med. **2015**, Article ID 954598 (2015)
3. Di Gangi, M., Lo Bosco, G., Rizzo, R., Deep learning architectures for prediction of nucleosome positioning from sequences data. BMC Bioinform. **19**, Article no. 418 (2018)

4. Di Taranto, M.D., et al.: Association of USF1 and APOA5 polymorphisms with familial combined hyperlipidemia in an Italian population. Mol. Cell. Probes **29**(1), 19–24 (2015)
5. Fiannaca, A., et al.: Deep learning models for bacteria taxonomic classification of metagenomic data. BMC Bioinform. **19**, Article no. 198 (2018)
6. Staiano, A., et al.: Investigation of single nucleotide polymorphisms associated to familial combined hyperlipidemia with random forests. In: Apolloni, B., Bassis, S., Esposito, A., Morabito, F. (eds.) Neural Nets and Surroundings, vol. 19, pp. 169–178. Springer, Heidelberg (2013). https://doi.org/10.1007/978-3-642-35467-0_18
7. Wang, A., Chen, Y., An, N., Yang, J., Li, L., Jiang, L.: Microarray missing value imputation: a regularized local learning method. IEEE/ACM Trans. Comput. Biol. Bioinform. **16**, 980–993 (2018)
8. Giancarlo, R., Bosco, G.L., Pinello, L., Utro, F.: The three steps of clustering in the post-genomic era: a synopsis. In: Rizzo, R., Lisboa, P.J.G. (eds.) CIBB 2010. LNCS, vol. 6685, pp. 13–30. Springer, Heidelberg (2011). https://doi.org/10.1007/978-3-642-21946-7_2
9. DeRisi, J.L., Iyer, V.R., Brown, P.O.: Exploring the metabolic and genetic control of gene expression on a genomic scale. Science **278**(5338), 680–686 (1997). PMID: 9381177
10. Candès, E.J., Wakin, M.B.: An introduction to compressive sampling. IEEE Signal Process. Mag. **25**(2), 21–30 (2008)
11. Ciaramella, A., Gianfico, M., Giunta, G.: Compressive sampling and adaptive dictionary learning for the packet loss recovery in audio multimedia streaming. Multimed. Tools Appl. **75**(24), 17375–17392 (2016)
12. Ciaramella, A., Giunta, G.: Packet loss recovery in audio multimedia streaming by using compressive sensing. IET Commun. **10**(4), 387–392 (2016)
13. Scholz, M., Fraunholz, M., Selbig, J.: Nonlinear principal component analysis: neural network models and applications. In: Gorban, A.N., Kégl, B., Wunsch, D.C., Zinovyev, A.Y. (eds.) Principal Manifolds for Data Visualization and Dimension Reduction. LNCSE, vol. 58, pp. 44–67. Springer, Heidelberg (2007). https://doi.org/10.1007/978-3-540-73750-6_2
14. Ciaramella, A., Longo, G., Staiano, A., Tagliaferri, R.: NEC: a hierarchical agglomerative clustering based on fisher and negentropy information. In: Apolloni, B., Marinaro, M., Nicosia, G., Tagliaferri, R. (eds.) NAIS/WIRN -2005. LNCS, vol. 3931, pp. 49–56. Springer, Heidelberg (2006). https://doi.org/10.1007/11731177_8
15. Nye, T.M., Lió, P., Gilks, W.R.: A novel algorithm and web-based tool for comparing two alternative phylogenetic trees. Bioinformatics **22**(1), 117–9 (2006)

Variational Inference in Probabilistic Single-cell RNA-seq Models

Pedro F. Ferreira[1(✉)] ⓘ, Alexandra M. Carvalho[1,2] ⓘ, and Susana Vinga[1,3] ⓘ

[1] Instituto Superior Técnico, ULisboa, Av. Rovisco Pais, 1049-001 Lisboa, Portugal
{pedro.fale,alexandra.carvalho,susanavinga}@tecnico.ulisboa.pt
[2] Instituto de Telecomunicações, Av. Rovisco Pais, 1049-001 Lisboa, Portugal
[3] INESC-ID, R. Alves Redol 9, 1000-029 Lisboa, Portugal

Abstract. Single-cell sequencing technology holds the promise of unravelling cell heterogeneities hidden in ubiquitous bulk-level analyses. However, limitations of current experimental methods also pose new obstacles that prevent accurate conclusions from being drawn. To overcome this, researchers have developed computational methods which aim at extracting the biological signal of interest from the noisy observations. In this paper we focus on probabilistic models designed for this task. Particularly, we describe how variational inference constitutes a powerful inference mechanism for different sample sizes, and critically review two recent scRNA-seq models which use it.

Keywords: scRNA-seq · Probabilistic modelling · Bayesian inference · Dimensionality reduction · Imputation

1 Scientific Background

Single-cell RNA-sequencing (scRNA-seq) has emerged in the last decade as a key technology in using gene expression to study cell heterogeneity [1]. With the obtained data, researchers can, for example, apply clustering algorithms to identify cell types and find genes which are differentially expressed between two conditions.

In scRNA-seq data, each observation is a cell and, for each cell, the expression of all detected genes is measured through the set of all RNA molecules present, i.e., its transcriptome. Specifically, each entry in the $N \times P$ data matrix, where N is the number of cells and P the number of genes, contains the number of mRNA molecules corresponding to gene p detected in cell n. Depending on the experimental protocol and quality control pipelines, data set sizes may vary from

Supported by the EU Horizon 2020 research and innovation program (grant No. 633974 – SOUND project), and the Portuguese Foundation for Science & Technology (FCT), through UID/EMS/50022/2019 (IDMEC,LAETA), UID/EEA/50008/2019 (IT), UID/CEC/50021/2019 (INESC-ID), PTDC/EMS-SIS/0642/2014, PTDC/CCI-CIF/29877/2017, PTDC/EEI-SII/1937/2014, IF/00653/2012, and by internal IT projects QBigData and RAPID.

hundreds to millions of cells [1] and from hundreds to tens of thousands of genes sequenced.

Although increasingly available, scRNA-seq data suffer from multiple confounding factors which may hide the biological signal of interest from analysis. These include varying sequencing depths and mRNA capture efficiency which lead to zero-inflated observations and library size (total number of mRNA molecules detected per cell) dispersion, as well as batch effects [1]. Because of this, applying generic computational methods for further downstream analyses, such as dimensionality reduction, clustering, or differential expression, yields spurious results. Indeed, while PCA may capture the existence of different clusters of cells in a data set, its principal components are highly correlated with technical factors [2]. Researchers have thus developed methods to extract only the biological signal of interest; the most commonly tackled issue is the unrealistic abundance of zero counts, termed "dropouts".

In this report, we focus on methods based on probabilistic models of scRNA-seq. Probabilistic modelling [3] constitutes a powerful framework for the disentanglement of multiple factors of variation in the stochastic generative process underlying the data. In general, probabilistic scRNA-seq models assume a lower-dimensional representation of the data, which is mapped into the observation space by some transformation, allowing for dimensionality reduction and a dropout-inducing process. This framework is particularly powerful because, by explicitly accounting for the different assumed factors of variation, it can be used for multiple downstream tasks after fitting to the data.

Probabilistic models define a joint probability distribution $p(\mathbf{X}, \mathbf{Z})$ over the observed data $\mathbf{X}$ and a set of latent variables $\mathbf{Z}$. These encode structure in the data, via some prior distribution $p(\mathbf{Z})$, and are related to the observations via the likelihood distribution $p(\mathbf{X}|\mathbf{Z})$. After defining such a model, inference of the latent variables is made via their probability distribution conditioned on the data, $p(\mathbf{Z}|\mathbf{X})$. This is called the posterior probability distribution and, according to Bayes' theorem, is given by

$$p(\mathbf{Z}|\mathbf{X}) = \frac{p(\mathbf{X}|\mathbf{Z})p(\mathbf{Z})}{p(\mathbf{X})}. \tag{1}$$

However, in general, for complex models, Eq. (1) can not be computed analytically. Approximating the posterior is thus the main computational challenge in probabilistic modelling. The common approach for this task is to use Markov Chain Monte Carlo sampling methods, with Gibbs sampling being the gold standard [3]. However, variational inference techniques are generally able to provide similar performance at a possibly lower computational cost, making them more suitable for large data sets.

In the following sections we describe two recent probabilistic models designed for scRNA-seq data which use variational inference to infer their latent variables: *Probabilistic Count Matrix Factorization* (pCMF) [4] and *Single Cell Variational Inference* (scVI) [5]. While both use the same inference engine, the modelling details of each allow for different techniques to be used, which we outline. The performance of both models is measured in terms of cell type separability in the lower-dimensional latent space, and dropout imputation error. As a baseline for comparison, we consider a state-of-the-art scRNA-seq probabilistic model, *Zero Inflated Factor Analysis* (ZIFA) [6].

2   Materials and Methods

We first describe variational inference. Then we describe pCMF and scVI. In particular, we aim at illustrating how the variational scheme allows for efficient inference of complex probabilistic models, both in small N, large P data sets, and the inverse.

2.1   Variational Inference

In variational inference the true posterior $p(\mathbf{Z}|\mathbf{X})$ defined in Eq. (1) is approximated via a distribution $q(\mathbf{Z}; \mu)$, which belongs to a certain family $\mathcal{Q}$, over the latent variables $\mathbf{Z}$ with free parameters μ [7]. These parameters are adjusted so as to minimize some distance between $q(\mathbf{Z})$ and $p(\mathbf{Z}|\mathbf{X})$. We thus turn inference into an optimization problem. The most commonly used distance metric between these distributions is the Kullback-Leibler (KL) divergence. In this case, the optimization problem becomes

$$q(\mathbf{Z}) = \operatorname*{argmin}_{q(\mathbf{Z}) \in \mathcal{Q}} \mathrm{KL}(q(\mathbf{Z}) \parallel p(\mathbf{Z}|\mathbf{X})). \tag{2}$$

The objective in Eq. (2) is not available because it depends on the posterior distribution which we aim at approximating. However, we can re-write the KL divergence in terms of a lower bound of $p(\mathbf{X})$ which we call the Evidence Lower BOund (ELBO). Minimizing the KL divergence is now achieved by maximizing the ELBO:

$$q(\mathbf{Z}) = \operatorname*{argmax}_{q(\mathbf{Z}) \in \mathcal{Q}} \mathrm{ELBO}(\mu) = \mathrm{E}_q\left[\log(p(\mathbf{X}, \mathbf{Z}))\right] - \mathrm{E}_q\left[\log(q(\mathbf{Z}; \mu))\right]. \tag{3}$$

This optimization is constrained not only by the family of distributions $\mathcal{Q}$ we choose, but also by the widely used mean-field approximation, where we assume each of the M latent variables to be independent from all the others and governed by their own variational density [7]. This makes the ELBO a non-convex function.

The most commonly used algorithm to find the μ that correspond to a local maximum of the ELBO is coordinate ascent, which we refer in the following sections as CAVI (Coordinate Ascent Variational Inference). CAVI algorithms can be easily derived for conditionally conjugate models. More recently, ELBO optimization has been generalized into the wider class of non-conditionally conjugate models, effectively allowing the design of more expressive models [7].

2.2 Probabilistic Count Matrix Factorization (pCMF)

This model consists of a Bayesian matrix factorization method for count data. Its latent variables are $\mathbf{U}$, $\mathbf{D}$ and $\mathbf{V}$.[1] $\mathbf{U}$ represents the cells in a lower-dimensional space of size $K < P$, $\mathbf{V}$ is the map from $\mathbf{U}$ to the observation space, and $\mathbf{D}$ models the occurrence of dropout in each observation.

By considering Gamma priors on $\mathbf{U}$ and $\mathbf{V}$, pCMF models the over-dispersion of the count data. $\mathbf{D}$ is given by a Bernoulli distribution. The model is represented graphically in Fig. 1 and defined by the generative process which Algorithm 1 outlines.[2]

Algorithm 1. Generative process for pCMF

For each cell n:
    For each k, sample a latent factor:
    $U_{nk} \sim \text{Gamma}\left(\alpha_{k1}, \alpha_{k2}\right).$
For each gene p:
    For each k, sample a factor load:
    $V_{pk} \sim \text{Gamma}\left(\beta_{k1}, \beta_{k2}\right).$
For each cell n and gene p,
    Sample dropout event:
    $D_{np} \sim \text{Bernoulli}\left(\pi_p\right).$
    Sample the observed count:
    $X_{np} \sim \text{Poisson}\left(\left(1 - D_{np}\right)\mathbf{U}_n\mathbf{V}_p^T\right).$

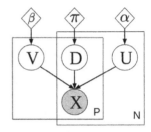

Fig. 1. Graphical representation of pCMF.

Because this model is conditionally conjugate (if we consider an auxiliary variable, see [4] for details), the posterior can be approximated via a CAVI algorithm.

While in traditional machine learning settings there are more observations than features, and thus an un-regularized point estimate of the global variables (commonly called "parameters") is enough, most initial scRNA-seq data sets, and the ones pCMF was designed for, are such that $P \gg N$. In these cases, the number of global variables (V_{pk}) is larger than the number of local variables (U_{nk}) in the model, which makes correct estimation of the global variables difficult. Thus the need for a Bayesian approach even for the global variables in this case.

However, CAVI requires the whole data set to compute each variational parameter update. As such, inference of pCMF on a data set with a much larger number of cells would imply a great computational effort, due to the need to infer a posterior over the global variables. In that case, it would suffice to use point estimates for the global variables instead, for example in an Expectation-Maximization algorithm.

---

[1] For brevity, here we do not consider the sparse loadings of the original model. In our experiments the resulting performance did not change significantly.

[2] $\alpha_{k1,2}$, $\beta_{k1,2}$ and π_p are fixed hyperparameters which can be estimated in an Expectation-Maximization scheme. See the original paper for details.

2.3 Single Cell Variational Inference (scVI)

scVI models the distribution of observed counts as conditioned on: $\mathbf{L}$, the variations due to capture efficiency and sequencing depth of each cell; $\mathbf{W}$, the normalized mean expressions; dropout events $\mathbf{D}$; θ for gene-specific dispersion and $\mathbf{Z}$, a lower-dimensional space where biological variability is encoded. It can also include the batch annotation of each cell in order to subtract batch effects from the biological signal.

Additionally, scVI utilizes neural networks to specify non-linear transformations between latent variables. Notably, it associates the cells' latent representations $\mathbf{Z}$ with the probability of dropout occurrence, encoded by a Bernoulli distribution on $\mathbf{D}$ whose parameter is given by a neural network f_D with output in the $[0,1]$ interval. Another neural network f_W is used to map from $\mathbf{Z}$ to the original-dimensional space containing $\mathbf{W}$, which is encoded by a Gamma-distributed random variable. The generative process is described in Algorithm 2 and Fig. 2 presents the corresponding graphical model.[3,4]

Algorithm 2. Generative process for scVI
For each cell n:
For each k, sample a latent factor:
$Z_{nk} \sim \text{Normal}\,(0,1)$.
Sample a cell-scaling factor:
$L_n \sim \text{LogNormal}\,\left(l_\mu, l_\sigma^2\right)$.
For each cell n and gene p,
Sample mean expression:
$W_{np} \sim \text{Gamma}\,\left(f_W\left(Z_n\right), \theta_p\right)$.
Sample dropout event:
$D_{np} \sim \text{Bernoulli}\,\left(f_D\left(Z_n\right)\right)$.
Sample the observed count:
$X_{np} \sim \text{Poisson}\,\left(\left(1 - D_{np}\right) L_n W_{np}\right)$.

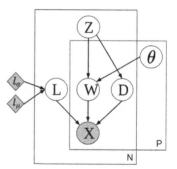

Fig. 2. Graphical representation of scVI.

Inference of scVI's latent variables is performed using neural networks specifying the approximate posterior distributions $q(\mathbf{Z})$ and $q(\mathbf{L})$ (the "inference networks" [8]), in which the other latent variables are integrated out. This allows for inference to be reduced to optimizing the weights of the four neural networks: f_W, f_D and $q(\mathbf{Z})$, $q(\mathbf{L})$ [5]. Unlike CAVI, this is a general mechanism possible for models without conditional conjugacy. It also makes inference amenable to stochastic optimization, meaning global variables can be estimated using small subsets of the data per iteration.

While the use of neural networks allows for great model expressiveness, the typically large number of parameters to fit may render them inadequate for small

---

[3] In these simplified descriptions we ignore the batch annotation observations, for brevity.

[4] l_μ and l_σ^2 are the observed log-library size mean and variance, respectively.

sample sizes. In addition, the ability to approximate the true posterior is limited by the flexibility of the inference networks.

3 Results

We test the methods described in Sect. 2 on real scRNA-seq data. For ZIFA and scVI, we use the implementations provided by the authors with the original publications. For pCMF we used our own implementation which allows for more flexibility in the inference scheme (i.e., inclusion of sparsity and hyperparameter estimation) than the one provided by the authors. We did, however, compare our implementation with the original one, and the same results were obtained. In our tests, we set the latent space dimensionality to $K = 10$ and apply all models to two real data sets, whose main characteristics are summarized in Table 1.

Table 1. Statistics of the considered experimental data sets.

Data set	# cells	# genes	# cell types	% zeros
Pollen [9]	249	6982	11	25.33
Zeisel [10]	3005	558	7	29.01

Figure 3 shows the evaluation of cluster separability in the latent space using different metrics: Average Silhouette Width (ASW), Adjusted Rand Index (ARI) and Normalized Mutual Information (NMI). For ARI and NMI, we used the K-means clustering method to obtain partitions. Factor Analysis (FA) and ZIFA perform similarly and always better than pCMF and scVI. pCMF is always worse than scVI.

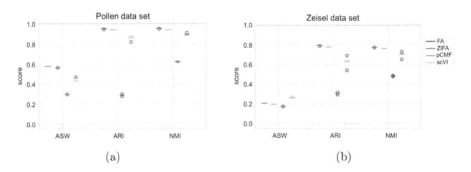

(a)   (b)

Fig. 3. Boxplots of ASW, ARI and NMI scores for each method for five repetitions on the (a) Pollen and (b) Zeisel data sets. Lines indicate no variation in the score for the five runs.

Following [5] we apply dropouts to 10% of the non-zero entries in each data set and compare the values imputed by the model with the original ones. Figure 4 shows the results for five repetitions of this process. In this sense, scVI is more sensitive than pCMF to the change of proportion between cells and genes.

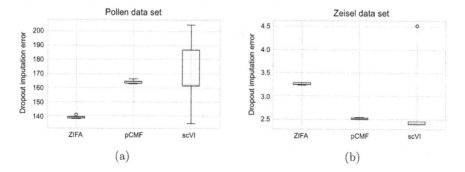

Fig. 4. Boxplots for the median L1 distance between imputed and original values for each scRNA-seq model fitted to five corrupted versions of the (a) Pollen and (b) Zeisel data sets.

Finally, we aim at understanding the effect of cell-specific scalings in the performance of scVI – its flexibility allows us to easily include them or not in the inferred model. Table 2 shows the results for 5-fold cross-validation (CV) on the Zeisel data set regarding imputation error and likelihood of held-out data. The results show that including cell-specific scalings to account for capture efficiency and sequencing depth improves the model fitness of scVI.

Table 2. Mean and standard deviation of scVI fitness metrics for a 5-fold CV on the Zeisel data set.

Scalings	Dropout imputation error	Predictive log-likelihood
No	4.224 (0.433)	−17245.777 (4750.590)
Yes	**2.338 (0.105)**	**−1515.624 (244.203)**

4 Conclusion

The results show that the separability of clusters in the latent space achieved by pCMF and scVI are not as good as the ones achieved by ZIFA or even FA. However, while pCMF and scVI may not provide better separations than FA and ZIFA, the explicit modelling of confounding factors guarantees that the structure they infer in the latent space is more related with actual biology rather than technical variability. In this light, scVI is more powerful than pCMF: not only does it account for more factors of variation, but it also achieves better cluster separability.

The expressiveness of scVI also allows for better imputation of dropouts in data sets with more cells than genes, such as Zeisel's. In the reverse case, due to scVI's complexity, it is expected to underfit the data, resulting in higher imputation errors (a behaviour also observed in [5] for a different data set). In this case, additional gene filtering must be used before applying scVI.

One of the main modelling issues that allow the good results of scVI is the use of cell-specific scalings. As shown in Table 2, their inclusion results in a large increase in dropout imputation error and the likelihood assigned to held-out data.

Additionally, leveraging the modelling power of scVI is easily done via the flexible inference process based on inference networks. This proves the versatility of this recent variational inference technique, which ultimately allows for models to be designed without worrying about the inference process, thus allowing the model designer to iterate faster over different model choices.

Acknowledgements. The authors thank Ghislain Durif for the helpful discussions about pCMF.

References

1. Kolodziejczyk, A.A., Kim, J.K., Svensson, V., Marioni, J.C., Teichmann, S.A.: The technology and biology of single-cell RNA sequencing. Mol. Cell **58**(4), 610–620 (2015)
2. Hicks, S.C., et al.: Missing data and technical variability in single-cell RNA-sequencing experiments. Biostatistics **19**, 562–578 (2017)
3. Murphy, K.: Machine Learning: A Probabilistic Approach. MIT Press, Cambridge (2012)
4. Durif, G., Modolo, L., Mold, J.E., Lambert-Lacroix, S., Picard, F.: Probabilistic count matrix factorization for single cell expression data analysis. arXiv (2018)
5. Lopez, R., Regier, J., Cole, M.B., Jordan, M., Yosef, N.: Bayesian inference for a generative model of transcriptome profiles from single-cell RNA sequencing. bioRxiv (2018)
6. Pierson, E., Yau, C.: ZIFA: dimensionality reduction for zero-inflated single-cell gene expression analysis. Genome Biol. **16**(1), 241 (2015)
7. Blei, D.M., Kucukelbir, A., McAuliffe, J.D.: Variational inference: a review for statisticians. J. Am. Stat. Assoc. **112**(518), 859–877 (2017)
8. Kingma, D., Welling, M.: Stochastic gradient VB and the variational auto-encoder. In: Second International Conference on Learning Representations, ICLR (2014)
9. Pollen, A.A., et al.: Low-coverage single-cell mRNA sequencing reveals cellular heterogeneity and activated signaling pathways in developing cerebral cortex. Nat. Biotechnol. **32**, 1053–1058 (2014)
10. Zeisel, A., et al.: Cell types in the mouse cortex and hippocampus revealed by single-cell RNA-seq. Science **347**(6226), 1138–1142 (2015)

Centrality Speeds the Subgraph Isomorphism Search Up in Target Aware Contexts

Vincenzo Bonnici$^{(\boxtimes)}$ (ID), Simone Caligola (ID), Antonino Aparo (ID), and Rosalba Giugno (ID)

Department of Computer Science, University of Verona,
Strada le Grazie 15, 37134 Verona, Italy
{vincenzo.bonnici,simone.caligola,antonino.aparo,rosalba.giugno}@univr.it

Abstract. The subgraph isomorphism (SubGI) problem is known to be a NP-Complete problem. Several methodologies use heuristic approaches to solve it, differing into the strategy to search the occurrences of a graph into another. This choice strongly influences their computational effort requirement. We investigate seven search strategies where global and local topological properties of the graphs are exploited by means of weighted graph centrality measures. Results on benchmarks of biological networks show the competitiveness of the proposed seven alternatives and that, among them, local strategies predominate on sparse target graphs, and closeness- and eigenvector-based strategies outperform on dense graphs.

Keywords: Subgraph isomorphism · Biological graphs · Variable orderings · Graph centrality · Label frequency

1 Introduction

The analysis of biological systems involves the study of large-scale networks from different *omics* such as genome, transcriptome, metabolome and proteome. Such biological networks are represented as graphs where, for example, graph vertices are molecular components and edges are the interactions among them. Vertices and edges are labeled with the information they carry with them. For example, in chemical structures, node labels are informative about the atom they represent and edge labels inform about the properties of chemical bonds. Understanding properties of biological networks involves the application of subgraph isomorphism algorithms (SubGI) allowing the search of small graphs, named patterns, into larger networks, named targets. SubGI algorithms are applied for the prediction of the biological activity of a molecule [1], for the analysis of network motifs (i.e. subgraphs) within complex biological networks [2] or for searching protein complexes in protein-protein interaction networks [3]. Subgraph isomorphism is a NP-complete problem, therefore the development of efficient heuristics is necessary to make the problem affordable.

M. Raposo et al. (Eds.): CIBB 2018, LNBI 11925, pp. 19–26, 2020.
https://doi.org/10.1007/978-3-030-34585-3_3

Several strategies have been proposed to speed the subgraph isomorphism up in the specific context of biological networks [4]. Algorithms rely on techniques that show different performance depending on the properties of the pattern and target graphs such as size, density and number of labels [5]. A crucial aspect regards also the strategy applied to drive the search process. In [4], authors report an approach, named RI, that takes advantage of local topological properties of the pattern graph. It shows very good performance compared with the state of art algorithms in SubGI [6,7], but it results in lowly suitable in subgraph isomorphism instances characterized by dense target graphs.

We define and investigate seven search strategies based on target-aware graph centrality measures in order to decrease the dependence of performance from local graph properties. Tests were run on a variegate benchmark of biological networks composed of dense protein contact maps, semi-dense protein inter-action networks, and sparse chemical structures. Results show that different combinations of the properties of pattern and target graphs lead the proposed strategies to differentially outperform. Results of this study are the base of new algorithms formulation for SubGI solutions.

2   Materials and Methods

2.1   Basic Notions

A graph is a pair $G = (V, E)$, where V is the set of vertices and E is the set of edges connecting them. Given two vertices $v, v' \in V$, the pair (v, v') represents an undirected edge. The neighbourhood $N(v) = \{v' \in V : (v, v') \in E\}$ of a vertex v is the set of the vertices connected to v, and the neighbourhood of a set of vertices $V' \subseteq V$ is given by $N(V') = \{\bigcup_{v \in V'} N(v) \setminus V'\}$. The degree of a vertex v is $|N(v)|$ i.e. the number of neighbours of v. A labeled graph is defined as a triple $G = (V, E, \alpha)$, where $\alpha : V \mapsto \Sigma_V$ is a injective function that maps vertices to the set of labels Σ_V. The density of a graph is defined as $D = |E|/((|V|)(|V| - 1))$.

Given a pattern graph $G_p = (V_p, E_p, \alpha_p)$ and a target graph $G_t = (V_t, E_t, \alpha_t)$, the subgraph isomorphism problem (SubGI) is to find a injective function $f : V_p \mapsto V_t$ that maps each vertex of G_p in G_t by preserving the pattern topology and vertex label compatibility. In particular, for each pattern vertex v, the label associated with it must be compatible to the label associated to the target vertex $f(v)$ with which it is matched, thus $\alpha_p(v) = \alpha_t(f(v))$. Pattern topology is preserved by forcing the existence of pattern edges among target vertices, thus for each edge $(v, v') \in E_p$ the edge $(f(v), f(v'))$ must resides in E_t.

2.2   Variable Ordering in Subgraph Isomorphism

Given an ordering $\mu_p = (u_p^1, u_p^2, \ldots, u_p^{|V_p|})$ of the vertices of V_p, the SubGI problem can be defined as the combinatorial problem of finding all possible rearrangements of size $|V_p|$ of the target vertices, $\mu_t = (u_t^1, u_t^2, \ldots, u_t^{|V_t|})$. A mapping can be represented as the ordered set of pairs $M = \{(u_p^1, u_t^1), (u_p^2, u_t^2), \ldots, (u_p^{|V_p|}, u_t^{|V_p|})\}$. The set of all possible mappings constitutes the search space of the problem. The space is commonly represented as a tree, where mapping solutions are branched together by the prefixing partial solution they share. A backtracking approach is used to visit the search-space tree (SST) in order to extend partial solutions is case of positive evaluation. Negative evaluations, that fail to satisfy SubGI conditions, are pruned and the algorithm backtracks to a parent partial solution.

Vertex ordering is also referred as *variable ordering* by analogy with the constraint satisfaction problem where values (target vertices) are assigned to variables (pattern vertices) and constraints are asserted [8]. Here constraints are the SubGI rules. The chosen preference to explore assignments of a specific variable w.r.t. other pattern vertices substantially reduces the set of values that can be assigned to forward variables.

A strategy for ordering variables may rely on features of both pattern and target graphs (e.g. topology, label properties). The frequency of the labels along the target graphs and the probability of pattern degree within the target, improve the power of the search strategy [7]. Moreover, properties of partial ordering are crucial, too [4]. A central concept adopted in several strategies is the *fail-first* principle, which states that variables having a high chance to produced pruning operations over the SST are queued early in the ordering. The concept is often implemented by giving preference to the *most constraint* variables according to the subgraph isomorphism rules.

Given a partial ordering $\mu_p^i = (u_p^1, u_p^2, \ldots, u_p^i)$, RI [4] applies a three-stages evaluation rules to choose the next vertex u_p^{i+1} to be queued to the ordering. At first, the vertex that maximizes the edges towards μ_p^i, namely the cardinality of the edge set $\{(u_p^{i+1}, u') \in E_p : u' \in \mu_p^i\}$, is preferred. If two or more vertices have the same number of connections, then their set of edges toward $N(\mu_p^i)$ are taken into account. The total vertex degree is take as final stage of discrimination between the vertices.

2.3   Ordering Strategies by Weighted Centralities

The RI approach does not explore global structures of the pattern graph, neither it relies on any target aware strategy. However, global structures and target information may improve effectiveness in evaluating future constraints [9]. Therefore, we investigated seven centrality graph measures that are commonly applied to the analysis of biological networks [10] and for each of them, we developed a modified version of the RI algorithm such that the centrality measure computed in the pattern graph is evaluated at first and, in case of equality in the variable ordering, the original RI ordering rules are applied.

The centrality measures can be divided in three groups:

– eigenvector-based measures (i.e., eigenvector centrality EIG, subgraph centrality SUB),
– local measures of the graphs (i.e., local-average-based centrality LAC, network centrality NC),

– centrality measures based on paths (i.e., information centrality INFO, closeness centrality CL, betweenness centrality BET).

We use weighted centrality measures, defining the weight of an edge of the
pattern according to the frequency of the pair of labels of its vertices in the
target graph. Given a pattern edge (v, v'), the weight assigned to it equals
$freq_t(\alpha_t(f(v)), \alpha_t(f(v')))$. The frequency of a pair of vertex labels, $l_1, l_2 \in
\Sigma_V$, linked by a given edge belonging to the target graph, is computed as:
$freq_t(l_1, l_2) = |\{(u, u') \in E_t : \alpha_t(u) = l_1 \text{ and } \alpha_t(u') = l_2\}|/|E_t|$. In this way,
strategies can take advantage of target properties, a behaviour that is missing
in the original formulation of RI.

Let $c(v)$ to be the centrality measure of a vertex $v \in G_p$, computed by one
the centrality measure listed above. The vertex u_p^{i+1} to be queued to the variable
ordering is the one, within the set $N(\mu_i^p)$, who maximizes the value $c(u_p^{i+1})$. In
case of vertices with equal maximal centrality value, the original RI's precedence
rules are applied.

2.4   Data

We evaluated performances of the investigated strategies over a widely used
benchmark of biological graphs [4]. We have taken into account three types of
biological graphs which differ in their structural and labeling properties.

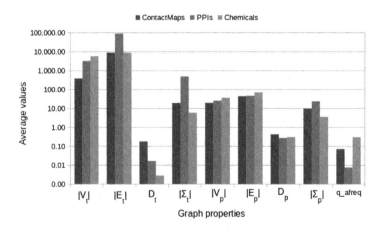

Fig. 1. Statistics regarding the properties of the analyzed datasets are shown in terms
of average number of vertices, edges and labels of target and pattern graphs, plus
densities and average frequency of pattern graph labels w.r.t. the corresponding target
graphs.

Protein contact maps are dense graphs representing proximity of protein
peptides, thus labels on vertices are the corresponding residues. Chemical protein structures represent chemical bonds between atoms (graph vertices labeled

with their atomic symbol) by means of sparse graphs. Protein-protein interaction (PPI) networks are physical interactions among proteins. PPIs are semi-sparse graphs where vertices represent proteins. PPI vertex labels are randomly assigned by varying the number of distinct labels from 1 to 2048.

Labels are provided in textual form by the input file format and converted to numeric identifiers in a preprocessing step. Each node is equipped with just one label. Pattern graphs are extracted from the benchmark by varying the number of pattern edges, from 8 to 256. An equal number of patterns has been extracted for each edge amount.

Figure 1 shows the compositional properties of the benchmark graphs. Pure structural properties are reported in terms of number of vertices, number of edges, graph density, and number of distinct labels. We also report an important feature in solving SubGI, mentioned as *q_afreq*, which reports the average frequency of the labels belonging to the pattern graph calculated within the target graph.

3   Results

Tests were performed by searching a pattern within the target from which it was extracted, and by putting a timeout of 10 min for the execution. We evaluated the proposed ordering strategies by calculating the number of times an approach is recorded as the fastest solution, and by summarizing execution times by means of clustering approaches.

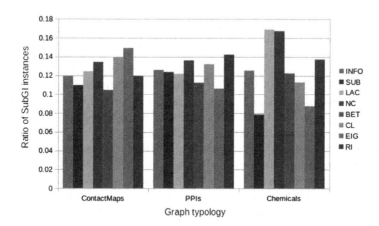

Fig. 2. The figure shows the ratio of SubGI instances in which each strategy has been the fastest solution. Instances are grouped by the type of graph.

Figure 2 shows the number of instances (expressed as ratio over the complete benchmark) for which each strategy outperformed all the others. Figure 3 details these results per graph type. In contact maps, the EIG strategy produces the

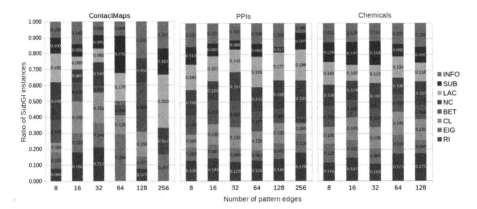

Fig. 3. The figure shows for each methodology the ratio of SubGI instances for which it has been registered has the fastest strategy (shown on top of the bars). Instances are grouped by graph typology and number of pattern edges (from 8 to 256). Patterns of 256 edges from chemical graphs were discarded because the sparsity of such graphs the extracted subgraphs tend to degenerate into simple paths.

ordering that performs at the best, followed by the closeness centrality (CL). The strategies (see Fig. 3) show stable trends on PPIs and chemical graphs. The RI and NC orderings are the most effective over PPI networks. Local strategies, such as LAC and NC, show clear predominance on chemical graphs. Thus, local orderings are more suitable in sparse and semi-sparse graphs, while global measurements provide more effective orderings for dense graphs. Unexpectedly, the BET and CL strategies have opposite trends compared to each other. In fact, the outcomes of the BET strategy increase with the graph sparsity, while CL tends to be more effective on dense graphs. Beside BET and CL, the EIG strategy shows clear dependence from the decreasing of graph density, and the other strategies do not seem to be linearly correlated with this graph property.

Performance similarities of related centralities emerge from two-samples statistical tests shown in Fig. 4 (left side). Clusters of statistical dependence clearly aggregate strategies depending on the type of centrality they are based on. Differences emerge for path-based centralities, in particular, the betweenness centrality seems to be more similar to local orderings rather than path-based strategies. Such a similarity is shown by the clustering performed in Fig. 4 (right side). Running times regarding each SubIG are firstly normalized by the time needed to solve it by the slowest strategy, then similarity between two strategies is computed by summing the absolute time difference along the whole benchmark.

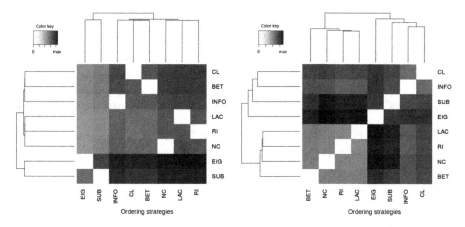

Fig. 4. On the left side, a heat map shows correlations between the investigated strategies by means of the Mann-Whitney U test applied to the running times. On the right side, similarities among strategies are computed by summing the absolute differences of their normalized running times. Normalization is performed by setting to 1 the slowest performance registered for each SubGI instance. U statistics and absolute time differences are used to perform hierarchical clusterings of the strategies.

4   Conclusion

We investigated alternatives to a well-established approach for solving the subgraph isomorphism problem (SubGI). The RI approach exploits a search strategy based on an effective ordering of the pattern graph in order to address the *fail first* and *most constrained* principles. The strategy is inherently local and does not explore wide properties of the pattern graph. In addition, the strategy does not take into account an important aspect for SubGI, namely how frequently labels on pattern vertices appear on the target graph. We evaluated both local strategies taking into account label frequency and strategies based on global centrality. The analysis was performed on three representative benchmarks of biological graphs to define, in the perspective of a multi-approach environment, the strategy to chose depending on the properties of the graphs involved. Local strategies predominate on sparse target graphs and closeness- and eigenvector-based strategies on dense graphs.

Acknowledgments. This work has been partially supported by the following projects: GNCS-INDAM, Fondo Sociale Europeo, and National Research Council Flagship Projects Interomics. This work has been partially supported by the project of the Italian Ministry of education, Universities and Research (MIUR) "Dipartimenti di Eccellenza 2018–2022".

References

1. Gifford, E., Johnson, M., Smith, D., Tsai, C.: Structure-reactivity maps as a tool for visualizing xenobiotic structure-reactivity. Netw. Sci. **2**, 1–33 (1996)

2. Milo, R., Shen-Orr, S., Itzkovitz, S., Kashtan, N., Chklovskii, D., Alon, U.: Network motifs: simple building blocks of complex networks. Science **298**(5594), 824–827 (2002)

3. Tian, Y., Mceachin, R.C., Santos, C., States, D.J., Patel, J.M.: SAGA: a subgraph matching tool for biological graphs. Bioinformatics **23**(2), 232–239 (2006)

4. Bonnici, V., Giugno, R.: On the variable ordering in subgraph isomorphism algorithms. IEEE/ACM Trans. Comput. Biol. Bioinform. (TCBB) **14**(1), 193–203 (2017)

5. McCreesh, C., Prosser, P., Solnon, C., Trimble, J.: When subgraph isomorphism is really hard, and why this matters for graph databases. J. Artif. Intell. Res. **61**, 723–759 (2018)

6. Aparo, A., et al.: Simple pattern-only heuristics lead to fast subgraph matching strategies on very large networks. In: Fdez-Riverola, F., Mohamad, M., Rocha, M., De Paz, J., González, P. (eds.) PACBB 2018. AISC, vol. 803, pp. 131–138. Springer, Heidelberg (2018). https://doi.org/10.1007/978-3-319-98702-6_16

7. Carletti, V., Foggia, P., Saggese, A., Vento, M.: Introducing VF3: a new algorithm for subgraph isomorphism. In: Foggia, P., Liu, C.L., Vento, M. (eds.) GbRPR 2017. LNCS, vol. 10310, pp. 128–139. Springer, Cham (2017). https://doi.org/10.1007/978-3-319-58961-9_12

8. Solnon, C.: AllDifferent-based filtering for subgraph isomorphism. Artif. Intell. **174**(12–13), 850–864 (2010)

9. Tarjan, R.E., Yannakakis, M.: Simple linear-time algorithms to test chordality of graphs, test acyclicity of hypergraphs, and selectively reduce acyclic hypergraphs. SIAM J. Comput. **13**(3), 566–579 (1984)

10. Tang, Y., Li, M., Wang, J., Pan, Y., Wu, F.X.: CytoNCA: a cytoscape plugin for centrality analysis and evaluation of protein interaction networks. Biosystems **127**, 67–72 (2015)

Structure-Based Antibody Paratope Prediction with 3D Zernike Descriptors and SVM

Sebastian Daberdaku$^{(\boxtimes)}$ (iD)

Department of Information Engineering, University of Padova,
Via Gradenigo 6/A, 35131 Padua, PD, Italy
sebastian.daberdaku@unipd.it

Abstract. Antibodies currently represent the most valuable category of biopharmaceuticals for both diagnostic and therapeutic applications. They are a class of Y-shaped proteins capable of specifically recognizing and binding to a virtually infinite number of antigens. Being able to identify the antigen-binding residues in an antibody structure is crucial for all antibody design methods and for shedding light on the complex mechanisms that govern antigen recognition and binding. This paper presents a method for antibody interface prediction from their experimentally-solved structures based on 3D Zernike Descriptors. Roto-translationally invariant descriptors are computed from circular patches of the antibody surface enriched with a chosen subset of physicochemical properties from the AAindex1 amino acid index set, and are used as samples for a binary classification problem. An SVM classifier is used to distinguish interface surface patches from non-interface ones. The proposed method outperforms other antigen-binding interface prediction software, namely Paratome, Antibody i-Patch and Parapred, on a novel dataset of experimentally-solved antibody–antigen complex structures.

Keywords: Antibody interface · Paratope · 3D Zernike descriptors · SVM · AAindex

1 Introduction

Antibodies (Abs), also known as immunoglobulins (Igs), are proteins that play a pivotal role in the adaptive immune response of vertebrates [1] as they bind to a potentially infinite number of antigens (Ags) with very high affinity and specificity. They are a class of Y-shaped, flexible proteins consisting of two heavy (H) and two light (L) chains, linked together by disulfide bonds (see Fig. 1). The light and heavy chains are composed of constant domains (C) that determine the functional properties of the antibody, and variable domains (V) that bind the antigens [2]. The heavy chain portion of the C domains can be used to categorise antibodies into five main classes or isotypes, each with its own biological properties, i.e. IgA, IgG, IgD, IgE and IgM. Moreover, IgG can be further split into

© Springer Nature Switzerland AG 2020
M. Raposo et al. (Eds.): CIBB 2018, LNBI 11925, pp. 27–49, 2020.
https://doi.org/10.1007/978-3-030-34585-3_4

four subclasses, i.e. IgG1, IgG2, IgG3, and IgG4; and IgA can similarly be split into IgA1 and IgA2 [3]. The isotype determines the antibody's capacity to reach a certain area (for instance, by diffusion or by engaging specific transporters that deliver them across epithelia), and initiate the adequate effector mechanism (i.e. trigger an immediate allergic reaction) [4]. In both H and L chains, the V domains are composed of three portions of variable sequence known as Complementarity Determining Regions (CDRs) and four portions of relatively constant sequence known as Framework Regions (FRs) [3].

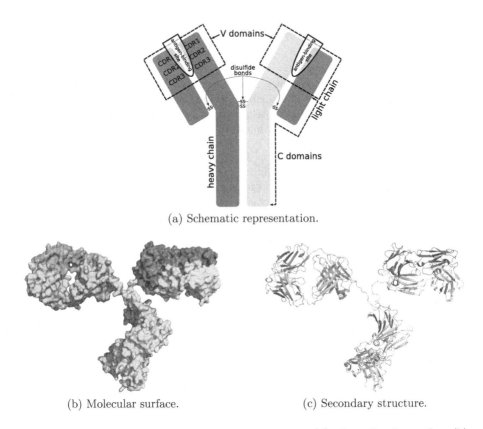

(a) Schematic representation.

(b) Molecular surface.   (c) Secondary structure.

Fig. 1. The figure depicts the schematic representation (a), the molecular surface (b) and the secondary structure (c) of an antibody (PDB entry *1igt*). The heavy chains are coloured in blue and yellow while the light chains are coloured in green and pink. (Color figure online)

Many antigens are rendered harmless simply by binding with a specific antibody, as in the case of some bacterial toxins or viral particles which are neutralised upon recognition. Antibodies can also mark their target pathogens for destruction by rendering them more susceptible to phagocytosis, or by activating the complement family of proteins which leads to cell lysis [2]. The large repertoire of V domains, in combination with the various isotypes (which can potentially be switched to allow altered effector function while maintaining antigen specificity), enables antibodies to bind to a potentially infinite number of antigens with very high affinity and specificity. The wide range of possible V domains results from a complex series of gene rearrangement events, and from somatic hypermutation after repeated exposures to the same antigen to allow affinity maturation.

Because of their peculiar antigen-binding properties, there has been an ever increasing interest in antibodies which has led to the development of several antibody-based drugs and antibody design methods [5–7]. Currently, antibodies are the most important class of biopharmaceuticals [8,9]. The most notable application is probably the development of Antibody-drug Conjugates (ADGs), i.e. monoclonal antibodies covalently linked to small drug molecules [10]. ADGs are designed to exploit the targeting ability of antibodies in order to precisely release the drug molecule on a very specific destination, such as cancer cells [7,11], in order to reduce systemic toxicity.

Early efforts to produce therapeutic antibodies consisted in raising antibodies against a target antigen in immunised mice. This methodology suffered from low clinical success rate as mouse antibodies were often highly immunogenic in humans [12]. Chimerisation was later introduced to overcome these limitations: "humanised" immunoglobulins were obtained by grafting the V domains of animal antibodies to the C domains of human ones, inducing considerably less response by the human immune system [13]. Currently, the most widely used technique known as phage display [14] can produce potent antibodies which are completely human. Despite all these advances in the antibody design methods, current experimental techniques exhibit several limitations, such as the inability to target a specific antigen or to employ rationally-engineered mutations without triggering immunogenic response [15]. On the other hand, computational methods could overcome these shortcomings and even provide a general methodology for antibody design. Several computational methods can aid in the design of antibodies: protein design algorithms as well as antibody-specific modelling techniques [16]. These methods have been successfully employed to improve antibody binding affinity [17–19], humanisation [20], homology modelling [21,22] and antibody *de novo* design [6,15].

This whole multitude of applications would greatly benefit from the ability to precisely pinpoint the antigen-binding residues (paratope) on a given antibody. This paper describes an Ab paratope prediction method based on an extended notion of local surface similarity that takes into account geometric information enriched with physicochemical properties. Circular patches of the antibody surface are classified by an SVM to distinguish interface surface patches from non-interface ones. This is an extension of the previous works on Ab paratope identification using 3D Zernike Descriptors and SVM [23,24]. Here, a new set of physicochemical features, specific to the antibody paratope prediction task, has been identified. A novel, larger set of representative and non-redundant bound antibody–antigen complex structures was employed to train the SVM model, and the resolution of the 3D Zernike descriptors was adjusted according to previous findings. The proposed method was experimentally validated on a novel testing set, composed of non-redundant Ab structures (153 Ab structures with protein antigens and 123 Ab structures with non-protein antigens) which have never been used to train any of the previously introduced predictors. The performance of the proposed method was compared with those of other antigen-binding interface prediction software, namely Paratome, Antibody i-Patch and Parapred.

1.1   Algorithm Overview

The workflow of the algorithm presented in this work is given in Fig. 2. The prediction method starts from the given experimentally-determined 3D structures (PDB files) of Abs. In order to effectively discriminate between interacting and non-interacting sites, a set of twenty amino acid indices of physicochemical and biochemical properties extracted from the AAindex1 dataset, specifically selected for the paratope prediction task, was employed. These indices are mapped onto the voxelised representation of the antibody surface, obtaining a geometrical representation of the latter enriched with the physicochemical and biochemical properties of the underlying residues. Spherical patches are then uniformly sampled from the antibody surface and, for each patch, a rotationally-invariant local descriptor based on 3D Zernike moments is computed. The 3D Zernike descriptors (3DZDs) possess several attractive features such as a compact representation, roto-translational invariance, and have been shown to adequately capture global and local protein surface shape [25–28] and to naturally represent physicochemical properties on the molecular surface [29].

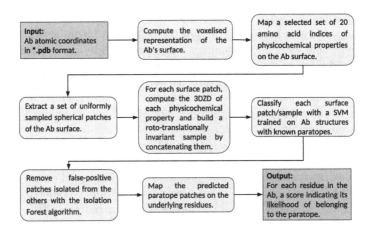

Fig. 2. Workflow of the Ab interface prediction algorithm.

3DZDs are employed to quickly evaluate the shape and physicochemical similarity of Local Surface Patches (LSPs), since similar patches have similar descriptors. The resulting descriptors are then used as samples for a binary classification problem: Support Vector Machines are used as a classifier to distinguish interface LSPs (surface patches belonging to the paratope) from non-interface ones. Since the antigen-binding regions in antibodies exhibit a certain spatial continuity, the Isolation Forest algorithm for outlier detection is used to discard false-positive patches isolated from the others. Finally, the predicted paratope patches are mapped on the underlying residues. Each residue is assigned a score indicating its likelihood of belonging to the binding region: a residue can be identified as belonging to the paratope only if its score reaches a minimum threshold value.

2   Scientific Background

The CDRs have been long considered as the locus of the paratope, so extensive research has been carried for their determination in the last decades [30–33]. Kabat et al. provided the means to systematically identify CDRs in antibodies with multiple sequence alignments (MSAs) [30,34]. This approach exploited the assumption that CDRs were composed of the most variable portions of the antibody sequence and thus could be identified by numbering the residues in the hypervariable regions according to the MSAs (i.e. the Kabat numbering system). Later, Chothia and co-workers included structural considerations to the Kabat numbering system, redefining the location of insertions and deletions sites in CDRs [31,33,35]. The boundaries of the CDR and FR regions were shown to adopt a restricted number of conformations depending on the presence of certain residues at key points in the CDRs and surrounding FRs. Lefranc et al. introduced the International Immunogenetics Information System (IMGT) database [36] which curates sequence information not only for antibodies, but also T-cell receptors, major histocompatibility complex (MHC) and related proteins of the immune system of human and other vertebrates. The IMGT numbering system combines the Kabat definition of FRs and CDRs with Chothia's characterization of the hypervariable loops and structural information.

Instead of region-wide annotations such as CDR definitions, in recent years there has been an increasing interest in developing methodologies that predict specific antigen-binding residues [8]. On average, CDRs contain between 40 and 50 residues, while the paratope is composed of only 18–19 residues [37], whose majority is located within the CDRs. Moreover, in [38,39], Kunik et al. have shown that CDRs contain on average only about 80% of the overall antigen-binding residues in the antibody and that the remaining ones are usually found in the immediate vicinity. They developed Paratome, a method for the identification of antigen-binding regions (ABRs) from the sequence or 3D structure of an antibody, which is based on the assumption that most of the antigen-binding residues are located in regions of structural consensus between antibodies. The algorithm derives the structural consensus regions from multiple structure alignment (MSTA) of a non-redundant set of antibody–antigen complexes to identify ABRs in other antibodies. It is worth noticing that the definition of ABR is not a refinement of the concept of CDR, but rather a complementary concept focusing on a different aspect of the antibody-binding site. Another recent method for the prediction of paratope residues is proABC [40]. ProABC is based on a machine-learning method (Random Forest) trained on sequence and sequence-derived features. Using the sequence alone as input, for each amino acid in the antibody, the algorithm estimates its interaction probability with the cognate antigen. Three different types of interaction are considered and predicted separately: hydrogen bond, hydrophobic and other non-bonded interactions.

In [41], Krawczyk et al. introduced Antibody i-Patch, a statistical method which relies on the antibody structure. The method annotates each residue with a binding likelihood score using antibody-specific statistics. This allows the users to differentiate between higher and lower confidence predictions, which can be used to trade off between higher precision or better coverage and potentially as a guide for introducing mutations to the CDR region. Antibody i-Patch was shown to work well also when only the homology model of the antibody is provided as input. In [42], the In-silico Molecular Biology Lab protein-protein interaction (ISMBLab-PPI) prediction algorithm, previously introduced in [43], was used to identify antigen-binding sites on the surfaces of antibodies using general protein–protein interaction principles. The physicochemical complementarity of protein–protein binding surfaces is described with several 3D probability density maps, each characterising the distribution probability of a protein atom type. Artificial neural network machine learning models were trained for each protein atom type, and the prediction of the antigen-binding site was obtained by combining the output of all models. Because its training set is not restrained only to antibody-antigen complexes, this method was shown to underperform if compared to specific antigen-binding prediction methods such as proABC. More recently, [44] developed the Parapred method, a sequence-based probabilistic machine learning algorithm for paratope prediction which relies on a deep-learning architecture to leverage features from both local residue neighbourhoods and across the entire sequence.

The general idea that the knowledge of the CDRs is sufficient for performing Ab rational design and engineering through targeted mutagenesis is probably the main reason why the field of paratope residue prediction has remained so underdeveloped [8]. Although small CDR regions with 6–8 residues or less can be targeted for saturation mutagenesis, for longer sequences this process becomes not tractable due to the limitation in the phage library sizes that may be generated and surveyed with confidence [45]. It is now clear that knowing the role played by each specific residue in the binding of the antigen is a fundamental aspect in antibody rational design [40]. The current available methods tend to predict almost all paratope residues (high recall) with low precision [46]. On the other hand, knowing fewer binding residues but with a higher precision will be beneficial for guiding mutations in antibody engineering, since minor mutations to the binding site can lead to significant changes to both the affinity and specificity of a given antibody [41]. In fact, numerous studies have revealed that, in most cases, amino acid substitutions in a few specific positions (typically no more than three or four) in the targeted CDR are sufficient to significantly improve affinity [45].

3   Materials and Methods

The methodology presented in this work was derived from a previous study on protein–protein interface prediction based on 3D Zernike Descriptors [47], where one of the main conclusions was that the protein interface prediction task should be addressed separately for each protein class, since the physicochemical properties that discriminate the protein interaction interface from the rest of the protein surface can vary widely from one protein class to another. Taking into account this result, the presented Ab interface prediction method was tailored to specifically address the paratope prediction task. A new set of physicochemical properties specific to the antibody interface identification task was identified, and a representative and non-redundant set of bound antibody–antigen complex structures was selected to train the SVM model. This section briefly describes the methodology, highlighting the differences with the previous study.

3.1   Antibody Surface Representation

The voxelised representation of the Solvent Excluded surface (SES) [48] was employed in this work. A voxel (volumetric pixel) represents a single, discrete data point on a regular grid in the 3D space. Voxelised surface representations can be used to store multiple data in each voxel in order to represent various properties of a certain portion of space in a simple and effective way.

The voxelised SES of antibodies were computed with the region-growing Euclidean distance transform methodology described in [49,50] at a resolution of 64 voxels per $Å^3$, using a 1.4 Å radius for the solvent probe. Patch centres were extracted from each protein surface uniformly, at a minimum separation of 1.8 Å, while LSPs were extracted using a sphere with a 6.0 Å radius [28,51,52] centred at each patch centre, which ensures plenty overlap among neighbouring patches.

3.2   Paratope Definition

The recognition of paratope regions can be seen as a classification problem, i.e., each LSP is assigned to one of the two classes: *interface* or *non-interface* surface patches. Consequently, the problem may be solved using machine learning techniques such as Support Vector Machines. Given an Ab–Ag complex with known 3D structure, an Ab residue is said to belong to the paratope if at least one of its heavy atoms is located within 4.5 Å from any Ag heavy atom (the same threshold was used in all previously developed predictors [38,41,44]). Since the selection of the distance cut-off value was shown not to significantly affect the prediction performance of paratope residues [38] and protein–protein interaction interface residues in general [8,53], in this work to the most commonly used threshold value is employed. The interface surface is defined as the portion of the Ab's SES corresponding to the paratope residues. Finally, a patch is an interface patch if its centre is located in the Ab's interface surface, otherwise it is categorised as a non-interface patch.

3.3   Datasets

In order to train and assess the prediction quality of the proposed model in comparison with other paratope prediction methods, the training and testing datasets were built as follows. First, the structures from the training datasets used to train the Paratome, Antibody i-Patch and Parapred predictors were collected. Because of the inability to determine the Abs used to train the ProABC predictor, it was excluded from the comparison. These sets were joined into one large training set. All the Ab–Ag complex structures present in the AbDb database [54] were selected to be used as testing set. AbDb has been recently introduced, and is the most comprehensive database of antibodies with known 3D structure to date. The Ab–Ag complexes present in AbDb (version 15/06/2018) are split into two categories depending on whether their antigen is a protein or not, and this separation was kept in the testing set. All the structures from the testing set having a sequence identity of over 95% with any Ab in the previously identified training set were removed using CD-HIT [55,56]. This ensures that no structure in the testing set was ever present in the training set of any of the paratope predictors under consideration. All structures with a resolution worse than $3.0\,\text{Å}$ were also removed, and the redundancy was further reduced so that no two Abs have more than 95% sequence identity, for both the training and testing sets. The training set was further split into two disjoint sets: a reduced training set and a development/validation set used to tune the various aspects of the predictive model. The reduced training set has 213 Ab structures, the development set contains 106 Ab structures, and the testing set contains 153 Ab structures with protein antigens and 123 Ab structures with non-protein antigens.

The paratope generally corresponds to a small portion of an Ab's surface, thus, a uniform sampling of the surface will result in a highly-imbalanced classification problem where the interface patches are the minority class. A combination of undersampling of the majority class and oversampling of the minority class were used in this work in order to balance the training set. The surface of each Ab in the training set was first sampled into LSPs with a minimum separation of $3.0\,\text{Å}$ between patch centres. Then, only the interface regions were sampled with a minimum separation of $1.0\,\text{Å}$ between patch centres. This procedure yields more balanced training sets and guarantees that both the interface and non-interface Ab surface regions are sampled in a fairly uniform fashion. For the training set, a total of 307 658 samples were obtained, 142 380 (46.3%) of which are positive (belonging to the paratope) and the remaining 165 278 (53.7%) are negative. The test and validation samples, on the other hand, were obtained by uniformly sampling the surfaces of the proteins in the test set with a minimum separation of $1.8\,\text{Å}$ between patch centres, thus retaining the native distribution of positive and negative samples. From the Abs of the test set with protein Ags, 250 382 (91.3%) negative samples were obtained and only 23 983 (8.7%) positive ones; from the Abs of the test set with non-protein Ags, 212 262 (96.1%) negative samples were obtained and only 8 711 (3.9%) positive ones. Similarly, from the validation set 182 065 (91.3%) negative samples and only 17 339 (8.7%) positive ones were obtained.

Since the proposed paratope prediction algorithm is structure based, and Ab structures might not be always available, a set of homology based modelled Ab structures was built in order to evaluate the prediction capabilities of the proposed method when only Ab sequence data is known. ABodyBuilder, a fully automated Ab structure prediction server [57], was employed to generate modelled structures of all the Abs in the test set using only their sequences. Since all the Ab structures in the datasets are already present in the PDB, for each Ab, all PDB IDs sharing a sequence identity of over 95% with it were excluded from the template search (including the query Ab). ABodyBuilder was chosen among other state-of-the-art Ab structure predictors because it gives the possibility to exclude any user-specified structures from the template search. The exclusion of all structures sharing too much sequence identity with the query Ab ensures that reasonably precise homology based modelled structures are obtained, as if their native structures were actually unavailable.

3.4 Feature Selection

In order to reliably predict paratope residues, the physicochemical characteristics (features) that can best discriminate between interacting and non-interacting sites must be identified [58]. Twenty physicochemical properties were selected in order to distinguish interface interaction regions from non-interface ones in antibodies. These properties were extracted from the AAindex [59]: a database of numerical indices representing various physicochemical and biochemical properties of residues and residue pairs derived from published literature. An amino acid index is a set of 20 numerical values representing any of the different properties of each amino acid: the AAindex1 section of the database is a collection of such indices. From the 566 indices in the AAindex v9.2, the ones that did not have a defined numerical value for all 20 amino acids were excluded, reducing the list to 545 entries. For all Abs in the development set, each residue was assigned a $+1$ label when identified as interface and a -1 label otherwise, together with the corresponding 545 physicochemical properties from the AAindex1. This way, 25 834 samples were obtained, 2 292 of which positive and the remaining 23 542 negative.

These samples were used as input to a feature selection method known as Randomised Logistic Regression [60] in order to reduce the number of features to a subset of relevant ones. This method works by sub-sampling the training data and fitting a L1-regularised Logistic Regression model where the penalty of a random subset of coefficients has been scaled. By performing this double randomization several times, the method assigns high scores to features that are repeatedly selected across randomizations. The implementation of the Randomised Logistic Regression method provided in scikit-learn was used. By setting the number of repetitions to 2000 and the minimum score threshold to 0.3, a set of 27 best-scoring indices was identified (see Fig. 3). This set was further reduced

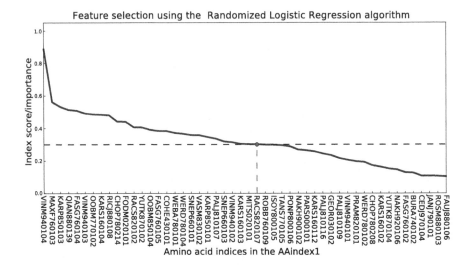

Fig. 3. Plot of the feature importance/score for the top 50 indices.

Table 1. The list of selected features ordered by decreasing importance.

AAindex ID	Description
VINM940104	Normalized flexibility parameters (B-values) for each residue surrounded by two rigid neighbours
MAXF760103	Normalized frequency of zeta R
KARP850103	Flexibility parameter for two rigid neighbors
QIAN880139	Weights for coil at the window position of 6
FASG760104	pK-N
OOBM770102	Short and medium range non-bonded energy per atom
RICJ880108	Relative preference value at N5
CHOP780214	Frequency of the 3rd residue in turn
FODM020101	Propensity of amino acids within pi-helices
RACS820102	Average relative fractional occurrence in AR(i)
YUTK870102	Unfolding Gibbs energy in water, pH9.0
OOBM850104	Optimized average non-bonded energy per atom
FASG760105	pK-C
WEBA780101	RF value in high salt chromatography
WERD780104	Free energy change of epsilon(i) to alpha(Rh)
SNEP660101	Principal component I
VASM830102	Relative population of conformational state C
KARP850101	Flexibility parameter for no rigid neighbors
PALJ810107	Normalized frequency of alpha-helix in all-alpha class
RACS820107	Average relative fractional occurrence in A0(i − 1)

to 20 indices by removing redundancy so that no two indices have a Pearson correlation coefficient of over 70% in absolute value: when selecting among highly correlated indices, the one which obtained the highest score during the Randomised Logistic Regression step was kept. The AAindex accession number and the description of each selected index is given in Table 1.

3.5 3D Zernike Descriptors

The 3D Zernike descriptors were first used as a representation of the protein surface shape in [25], and have since been employed in several bioinformatics and computational biology tasks given their advantages over other surface representations. 3DZDs can represent protein surfaces very compactly as a vector of numbers and are invariant to roto-translations. Because of this property, time-consuming spatial alignments are not required and the descriptors can be pre-computed and stored. They can also represent physicochemical properties on the molecular surface [29,61]. Lastly, by changing the order of the series expansion, the resolution of the surface representation can be easily controlled.

In what follows, a brief description of the 3DZD is provided. Refer to [62] for the exhaustive mathematical derivation and to [63] for the implementation details. The 3D Zernike functions Z_{nl}^m of order n and repetition m are defined as:

$$Z_{nl}^m(r, \theta, \phi) = R_{nl}(r) \cdot Y_l^m(\theta, \phi). \tag{1}$$

$Y_l^m(\theta, \phi)$ are the spherical harmonics in polar coordinates of lth degree, where $l \leq n$, $-l \leq m \leq l$, with $n - l$ an even number. $R_{nl}(r)$ are the radial polynomials of radius r which guarantee the orthonormality of the $Z_{nl}^m(r, \theta, \phi)$ polynomials in Cartesian coordinates. The expression of Z_{nl}^m can be rewritten in Cartesian coordinates in a compact form as a linear combination of monomials of order up to n:

$$Z_{nl}^m(\boldsymbol{x}) = \sum_{r+s+t \leq n} \chi_{nlm}^{rst} \cdot x^r y^s z^t. \tag{2}$$

The 3D Zernike moments Ω_{nl}^m of function $f(\boldsymbol{x}), \boldsymbol{x} \in \mathbb{R}^3$ are defined as:

$$\Omega_{nl}^m := \frac{3}{4\pi} \int_{|\boldsymbol{x}| \leq 1} f(\boldsymbol{x}) \overline{Z_{nl}^m(\boldsymbol{x})} d\boldsymbol{x}. \tag{3}$$

Using Eq. 2, the 3D Zernike moments Ω_{nl}^m of an object can be written as a linear combination of geometric moments of order up to n:

$$\Omega_{nl}^m = \frac{3}{4\pi} \cdot \sum_{r+s+t \leq n} \overline{\chi_{nlm}^{rst}} \cdot M_{rst}, \tag{4}$$

where M_{rst} is the geometric moment of the object scaled to fit in the unit ball. The 3D Zernike moments Ω_{nl}^m are not invariant under rotations. In order to achieve invariance, moments are collected into $(2l + 1)$-dimensional vectors

$\boldsymbol{\Omega}_{nl} = \left(\Omega_{nl}^{l}, \Omega_{nl}^{l-1}, \Omega_{nl}^{l-2}, \ldots, \Omega_{nl}^{-l}\right)^{\top}$, and the rotationally invariant 3D Zernike descriptors F_{nl} are defined as norms of vectors $\boldsymbol{\Omega}_{nl}$:

$$F_{nl} := \|\boldsymbol{\Omega}_{nl}\|. \tag{5}$$

Given the maximum moment order N, the number of 3D Zernike descriptors can be easily determined by using the following formula:

$$\text{No. 3DZDs} = \begin{cases} \left(\frac{N+2}{2}\right)^2, & \text{if } N \text{ is even} \\ \frac{(N+1)(N+3)}{4}, & \text{if } N \text{ is odd.} \end{cases} \tag{6}$$

3.6   Patch Representation Using 3D Zernike Descriptors

The selected set of twenty physicochemical and biochemical properties are mapped on the voxelised representation of the Ab's SES. Let A_{Ab} be the set of atoms in the current Ab, and let $\Phi_i : A_{Ab} \to \mathbb{R}$ be the function which assigns to each atom the numeric value of the corresponding amino acid index i. Then, for a given index i, the corresponding property is mapped on the $SES(Ab)$ according to the following 3D function:

$$f_i(\boldsymbol{v}) = \begin{cases} \displaystyle\sum_{a \in A_{Ab}} \frac{\Phi_i(a)}{r_a} \mathbb{1}_a(\boldsymbol{v}), & \text{if } \boldsymbol{v} \in SES(Ab) \\ 0, & \text{if } \boldsymbol{v} \notin SES(Ab), \end{cases} \tag{7}$$

where r_a is the radius of atom a, and $\mathbb{1}_a(\boldsymbol{v})$ is the indicator function for atom a, which is 1 if voxel $\boldsymbol{v} \in a$ and 0 if $\boldsymbol{v} \notin a$.

Zernike descriptors cannot distinguish between positive and negative valued functions, as any positive valued function $f(\boldsymbol{x})$ will have the same 3DZDs as $-f(\boldsymbol{x})$. To correctly handle these situations, a 3D function $f(\boldsymbol{x})$ can be expressed as the difference of its positive $f^+(\boldsymbol{x}) = \max(f(\boldsymbol{x}), 0)$ and negative parts $f^-(\boldsymbol{x}) = -\min(f(\boldsymbol{x}), 0)$, i.e. $f(\boldsymbol{x}) = f^+(\boldsymbol{x}) - f^-(\boldsymbol{x})$, and by computing the 3DZDs of these two functions separately. Three of the twenty selected amino acid indices can assume both positive and negative values. The positive and negative parts were considered separately for these three indices, yielding a total of 23 3DZDs for each LSP. In [47], it can be observed that when using a high 3DZD order (i.e. 20, which corresponds to 121 descriptors for each represented 3D function), only a small subset of the resulting features were actually selected and used by the learning algorithm (about 10% of the total). For this reason, in this work the maximum 3DZD order was lowered to 5, which gives 12 descriptors for each 3D function (according to Eq. 6), yielding a total of $23 \times 12 = 276$ features for each LSP.

3.7   Support Vector Machine

Support vector machine (SVM) is a binary classification technique introduced in [64] that minimizes the upper bound of the generalization error by maximizing the margin between the separating hyperplane and the data, abiding to the structure risk minimization principle for model selection. A binary classification problem usually involves separating data into training and testing sets. The instances (samples) of the training set are the pairs (x_i, y_i), where x_i is a vector representing the features or attributes of the given sample and $y_i \in \{-1, +1\}$ is the corresponding class label. The goal of SVM is to produce a model based on the training data which predicts the class labels of the test data given only the feature vectors of the test data. SVM can perform non-linear classification in the feature space with the so called kernel trick by implicitly finding a separating hyperplane with maximal margin in a higher dimensional space. This is achieved by using a kernel function: the most common ones are (1) the linear kernel, (2) the polynomial kernel, (3) the radial basis function (RBF) kernel and (4) the sigmoid kernel.

In this work, the SVM implementation provided in the scikit-learn Python module for machine learning version 0.19.1 (http://scikit-learn.org) was employed to classify LSPs of the Ab surface into interface and non-interface ones. Instead of a binary classification result for each sample, the SVM can also output a signed real number b which tells on which side of the separating hyperplane the current sample is on and the distance from it. By thresholding on this value (the default threshold value is 0), the user can increase or decrease the selectivity of the classifier.

3.8   SVM Model Selection

Choosing an appropriate kernel function with the corresponding best hyper-parameters is critical for achieving good classification performance with SVMs. Hyper-parameters include the penalty C and the kernel parameters (which depend on the choice of the kernel function). In this work, a randomised search over the hyper-parameters was performed for each the following four kernel functions: (1) linear, (2) polynomial, (3) radial basis function (RBF) and (4) sigmoid. The penalty parameter C was sampled from the continuous random variable 2^X, where $X \sim \mathcal{U}(-5, 16)$ for all kernel functions. The γ parameter was sampled from the continuous random variable 2^Y, where $Y \sim \mathcal{U}(-15, 4)$ for the polynomial, RBF and sigmoid kernel functions. The degree d parameter of the polynomial kernel was sampled from the discrete uniform distribution $\mathcal{U}\{2, 10\}$ (the polynomial kernel of degree 1 is actually the linear kernel), while the r parameter of the polynomial and sigmoid kernels was sampled from the continuous uniform distribution $\mathcal{U}(-2, 2)$. The computation budget, i.e. the total number of sampled candidates or sampling iterations, was set to 200 iterations for each kernel function.

A 10-fold cross-validation (CV) was used (at the Ab level) on the training set for SVM kernel and hyperparameter selection. The training set was randomly split into 10 roughly-equal subsets, making sure that the samples of a given Ab were never split among different subsets. For a given combination of kernel type and hyperparameters to test, in turn, each subset was selected as testing set while the remaining 9 subsets were used to train the model. The Receiver

Operating Characteristic area under the curve (ROC-AUC) was employed as a performance measure throughout all experiments, and the combination of kernel and hyperparameters that yielded the best average ROC-AUC over the 10 folds was selected. The best average ROC-AUC of 0.895 (with a standard deviation of 0.020) was obtained with the RBF kernel with parameters $C = 540.2$ and $\gamma = 7.98e-3$.

3.9   Post-processing

The default $t = 0$ threshold used by the SVM classifier does not yield optimal results since the employed training set is balanced and does not reflect the natural distribution of interface and non-interface patches. The best threshold value $t_{\text{best}} = 0.6232$ was selected as the one that maximised the average F_1 score (58.02%) of LSP classification on the development set (see Fig. 4a). So, samples with $b \geq t_{\text{best}}$ are classified as positive while the ones with $b < t_{\text{best}}$ are classified as negative.

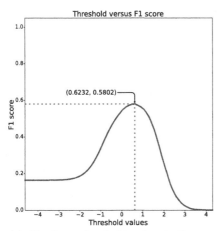
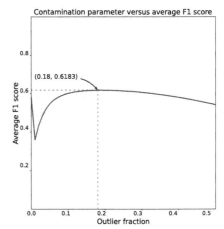

(a) The best threshold value is the one that maximises the average F_1 score on the development set.

(b) The best contamination parameter value is the one that maximises the average F_1 score on the development set over 100 runs for each Ab structure.

Fig. 4. Best threshold and contamination parameter values.

The Isolation Forest (IF) algorithm for outlier detection [65] is then used to reduce the number of spatially-isolated false positive LSPs. Paratope regions are composed of contiguous surface patches, thus isolated patches marked as positive by the SVM classifier (i.e. false positives) can be safely discarded. Similarly, patches classified as negative which are surrounded by many positive ones are most likely false negatives. For each antibody, an IF classifier is trained on the coordinates of the LSPs identified as interface patches by the SVM classifier, using their distances from the separating hyperplane (b values) as weights. Then, the IF classifier is used on the whole set of LSPs of the Ab to identify the ones belonging to the paratope. A contamination parameter of 0.18 was used in the testing phase. All contamination values ranging from 0.01 to 0.5 with a constant increment of 0.01 were tested, and the one that yielded the best average F_1 score (61.83%) on the development set over 100 runs for each Ab structure was selected (see Fig. 4b), since IF for outlier detection is a randomised algorithm.

3.10   Identifying Paratope Residues

The following procedure is used to map the predicted paratope LSPs on the underlying residues. Each residue in the Ab is assigned an initial score of 0. Then, for each LSP predicted as belonging to the interface, all the residues with at least one atom within 6.0 Å from its centre are identified as *covered*. The score of each covered residue is incremented by $1/(1 + d)$, where d is the minimum distance from its atoms to the current LSP's centre. At the end of the procedure, each residue in the Ab will have a final score which indicates its likelihood of belonging to the paratope. Since the algorithm contains a random step (the IF post-processing), the final residue scores are obtained as the average scores over 100 runs of this procedure. Each residue in the Ab can then be classified as Ag-binding or non-Ag-binding by thresholding on its final score.

4   Experimental Results

The proposed method was compared to three other paratope predictors, namely Paratome, Antibody i-Patch and Parapred. The Paratome web server accepts as input either the heavy (H) and light (L) chain sequences of an Ab in FASTA format or its PDB structure, and returns a binary prediction for each residue in the query Ab (interface/non-interface). Since Paratome is a structurally derived paratope identification tool, the Ab PDB structures were provided as input. The Antibody i-Patch server requires the structures of both the Ab and its possible binding partner as input in order to predict the paratope residues. For each residue in the input Ab structure, Antibody i-Patch returns a score which describes the residue's likelihood to belong to the interaction interface instead of a simple binary prediction. The Parapred server implements a purely sequence based Ab interface prediction method, and takes as inputs either the variable domain sequences or the full H and L chain sequences, and returns a score indicating the likelihood of belonging to the paratope of each residue in the CDRs. The full H and L chain sequences of the Abs in the test set were provided as input to this predictor.

The method comparison was carried out on the test set described in Sect. 3.3 which is composed of two disjoint subsets: a set of Abs complexed with protein Ags and a set of Abs complexed with non-protein Ags. The prediction performance comparison for each method in given terms of (macro) average Receiver Operating Characteristic (ROC) and (macro) average Precision-Recall

(PR) curves (Figs. 5, 6 and 7), and was carried out on the complete test set, as well as on the two disjoint subsets separately. The curves were obtained by computing the average precision and recall (for the PR curve) and the average true positive and false positive rates (for the ROC curve) at the Ab level for several classification threshold values. Usually, PR curves start from $(0, 1)$, based on the premise that when no positive samples are retrieved, i.e. the recall is 0, there are no false positives so the precision is conventionally set to 1. When computing the average PR curve however, this convention may results in an overestimation of the precision of a classifier. Lets suppose that we have chosen a high threshold value for the residue classification task, and the recall is greater than 0 (we are retrieving some positive samples, i.e. some interface residues are classified correctly as such) only for a certain number of Abs while being still 0 for the remaining ones in the dataset (the current threshold value could be larger than all the residue scores in these Abs). The precision for these Abs will be 1 by convention, thus the resulting average precision could be overestimated. For this reason, the average PR curves in this work start from $(0, 0)$.

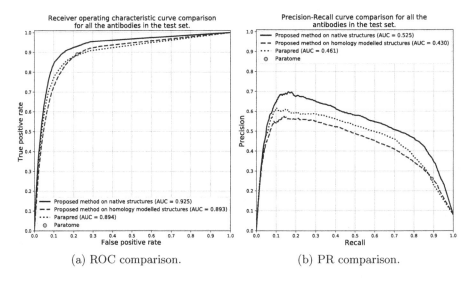

(a) ROC comparison.   (b) PR comparison.

Fig. 5. Average ROC and PR curve comparison of the different paratope prediction methods on the complete test set.

Antibody i-Patch was unable to make any prediction in the dataset of Abs with non-protein Ags: although the PDB structures of both the Ab and the Ag were correctly provided, a zero score was returned for all Ab residues, and thus its ROC and PR curve plots are provided only for the protein-binding Abs test set. Also, the results of the Paratome method are represented as a single point in the ROC and PR curve comparison plots since this predictor can only provide a binary classification output (all the other methods provide a classification

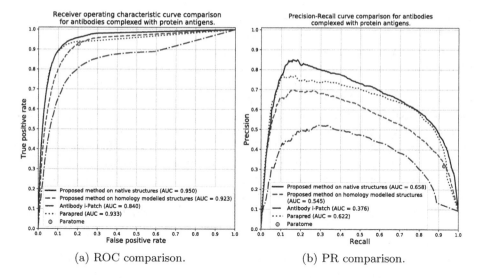

Fig. 6. Average ROC and PR curve comparison of the different paratope prediction methods on protein-binding Abs.

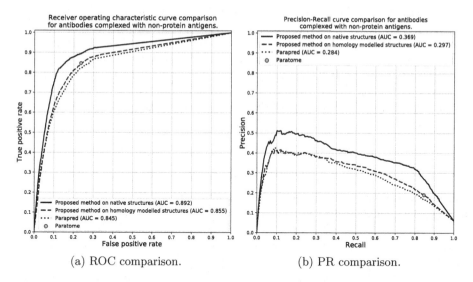

Fig. 7. Average ROC and PR curve comparison of the different paratope prediction methods on non-protein-binding Abs.

score which can be thresholded). The proposed method was also tested on the homology-based modelled structures of the Abs in the test set: the performance results are included in the ROC and PR plots.

By inspecting the ROC and PR curves it is clear that the proposed methodology significantly outperforms the Antibody i-Patch and Paratome methods.

When the paratope prediction is made starting from the native structures of the Abs, the proposed method also outperforms Parapred. In the test set of Abs with protein Ags, the proposed prediction method performs only slightly better than Parapred in terms of ROC-AUC and PR-AUC. The Wilcoxon signed-rank test (with $\alpha = 0.05$) was performed to verify that the performance difference between the proposed method and Parapred in terms of ROC-AUC and PR-AUC is statistically significant. A (two-tailed) p-value of $2.233e-8$ was obtained for the statistical test on the ROC-AUC and a (two-tailed) p-value of 0.0296 was obtained for the statistical test on the PR-AUC: since both p-values are smaller than α, we can assert that the difference in performance (in terms of ROC-AUC and PR-AUC) between the proposed method and Parapred is statistically significant at the 0.05 significance level. This performance difference becomes more pronounced in the test set of Abs with non-protein Ags. All paratope prediction methods obtain significantly worse results on the test set of Abs complexed with non-protein Ags (the proposed method remains the better performing). This is most likely due to the fact that there are very few instances of Abs with non-protein Ags in the training sets of the available paratope predictors (including the training set employed in this work).

When using homology modelled structures instead of native ones, the prediction performance of the proposed method is significantly reduced. By using the homology modelled structures of Abs with protein Ags, the proposed method is outperformed by Parapred. Still, it achieves similar results to Paratome, while remaining significantly superior to Antibody i-Patch. When predicting the interfaces of homology modelled structures of Abs with non-protein Ags, the proposed method obtains comparable results with Parapred (there is no statistically significant difference with Parapred both in terms of ROC-AUC and PR-AUC) and Paratome.

5   Conclusions

This paper presents a structure-based method for the prediction of the paratope in Abs. The reliable identification of Ag-binding residues is fundamental for understanding the biological mechanisms of Ab–Ag interactions and for immunological research aimed at Ab design and manipulation. The proposed method was shown to outperform its competitors on two separate test set of Ab structures: a set of Ab structures complexed with protein Ags and a set of Ab structures complexed with non-protein Ags. The performance of all the tested paratope predictors was lower in the second test set, probably due to the few instances of Abs complexed with non-protein Ags in the training sets. Apparently, the physicochemical properties that determine the Ag binding interface in Ab structures and/or their spatial distribution can vary depending on the Ag type. The fact that the proposed prediction method performs better than the others in this dataset is most likely due to the fact that it is purely structure-based, and thus is able to exploit the rather regular geometrical shape of antibodies when trying to locate the paratope. As the number of available Ab structures complexed with various Ag types in the PDB and Ab dedicated databases will increase, the predictors should become able to correctly identify non-protein binding paratopes as well.

When predicting homology modelled structures the performance of the proposed method drops sensibly, although the prediction quality remains reasonably good. In the test set of Abs complexed with non-protein Ags, the predictions on homology modelled structures made with the proposed method are very similar to the ones made by Parapred, which is a purely sequence-based method. With the development of ever improving new Ab structure prediction techniques and the increase in Ab structures in the PDB, the proposed method is expected to be applicable even when only the Ab sequence information is known.

The proposed methodology could aid rational Ab design applications. In the natural immune response, the binding residues of Abs are often mutated to increase the Ag binding affinity and specificity in a process known as *affinity maturation*. If Ab–Ag contact residues are known, one can computationally mutate them to search for residues that improve the binding affinity and specificity of the Ab. This computational process is usually comprised of three steps: paratope prediction, epitope prediction, and Ab–Ag docking [66]. The correct prediction of the paratope is very important because it can be used to limit the number of residues that need to be mutated for affinity maturation. Since the proposed method returns a score for each residue in the Ab structure indicating its likelihood to bind the Ag, it could be used to select the residues for mutagenesis. Moreover, the predictions at the local surface patch level can be readily used to guide Ab–Ag docking algorithms by restraining the conformational search space to the sole surface patches belonging to the paratope. Docking algorithms based on local surface descriptor matching could greatly benefit from the proposed approach, since this would sensibly limit the number of candidate patch pairs to be evaluated.

Although antigen specificity was not considered in this work, knowledge of the specific epitope residues could improve paratope prediction. The proposed methodology could be extended to consider binding-partner specificity by classifying pairs of LSPs from the surfaces of the interacting units instead of single patches. However, particular care should be taken in handling the resulting class imbalance and in selecting a proper set of physicochemical properties that characterise the Ag's binding interface, which could be different from the ones used to characterise the paratope.

The selected set of 20 amino acid indices of physicochemical and biochemical properties have demonstrated very good discriminative capabilities for the interface recognition of antibodies. Other sets of amino acid indexes can be similarly mapped on the voxelised protein surface and represented by 3DZDs, and the proposed methodology could also be extended to specifically address the interface prediction for other protein classes altogether. These issues will be investigated in future research work.

Availability of Data and Materials

Linux binaries, Python scripts and the datasets generated/analysed during the current study are available at https://doi.org/10.6084/m9.figshare.7797563.

Funding

This research has been partially supported by the University of Padova project CPDR150813/15 "Models and Algorithms for Protein–Protein Docking".

References

1. Frank, S.A.: Vertebrate immunity. In: Immunology and Evolution of Infectious Disease. Princeton University Press, Princeton (2002)
2. Moser, M., Leo, O.: Key concepts in immunology. Vaccine **28**, C2–C13 (2010)
3. Schroeder, H.W., Cavacini, L.: Structure and function of immunoglobulins. J. Allergy Clin. Immunol. **125**(2), S41–S52 (2010)
4. Murphy, K., Weaver, C.: The humoral immune response: the distribution and functions of immunoglobulin isotypes. In: Janeway's Immunobiology, 9th edn, chap. 10, pp. 423–432. Garland Science, Taylor & Francis Group (2017)
5. Griffiths, K., et al.: I-bodies, human single domain antibodies that antagonize chemokine receptor CXCR4. J. Biol. Chem. **291**(24), 12641–12657 (2016)
6. Poosarla, V.G., Li, T., Goh, B.C., Schulten, K., Wood, T.K., Maranas, C.D.: Computational de novo design of antibodies binding to a peptide with high affinity. Biotechnol. Bioeng. **114**(6), 1331–1342 (2017)
7. Trail, P.A., Dubowchik, G.M., Lowinger, T.B.: Antibody drug conjugates for treatment of breast cancer: novel targets and diverse approaches in ADC design. Pharmacol. Ther. **181**, 126–142 (2017)
8. Esmaielbeiki, R., Krawczyk, K., Knapp, B., Nebel, J.C., Deane, C.M.: Progress and challenges in predicting protein interfaces. Brief. Bioinform. **17**(1), 117–131 (2015)
9. Reichert, J.M.: Antibodies to watch in 2017. mAbs **9**(2), 167–181 (2017)
10. Beck, A., Goetsch, L., Dumontet, C., Corvaïa, N.: Strategies and challenges for the next generation of antibody-drug conjugates. Nat. Rev. Drug Discov. **16**(5), 315–337 (2017)
11. Prendergast, J.M., et al.: Novel anti-Sialyl-Tn monoclonal antibodies and antibody-drug conjugates demonstrate tumor specificity and anti-tumor activity. mAbs **9**(4), 615–627 (2017)
12. Carter, P.J.: Potent antibody therapeutics by design. Nat. Rev. Immunol. **6**(5), 343 (2006)
13. Owens, S.M.: Monoclonal antibodies. In: Montoya, I.D. (ed.) Biologics to Treat Substance Use Disorders: Vaccines, Monoclonal Antibodies, and Enzymes, pp. 107–108. Springer, Cham (2016). https://doi.org/10.1007/978-3-319-23150-1
14. Hammers, C.M., Stanley, J.R.: Antibody phage display: technique and applications. J. Invest. Dermatol. **134**(2), e17 (2014)
15. Li, T., Pantazes, R.J., Maranas, C.D.: OptMAVEn – a new framework for the de novo design of antibody variable region models targeting specific antigen epitopes. PLoS ONE **9**(8), e105954 (2014)

16. Kuroda, D., Shirai, H., Jacobson, M.P., Nakamura, H.: Computer-aided antibody design. Protein Eng. Des. Sel. **25**(10), 507–522 (2012)
17. Lippow, S.M., Wittrup, K.D., Tidor, B.: Computational design of antibody affinity improvement beyond in vivo maturation. Nat. Biotechnol. **25**(10), 1171 (2007)
18. Karanicolas, J., Kuhlman, B.: Computational design of affinity and specificity at protein-protein interfaces. Curr. Opin. Struct. Biol. **19**(4), 458–463 (2009)
19. Kiyoshi, M., et al.: Affinity improvement of a therapeutic antibody by structure-based computational design: generation of electrostatic interactions in the transition state stabilizes the antibody-antigen complex. PLoS ONE **9**(1), e87099 (2014)
20. Margreitter, C., Mayrhofer, P., Kunert, R., Oostenbrink, C.: Antibody humanization by molecular dynamics simulations-in-silico guided selection of critical back-mutations. J. Mol. Recognit. **29**(6), 266–275 (2016)
21. Weitzner, B.D., et al.: Modeling and docking antibody structures with Rosetta. Nat. Protoc. **12**(2), 401–416 (2016)
22. Sefid, F., Rasooli, I., Payandeh, Z.: Homology modeling of a Camelid antibody fragment against a conserved region of Acinetobacter baumannii biofilm associated protein (Bap). J. Theor. Biol. **397**, 43–51 (2016)
23. Daberdaku, S.: Paratope identification by classification of local antibody surface patches enriched with eight physicochemical properties. In: 15th International Conference on Computational Intelligence Methods for Bioinformatics and Biostatistics (CIBB 2018), Caparica, September 2018
24. Daberdaku, S., Ferrari, C.: Antibody interface prediction with 3D Zernike descriptors and SVM. Bioinformatics **35**(11), 1870–1876 (2018)
25. Sael, L., et al.: Fast protein tertiary structure retrieval based on global surface shape similarity. Proteins Struct. Funct. Bioinform. **72**(4), 1259–1273 (2008)
26. La, D., et al.: 3D-SURFER: software for high-throughput protein surface comparison and analysis. Bioinformatics **25**(21), 2843–2844 (2009)
27. Venkatraman, V., Sael, L., Kihara, D.: Potential for protein surface shape analysis using spherical harmonics and 3D Zernike descriptors. Cell Biochem. Biophys. **54**(1–3), 23–32 (2009)
28. Venkatraman, V., Yang, Y.D., Sael, L., Kihara, D.: Protein-protein docking using region-based 3D Zernike descriptors. BMC Bioinform. **10**(1), 407 (2009)
29. Sael, L., La, D., Li, B., Rustamov, R., Kihara, D.: Rapid comparison of properties on protein surface. Proteins **73**(1), 1–10 (2008)
30. Wu, T.T., Kabat, E.A.: An analysis of the sequences of the variable regions of Bence Jones proteins and myeloma light chains and their implications for antibody complementarity. J. Exp. Med. **132**(2), 211–250 (1970)
31. Chothia, C., Lesk, A.M.: Canonical structures for the hypervariable regions of immunoglobulins. J. Mol. Biol. **196**(4), 901–917 (1987)
32. MacCallum, R.M., Martin, A.C., Thornton, J.M.: Antibody-antigen interactions: contact analysis and binding site topography. J. Mol. Biol. **262**(5), 732–745 (1996)
33. Al-Lazikani, B., Lesk, A.M., Chothia, C.: Standard conformations for the canonical structures of immunoglobulins. J. Mol. Biol. **273**(4), 927–948 (1997)
34. Kabat, E.A., Te Wu, T., Perry, H.M., Gottesman, K.S., Foeller, C.: Sequences of Proteins of Immunological Interest, 5th edn. DIANE Publishing, Collingdale (1991)
35. Chothia, C., et al.: Conformations of immunoglobulin hypervariable regions. Nature **342**(6252), 877–883 (1989)
36. Lefranc, M.P., et al.: IMGT unique numbering for immunoglobulin and T cell receptor variable domains and Ig superfamily V-like domains. Dev. Comp. Immunol. **27**(1), 55–77 (2003)

37. Stave, J.W., Lindpaintner, K.: Antibody and antigen contact residues define epitope and paratope size and structure. J. Immunol. **191**(3), 1428–1435 (2013)
38. Kunik, V., Peters, B., Ofran, Y.: Structural consensus among antibodies defines the antigen binding site. PLoS Comput. Biol. **8**(2), e1002388 (2012)
39. Kunik, V., Ashkenazi, S., Ofran, Y.: Paratome: an online tool for systematic identification of antigen-binding regions in antibodies based on sequence or structure. Nucleic Acids Res. **40**(W1), W521–W524 (2012)
40. Olimpieri, P.P., Chailyan, A., Tramontano, A., Marcatili, P.: Prediction of site-specific interactions in antibody-antigen complexes: the proABC method and server. Method. Biochem. Anal. **29**(18), 2285–2291 (2013)
41. Krawczyk, K., Baker, T., Shi, J., Deane, C.M.: Antibody i-Patch prediction of the antibody binding site improves rigid local antibody-antigen docking. Protein Eng. Des. Sel. **26**(10), 621–629 (2013)
42. Peng, H.P., Lee, K.H., Jian, J.W., Yang, A.S.: Origins of specificity and affinity in antibody-protein interactions. Proc. Nat. Acad. Sci. **111**(26), E2656–E2665 (2014)
43. Chen, C.T., et al.: Protein-protein interaction site predictions with three-dimensional probability distributions of interacting atoms on protein surfaces. PLoS ONE **7**(6), e37706 (2012)
44. Veličković, P., Liò, P., Liberis, E., Vendruscolo, M., Sormanni, P.: Parapred: antibody paratope prediction using convolutional and recurrent neural networks. Bioinformatics **34**(17), 2944–2950 (2018). https://doi.org/10.1093/bioinformatics/bty305
45. Hu, D., et al.: Effective optimization of antibody affinity by phage display integrated with high-throughput DNA synthesis and sequencing technologies. PLoS ONE **10**(6), e0129125 (2015)
46. Dunbar, J., et al.: SAbPred: a structure-based antibody prediction server. Nucleic Acids Res. **44**(W1), W474–W478 (2016)
47. Daberdaku, S., Ferrari, C.: Exploring the potential of 3D Zernike descriptors and SVM for protein-protein interface prediction. BMC Bioinform. **19**(1), 35 (2018)
48. Connolly, M.L.: Analytical molecular surface calculation. J. Appl. Crystallogr. **16**(5), 548–558 (1983)
49. Daberdaku, S., Ferrari, C.: Computing discrete fine-grained representations of protein surfaces. In: Angelini, C., Rancoita, P.M., Rovetta, S. (eds.) CIBB 2015. LNCS, vol. 9874, pp. 180–195. Springer, Cham (2016). https://doi.org/10.1007/978-3-319-44332-4_14
50. Daberdaku, S., Ferrari, C.: Computing voxelised representations of macromolecular surfaces: a parallel approach. Int. J. High Perform. Comput. Appl. **32**(3), 407–432 (2016)
51. Duhovny, D., Nussinov, R., Wolfson, H.J.: Efficient unbound docking of rigid molecules. In: Guigó, R., Gusfield, D. (eds.) WABI 2002. LNCS, vol. 2452, pp. 185–200. Springer, Heidelberg (2002). https://doi.org/10.1007/3-540-45784-4_14
52. Schneidman-Duhovny, D., et al.: Taking geometry to its edge: fast unbound rigid (and hinge-bent) docking. Proteins: Struct. Funct. Bioinform. **52**(1), 107–112 (2003)
53. Valencia, A., Ezkurdia, I., Bartoli, L., Tress, M.L., Fariselli, P., Casadio, R.: Progress and challenges in predicting protein-protein interaction sites. Brief. Bioinform. **10**(3), 233–246 (2009)
54. Ferdous, S., Martin, A.C.R.: AbDb: antibody structure database – a database of PDB-derived antibody structures. Database **2018** (2018). https://doi.org/10.1093/database/bay040

55. Li, W., Godzik, A.: CD-HIT: a fast program for clustering and comparing large sets of protein or nucleotide sequences. Bioinformatics **22**(13), 1658–1659 (2006)
56. Niu, B., Fu, L., Wu, S., Li, W., Zhu, Z.: CD-HIT: accelerated for clustering the next-generation sequencing data. Bioinformatics **28**(23), 3150–3152 (2012)
57. Leem, J., Dunbar, J., Georges, G., Shi, J., Deane, C.M.: ABodyBuilder: automated antibody structure prediction with data-driven accuracy estimation. mAbs **8**(7), 1259–1268 (2016)
58. Xue, L.C., Dobbs, D., Bonvin, A.M., Honavar, V.: Computational prediction of protein interfaces: a review of data driven methods. FEBS Lett. **589**(23), 3516–3526 (2015)
59. Kawashima, S., Pokarowski, P., Pokarowska, M., Kolinski, A., Katayama, T., Kanehisa, M.: AAindex: amino acid index database, progress report 2008. Nucleic Acids Res. **36**(Suppl. 1), D202–D205 (2008)
60. Meinshausen, N., Bühlmann, P.: Stability selection. J. Roy. Stat. Soc. Ser. B (Stat. Methodol.) **72**(4), 417–473 (2010)
61. Zhu, X., Xiong, Y., Kihara, D.: Large-scale binding ligand prediction by improved patch-based method Patch-Surfer2.0. Bioinformatics **31**(5), 707–713 (2015)
62. Canterakis, N.: 3D Zernike moments and Zernike affine invariants for 3D image analysis and recognition. In: 11th Scandinavian Conference on Image Analysis, pp. 85–93 (1999)
63. Novotni, M., Klein, R.: Shape retrieval using 3D Zernike descriptors. Comput. Aided. Des. **36**(11), 1047–1062 (2004)
64. Boser, B.E., Guyon, I.M., Vapnik, V.N.: A training algorithm for optimal margin classifiers. In: The Fifth Annual Workshop on Computational Learning Theory, COLT 1992, pp. 144–152. ACM, New York (1992)
65. Liu, F.T., Ting, K.M., Zhou, Z.H.: Isolation-based anomaly detection. ACM Trans. Knowl. Discov. Data **6**(1), 3 (2012)
66. Roy, A., Nair, S., Sen, N., Soni, N., Madhusudhan, M.: In silico methods for design of biological therapeutics. Methods **131**, 33–65 (2017). Systems Approaches for Identifying Disease Genes and Drug Targets

Simultaneous Phasing of Multiple Polyploids

Laxmi Parida$^{(\boxtimes)}$ (iD) and Filippo Utro (iD)

Computational Biology Center, IBM T. J. Watson Research,
Yorktown Heights, NY 10598, USA
{parida,futro}@us.ibm.com

Abstract. We address the problem of phasing polyploids specifically with polyploidy larger than two. We consider the scenario where the input is the genotype of samples along a genic chromosomal segment. In this setting, instead of NGS reads of the segments of a sample, genotype data from multiple individuals is available for simultaneous phasing. For this mathematically interesting problem, with application in plant genomics, we design and test two algorithms under a parsimony model. The first is a linear time greedy algorithm and the second is a more carefully crafted algebraic algorithm. We show that both the methods work reasonably well (with accuracy on an average larger than 80%). The former is very time-efficient and the latter improves the accuracy further.

Keywords: Polyploids · Phasing · Genotypes · Cross-overs · Plant genetics · Algebraic method

1 Introduction

Polyploidy, an important mechanism of adaptation, is common in cultivated plants. This was perhaps introduced in the wild or during the long history of human-assisted agriculture. Nevertheless, different cultivars of the same species (say potatoes, sugarcane, mangoes) may differ in taste, vigor, yield or diseases-resistance to mention a few phenotype attributes or traits. However, it is unknown what parts of the genome are responsible for the favorable traits. The task of trait mapping is to uncover this connection between he genotype and the phenotype. The phasing of the genotypes, i.e., separating into the constituent haplotypes aids in the trait mapping. This informs which alleles travel together and can be a powerful piece of information in delineating the primary markers of the trait from the travel-along markers. In this paper, we focus on the phasing problem, which can be set up as an elegant mathematical problem. When the ploidy of an organism goes up from two, the problem of phasing becomes considerably difficult. The phasing problem under various settings is NP-complete, even for diploids [4]. Here we propose two approaches to phase polyploidy genotypes at a set of polymorphic markers. In our problem scenario the absence of Next Generation Sequencing (NGS) reads, which potentially informs the phasing in a local neighborhood, is compensated by the availability of multiple individuals

M. Raposo et al. (Eds.): CIBB 2018, LNBI 11925, pp. 50–68, 2020.
https://doi.org/10.1007/978-3-030-34585-3_5

for simultaneous phasing. One way of relating multiple samples is to reconstruct the common ancestral haplotypes.

Furthermore, the available markers are in the genic regions. Note that crossovers in viable offsprings occur mostly in the intergenic regions. Thus the assumption of absence of recombinations in the phased haplotypes of the samples is reasonable and realistic. This also makes the underlying computational problem tighter and more well-defined.

2   Scientific Background

In this paper we address the phasing problem in a setting that differs from the others where NGS reads are used to phase at a single sample level [1,3,6,8]. Many legacy databases exist with seed companies and research facilities where the only available data is that of genotypes at a set of polymorphic markers (private communication). Due to "cost or lost" germplasms, extraction of NGS data is not practical or possible.

3   Materials and Methods

For this paper, the input to the phasing problem is defined as follows. The n samples are given as a $n \times m$ matrix D of genotypes, where the rows of D denote the samples and the m columns denote the m markers. For example, for a tetraploid the genotype may be coded as 0001 representing the four alleles of the Single Nucleotide Polymorphism (SNP) observed at the position. The SNPs are assumed to be bi-allelic with the nucleic acid base coded as 0 or 1. Without loss of generality the alternate allele is denoted by 1 and the reference or ancestral allele by 0. We begin by making the following observation.

Lemma 1. *Any $(2k+1)$-ploid with biallelic alleles (i.e., value 0 or 1) can be resolved into at most $k+1$ haplotypes.*

We first present a greedy algorithm to solve the problem and then develop an algebra of the genotypes to design the second algorithm.

3.1   Greedy Algorithm

Let $H_1, H_2, ..,$ be the haplotypes and let f_{H_i} be the frequency with which haplotype H_i is seen in the phased samples. We define the entropy E as

$$E = \Sigma_i f_{H_i}^2. \tag{1}$$

The cost function is defined as simultaneously minimizing both the following: (OPM I) the number of haplotypes (h) and (OPM II) the entropy (E). A lower entropy solution avoids peaks at some haplotypes, that could be artifacts of the algorithm. Usually, bi-allelic SNPs pull a solution configuration towards sharp peaks in the haplotype frequency distributions and minimizing the number of haplotypes is no guarantee for the absence of these artifacts. On the other hand, lowering the entropy in the solution discourages these types of artifacts. For convenience of the reader, we report here the guiding principles and the outline of the approach, while the data structure and concrete example are presented in the Supplement.

The two guiding principles are as follows:

AD I (One-Choice): We use the homozygous SNPs in a sample to make the "inevitable" one-choice decisions. Further, we make these choices as early as possible, both in the iterations and within each iteration, to mitigate error propagation.

AD II (Multiple-Choice): If the multiple options are not equally favorable, based on the cost function, we appropriately order the decision making. Further, for equally favorable multiple options, the implementation makes a random call, that could be different in each run of the algorithm. This randomness can be used to generate multiple possible solutions with each run to get confidence intervals.

Recall that the input is a genotype $n \times m$ matrix where the number of SNPs (or columns) is m and the number of samples (or rows) is n. A column C_j, $j = 1, 2, .., m$, refers to the alleles in SNP j and $C_j[i]$, $1 \leq i \leq n$, refers to the jth genotype in the ith sample, which is a set of p binary values. Each of these is accessed as $G[i][j][k]$. Let

$$U = \{(i, k) \mid 1 \leq i \leq n, 1 \leq k \leq p\}.$$

A trie $\mathcal{T}$ is initialized with a single node (also the root). Every node v on $\mathcal{T}$ has exactly one incoming edge and at most two outgoing edges, with the following exception: a *root* has no incoming edge and a *leaf* has no outgoing edges. If e is an incoming edge from v_1 to v_2 then v_2 is the *child* of v_1 and v_1 is the *parent* of v_2. Each node v is labeled with a pair denoted as

$$(x_v, \mathcal{L}_v) \qquad \text{where } x_v \in \{0, 1\} \text{ and } \mathcal{L}_v \subset U.$$

The root node is an exception and in the initialization, the root is assigned the label $(\emptyset, U)$. The root node is denoted as v_0. x_v is called the SNP label of v and $\mathcal{L}_v$ is the list label of v. The depth of a node v is defined as the distance, in terms of edges in the path, from the root v_0, Thus depth(v_0) is 0. We maintain the following invariance at each iteration d:

$$\text{depth}(v) = d, \quad \text{where } v \text{ is a leaf node}, \tag{2}$$

$$\mathcal{L}_v = \bigcup_{u \text{ is a child of } v} \mathcal{L}_u, \tag{3}$$

$$U = \bigcup_{v \text{ is a leaf}} \mathcal{L}_v. \tag{4}$$

Lemma 2. *If h is the number of leaves at depth m in $\mathcal{T}$, then h is the number of distinct haplotypes for the given instance of problem. Each leaf corresponds to a distinct haplotype and is obtained by reading off the 0/1 label of the nodes in the path from the root node to the leaf.*

Greedy Algorithm Outline. *Initialization:* Let $\mathcal{I} = \{1, 2, .., n\}$. Initialize each $C_0[i], i \in \mathcal{I}$ to be heterozygous. *Iteration: d* is set to 0.
REPEAT

1. (Book keeping)
   Increment d by 1. Consider the next column j such that the number of entries in C_0 or C_j (could be both) that are homozygous is maximized. Further,

$$I^{\text{hom}} = \{i \mid C_0[i] \text{ OR } C_j[i] \text{ is homozygous}\} ; I^{\text{het}} = \mathcal{I} \setminus I^{\text{hom}}.$$

2. (HOM: new nodes induced by $i \in I^{\text{hom}}$; AD I nodes and labels)
   For each $i \in I^{\text{hom}}$:
   Locate the node(s) in $\mathcal{T}$ at depth d that have i in the list of labels and introduce a new outgoing edge on a node -at level d- with label 1 if the genotype value is 1. Similarly if the value is 0. Update the node labels appropriately.
3. (HET: global optimization for AD II nodes and labels)
   Use a temporary matrix to make the optimal choices.
4. (Book-keeping)
   Update C_0 as: If $\{(i, 1), (i, 2), .., (i, p)\} \subset \mathcal{L}_v$ for some leaf node v of $\mathcal{T}$, then $C_0[i]$ is set to homozygous; else set to heterozygous.

UNTIL $d = m$. It is easy to verify that the algorithm runs in $O(nm)$ time and we present its accuracy in Sect. 4.

3.2   Algebraic Method

In iXoRA [9] the algebraic approach is used to address the phasing problem for diploid F1 populations, but with possible recombinations in the chromosomal segment. The problem was solved under a parsimony model, i.e., with the smallest number of recombinations across the individuals. As the population is F1, the number of ancestral haplotypes is four (two from each parent). We generalize this approach to the parsimony model on polyploids. Due to space constraints the development of the algebra of the genotypes is presented in the Supplement.

The Algebraic Algorithm Outline. Recall that the input is a matrix D where each row represents a sample and the ordered columns represent the ordered SNPs. Each entry of the matrix is a genotype of ploid k, i.e., a binary set of size k. The task is to resolve the D into the smallest number of haplotypes (or rows of ploidy 1).

Initialize by setting the ploidy of each row to be k, the input ploidy and for each ith row, $\langle X \rangle$, the set S_x is initialized to the sample represented by that row with multiplicity k. The heuristics are applied in this order, using the algebra of genotypes, until all the rows have ploidy 1.

1. H-Ia (homozygous SNPs). We target the homozygous genotypes (shown in red in the example). We associate a weight with each ith row as i_{wt} which is computed at every iteration, computed as the product of the ploidy i_p of the row and the number of homozygous genotypes in the row say i_h. Thus $i_{\text{wt}} = i_p i_h$. The rows are sorted in decreasing order of the weight and considered for the row operations.

2. H-Ib (fully homozygous row). If *all* the SNPs of a row are homozygous, then replace row i with i_p monoploid rows.
3. H-IIa (large ploidy from row-row operation $\cap$). For a pair of rows, carry out the row-row intersection operation $\cap$. Note that for this the ploidy of both the rows is greater than 1. Further, a larger ploidy in the resulting row-row operation is preferred over a smaller ploidy. If this fails, then resolve variables using the operation $\cap_f^1$ (i.e., $t = 1$).
4. H-IIb (large ploidy from row-row operation $\setminus$). For a pair of rows, carry out the row-row difference operation $\setminus$. Further, a larger ploidy in the resulting row-row operation is preferred over a smaller ploidy. If this fails, then resolve variables using the operation $\setminus_f$.
5. H-III (resolve variables using input haplotype-constraint). Resolve variables that best fits the input haplotype constraint (coming from the read-map data). The preference is to resolve as few variables as possible.
6. H-IV (homozygous rows). If none of the above apply, then scan the rows in decreasing order of weight and pick say the ith row $\langle X \rangle$ with ploidy $i_p > 2$. The $i$row is replaced by a pair of rows:
   (a) Extract a (random) homozygous diploid row $\langle H \rangle$ from $\langle X \rangle$. $S_h \leftarrow S_x$ and $\langle H \rangle_p \leftarrow 2$.
   (b) Let $\langle Z \rangle$ and $\langle X \rangle \setminus \langle H \rangle$. Then $\langle Z \rangle_p \leftarrow \langle X \rangle_p - 2$ and $S_z \leftarrow S_x$, each with multiplicity 2.

We present a simple illustrative example here:

Example 1. *The input matrix is D with 4 samples and 5 alleles.*

D

a 4	0000	0111	0000	0001	0011
b 4	0000	1111	0001	0011	0111
c 4	1111	0011	1111	0111	0001
d 4	1111	0001	1111	0111	0001

$\Rightarrow$

clst

a 1	0	0	0	0	0
b 1	0	1	1	1	1
a,b 3	000	111	000	001	011
c,d 3	111	001	111	11x	00y
c 1	1	1	1	$\bar{x}$	$\bar{y}$
d 1	1	0	1	$\bar{x}$	$\bar{y}$

Step 1: $a \ X \ b$; $c \ X \ d$

....

$\Rightarrow$

clst

a 1	0	0	0	0	0
b 1	0	1	1	1	1
a,b 1	0	1	0	1	0
a,b 2	00	11	00	00	11
c,d 2	11	00	11	11	00
c,d 1	1	1	1	x	y
c 1	1	1	1	$\bar{x}$	$\bar{y}$
d 1	1	0	1	$\bar{x}$	$\bar{y}$

Step 2: *Using Lemma 1*

The optimal solution has 8 distinct haplotypes (from a possible potential 16). Corresponding to the possible values of x and y; there are $2 \times 2 = 4$ distinct optimal configurations.

Each of a, b, c and d samples have the two identical haplotypes. Further, a and b have common haplotypes and so do c and d.

4   Experimental Results

The difficulty with the real data is that the true (gold) answers are unknown. Hence we use simulated data to evaluate the accuracy of the results.

4.1   Data Simulation

Using SimBA-hap [7], we generate different scenario datasets. In particular, a set of tetraploids with increasing number of samples (20, 30, 40) and number of markers (10, 20, 30, 40) was generated. For each setting, five independent simulations were carried out.

4.2   Measuring Accuracy

Two solutions may have differing number of haplotypes and the need is for a fair comparison between the two solutions. Furthermore, a switch error could unfairly penalize entire segments. For each sample we compute the Hamming distance between the estimated haplotype and the gold solutions [2]. To assign a unique haplotype from the gold solution to the computed haplotype, we pick the one from the gold which is at the shortest distance from the candidate haplotype. Then accuracy a is computed as:

$$a = \frac{p * m - D}{p * m} \times 100, \text{ where } D = \sum_i \left(\min_j \{\text{Ham}(H_i, G_j)\} \right).$$

p is the ploidy (in our case $p = 4$) and m is the number of markers.

4.3   Comparing the Performance of the Methods

See Fig. 1 for a summary of results on a set of experiments. For the algebraic method, if any variable is unresolved at the end of the algorithm execution, a random value is assigned for the purposes of measuring accuracy. The results show that the accuracy is more or less invariant over the number of samples, and improves with decrease in the number of markers. Both the methods reach an accuracy larger than 0.8. The algorithms cannot be compared nose-to-nose with other methods since the problem setting is different. We informally compared with methods that used simulated NGS reads and observe that the algebraic method (average accuracy 0.93) is almost comparable with that of [5] and marginally worse than a NGS based algorithm (average accuracy 0.95) [8].

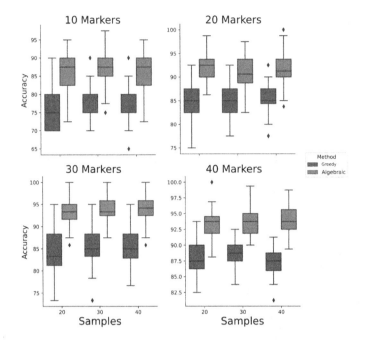

Fig. 1. Performance for the two methods in terms of accuracy for the different datasets.

5   Conclusion

We have presented two methods for simultaneously phasing multiple polyploids: one a straightforward greedy algorithm. For the second, we developed an algebra of the genotypes and used that to design the algorithm. Using only limited information, i.e., the genotypes and not the NGS read information, we show that the problem can be solved reasonably well by utilizing the information from multiple samples. This also indicates that methods that use the NGS reads can benefit further from multiple samples.

6   Supplement

6.1   On the Greedy Method

The Greedy Algorithm: Concrete Examples. Refer to Fig. 2 for the example input matrix G and the trie T that is constructed. The nodes in the Trie are labeled by the column identifier on top and then following vertically down from that column. For example, $v_{5,3}$ refers to the third node from top along the column marked 5. It is labeled by SNP 0 (i.e., hollow circle) and the cardinality of the label set is 4.

```
 1. 0000 0000 0000 1000 0000 0000 0010 0000
 2. 0100 0000 1010 0000 0100 1011 0000 0000
 3. 0011 0000 0000 0000 0011 0000 0000 0000
 4. 1000 0000 0000 0000 1000 0000 0000 0100
 5. 0001 0001 0000 1000 0000 0000 0001 0000
 6. 0000 0100 0000 0000 0001 0000 0000 0000
 7. 0000 0000 0001 0000 0000 0000 0000 0000
 8. 0100 0100 1000 0001 0000 1000 0100 0000
 9. 0000 0000 0000 0000 0000 0000 1000 0000
10. 0000 0000 0000 0000 0000 0000 0000 0000
```

00000000 $\mathcal{L}_{v_{1,1}} = \{(1,2),(1,4),(3,1),(3,2),(4,3),(4,4),(5,2),(6,1),(6,2),(7,1),$
$(7,2),(7,3),(9,2),(9,3),(9,4),(10,1),(10,2),(10,3),(10,4)\},$

00000010 $\mathcal{L}_{v_{1,2}} = \{(5,3),(6,3),(8,3)\},$

00000101 $\mathcal{L}_{v_{1,3}} = \{(5,1),(8,4)\},$

00000100 $\mathcal{L}_{v_{1,4}} = \{(1,1)\},$

00001000 $\mathcal{L}_{v_{1,5}} = \{(2,4)\},$

00011000 $\mathcal{L}_{v_{1,6}} = \{(2,1),(2,3),(8,1)\},$

00010000 $\mathcal{L}_{v_{1,7}} = \{(7,4)\},$

00100001 $\mathcal{L}_{v_{1,8}} = \{(2,2),(3,3),(3,4),(4,1)\},$

00100000 $\mathcal{L}_{v_{1,9}} = \{(6,4)\},$

01000000 $\mathcal{L}_{v_{1,10}} = \{(1,3),(5,4),(8,2),(9,1)\},$

10000000 $\mathcal{L}_{v_{1,11}} = \{(4,2)\}.$

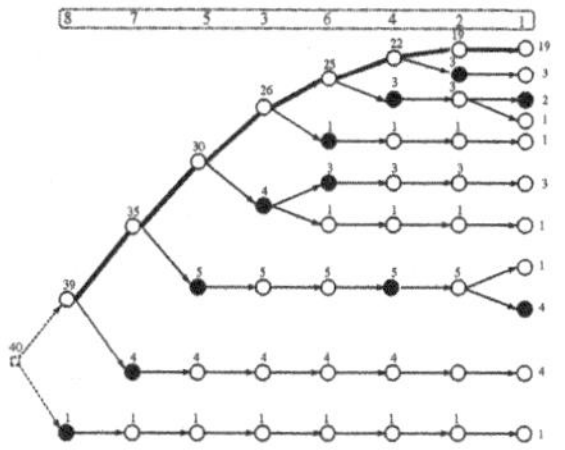

Fig. 2. An input genotype matrix G (70% homozygous) with $n = 10$ samples and $m = 8$ SNPs and $p = 4$. The trie $\mathcal{T}$ constructed by the PiXora algorithm. The root is the leftmost rectangular node. Each hollow circle has a SNP label of 0 and a solid circle a SNP label of 1. The columns are considered (left to right) in the order 8, 7, .., 2, 1. To avoid clutter we only give the cardinality of the list label of each node, which is also reflected in the thickness the incident in-coming edge on the vertex. The solution has 11 distinct haplotypes corresponding to each of the 11 leafnodes. The gold solution (simulated data) has 11 haplotypes with 5 unique occurrences; 2 twice, 2 thrice, 1 four times and 1 occurring twenty-one times in G. The PiXora solution is shown above.

d	j	I^{hom}	I^{het}	HOM	HET
1	8	$\mathcal{I}$	$\emptyset$	$v_{8,1}, v_{8,2}$	—
2	7	$\mathcal{I}$	$\emptyset$	$v_{7,1},$ $v_{7,2}, v_{7,3}$	—
3	5	$\mathcal{I}\setminus I^{\mathrm{het}}$	$\{4\}$	$v_{5,1},$ $v_{5,2}, v_{5,3}$	(see HET 3 below)
4	3	$\mathcal{I}\setminus I^{\mathrm{het}}$	$\{2,8\}$	$v_{3,1}, v_{3,2},$ $v_{3,3},$ $v_{3,4}, v_{3,5}$	(see HET 4 below)
5	6	$\mathcal{I}\setminus I^{\mathrm{het}}$	$\{2,8\}$	$v_{6,1}, v_{6,2},$ $v_{6,4}, v_{6,5},$ $v_{6,6}, v_{6,7}$	(see HET 5 below)
6	4	$\mathcal{I}\setminus I^{\mathrm{het}}$	$\{1,5,8\}$	$v_{4,1}, v_{4,3},$ $v_{4,4}, v_{4,5},$ $v_{4,6},$ $v_{4,7}, v_{4,8}$	(see HET 6 below)
7	2	$\mathcal{I}\setminus I^{\mathrm{het}}$	$\{5,6,8\}$	$v_{2,1}, v_{2,3},$ $v_{2,4}, v_{2,5},$ $v_{2,6}, v_{2,7},$ $v_{2,8}, v_{2,9}$	(see HET 7 below)
8	1	$\mathcal{I}\setminus I^{\mathrm{het}}$	$\{2,3,4,5,8\}$	$v_{1,1}, v_{1,2}$ $v_{1,4}, v_{1,7}$ $v_{1,9}, v_{1,10}$	see below

HET ($d=3$):

i	$v_{7,1}$	$v_{7,3}$	
	28/4	0/0	$G[i][5]$
target	24 1's	*	
4	3	1	0001
⇓			
4	001	0	—
	30/5	1/0	

HET ($d=4$):

i	$v_{5,1}$	$v_{5,2}$	$v_{5,3}$	
	23/1	4/0	3/0	$G[i][3]$
target	22 1's	0's	0's	
2	3	1	×	0011
8	3	×	1	0001
⇓				
2	011	0	×	—
8	001	×	0	—
	26/4	5/0	4/0	

HET ($d=5$):

i	$v_{3,1}$	$v_{3,2}$	$v_{3,3}$	$v_{3,4}$	
	23/0	1/0	4/0	3/0	$G[i][6]$
target	0's	0's	0's	0's	
2	1	2	1	×	0111
8	2	1	×	1	0001
⇓					
2	1	11	0	×	—
8	00	1	×	0	—
	25/1	1/3	5/0	4/0	

HET ($d=6$):

i	$v_{6,1}$	$v_{6,3}$	$v_{3,6}$	
	17/0	2/0	1/0	$G[i][4]$
target	0's	0's	0's	
1	3	×	1	0001
5	3	×	1	0001
8	2	1	1	0001
⇓				
1	001	×	0	—
5	001	×	0	—
8	01	0	0	—
	22/3	3/0	4/0	

HET ($d=7$):

i	$v_{4,1}$	$v_{4,2}$	$v_{4,5}$	$v_{4,6}$	$v_{4,7}$	
	16/0	1/0	0/0	4/0	2/0	$G[i][2]$
target	0's	0's	0's	0's	0's	
5	2	1	×	×	1	0001
6	3	×	×	1	×	0001
8	1	1	1	×	1	0001
⇓						
5	01	0	×	×	0	—
6	001	×	×	0	×	—
8	1	0	0	×	0	—
	19/3	3/0	1/0	5/0	4/0	

Step 3. HET (Ad II) nodes and labels based on global optimization

i	$v_{2,4}$ 0/0 *	$v_{2,5}$ 0/0 *	$v_{2,9}$ 0/0 *	$v_{2,1}$ 14/0 0's	$v_{2,2}$ 1/0 0's	$v_{2,3}$ 1/0 0's	$v_{2,7}$ 1/0 0's	$v_{2,8}$ 2/0 0's	$G[i][1]$ genotype
target	*	*	*	0's	0's	0's	0's	0's	genotype
2	1	2	×	×	×	×	1	×	0001
3	×	×	×	2	×	×	2	×	0011
4	×	×	1	2	×	×	1	×	0001
5	×	×	×	1	1	1	×	1	0001
8	×	1	×	×	1	1	×	1	0001

Initialize all to 0's ⇓

2	0	00	×	×	×	×	0	×	0001
3	×	×	×	00	×	×	00	×	0011
4	×	×	0	00	×	×	0	×	0001
5	×	×	×	0	0	0	×	0	0001
8	×	0	×	×	0	0	×	0	0001

Aim is to align all the 1's, genotypes permitting, along columns

Phase 1: a. For cols with target 1's; place 1's, $G[i][j]$ permitting; update genotype

b. If homozygous in 0's remove the row & update targets ⇓

2	1	00	×	×	×	×	0	×	000
3	×	×	×	00	×	×	00	×	0011
4	×	×	1	00	×	×	0	×	000
5	×	×	×	0	0	0	×	0	0001
8	×	0	×	×	0	0	×	0	0001

⇓ b.

target	×	0's	×	0's	0's	0's	0's	0's	
3	×	×	×	00	×	×	00	×	0011
5	×	×	×	0	0	0	×	0	0001
8	×	0	×	×	0	0	×	0	0001

Phase 2: Repeat a-c until all genotypes empty:

a. (OPM I) Solve *Min Set Cover* i.e., min number of cols covering all rows

(OPM II) Choose from multiple solns to *Min Set Cover* solution: $v_{2,7}$, $v_{2,8}$

– Solution that does NOT generate a new sibling preferred

– Solution that minimizes the sum of the square of the gap between siblings ⇓

b. Assign the 1's in cols $v_{2,7}$, $v_{2,8}$, $G[i][j]$ permitting

c. Update the genotype; If homozygous in 0's; remove the row

3	×	×	×	00	×	×	11	×	−
5	×	×	×	0	0	0	×	1	−
8	×	0	×	×	0	0	×	1	−
	0/1	3/0	0/1	5/0	2/0	3/0	2/2	0/2	
				19/0	3/0	3/0	3/2	2/2	
	$\bar{v}_{1,5}$	$\bar{v}_{1,6}$	$\bar{v}_{1,11}$				v_{new}	v_{new}	

Greedy Min Set cover algorithm:

1. Associate a Weight to the columns by the number of rows they cover (each multiplied by the number in that row).
2. Sort the columns by this weight.
3. Solution: Traverse down the sorted list in descending order till all rows are covered.

A node is *untouchable* if and only if all its ancestral nodes are produced by homozygosity. Every other node is *touchable*.

Priority to fix the matrix, after the 1's have been assigned:

1. Eliminate singletons.
   (a) Mark a singleton column as 1 if you need to remove the 1 and −1 if you need to remove a 0.
   (b) Then, search for pairs of −1 marked columns and 1 marked columns where the exchange can happen, i.e., there exists at least one row in the matrix such that both are not marked X for these 2 columns. Make the exchange.

(c) If columns are still marked; then check if they can be moved without generating new singletons.

(d) If all fails then leave the singletons as they are.

2. Reduce gaps between siblings.

(a) Mark a column as 1 if you need to remove the 1 to get a balance. Mark it as -1 if you need to move the 0 to balance it. A balanced column is marked 0.

(b) Then, search for pairs of -1 marked columns and 1 marked columns where the exchange can happen, i.e., there exists at least one row in the matrix such that both are not marked X for these 2 columns. Make the exchange.

The Heuristics proposed are: (Heuristic I): Using set cover (explained earlier); (Heuristic II) controlling the MAF allele (1's) along each path of the tree or haplotype; (Heuristic III (backtracking)) Collapsing two isomorphic sub-trees:

1. they must have the same values, i.e., 0 or 1 and
2. there must exist a column in the matrix where they can get aligned, i.e., collapsed.

and (Heuristic IV (Trie-shaking)) See Fig. 3.

d	j	I^{hom}	I^{het}	HOM	HET I	HET II
1	8	I	$\emptyset$	$v_{8,1}, v_{8,2}$	−	−
2	7	I	$\emptyset$	$v_{7,1}, v_{7,2}, v_{7,3}$	−	−
3	5	$5\ I \setminus I^{\mathrm{het}}$	$\{4\}$	$v_{5,1}, v_{5,2}, v_{5,3}$	(a) $v_{5,4}$	see sub-table (3)
4	3	$3\ I \setminus I^{\mathrm{het}}$	$\{2,8\}$	$v_{3,1}, v_{3,2}, v_{3,3}, v_{3,4}, v_{3,5}$	−	see sub-table (4)
5	6	$6\ I \setminus I^{\mathrm{het}}$	$\{2,8\}$	$v_{6,1}, v_{6,2}, v_{6,4}, v_{6,5}, v_{6,6}, v_{6,7}$	(b) $v_{6,3}$	see sub-table (5)
6	4	$4\ I \setminus I^{\mathrm{het}}$	$\{1,5,8\}$	$v_{4,1}, v_{4,3}, v_{4,4}, v_{4,5}, v_{4,6}, v_{4,7}, v_{4,8}$	(b) $v_{4,2}$	see sub-table (6)
7	2	$2\ I \setminus I^{\mathrm{het}}$	$\{5,6,8\}$	$v_{2,1}, v_{2,3}, v_{2,4}, v_{2,5}, v_{2,6}, v_{2,7}, v_{2,8}, v_{2,9}$	(b) $v_{2,2}$	see sub-table (7)
8	1	$1\ I \setminus I^{\mathrm{het}}$	$\{2,3,4,5,8\}$	$v_{1,1}, v_{1,2}, v_{1,4}, v_{1,7}, v_{1,9}, v_{1,10}$	(b) $v_{1,3}, v_{1,8}$	see below

HET II sub-table (3):

i	$v_{7,1}$ 28/4	$v_{7,3}$ 0/0	$G[i][5]$
target	24 1's	*	
4	3	1	0001
⇓			
4	001	0	−
	30/5	1/0	

HET II sub-table (4):

i	$v_{5,1}$ 23/1	$v_{5,2}$ 4/0	$v_{5,3}$ 3/0	$G[i][3]$
target	22 1's	0's	0's	
2	3	1	×	0011
8	3	×	1	0001
⇓				
2	011	0	×	−
8	001	×	0	−
	26/4	5/0	4/0	

HET II sub-table (5):

i	$v_{3,1}$ 23/0	$v_{3,2}$ 1/0	$v_{3,3}$ 4/0	$v_{3,4}$ 3/0	$G[i][6]$
target	0's	0's	0's	0's	
2	1	2	1	×	0111
8	2	1	×	1	0001
⇓					
2	1	11	0	×	−
8	00	1	×	0	−
	25/1	1/3	5/0	4/0	

HET II sub-table (6):

i	$v_{6,1}$ 17/0	$v_{6,3}$ 2/0	$v_{3,6}$ 1/0	$G[i][4]$
target	0's	0's	0's	
1	3	×	1	0001
5	3	×	1	0001
8	2	1	1	0001
⇓				
1	001	×	0	−
5	001	×	0	−
8	01	0	0	−
	22/3	3/0	4/0	

HET II sub-table (7):

i	$v_{4,1}$ 16/0	$v_{4,2}$ 1/0	$v_{4,5}$ 0/0	$v_{4,6}$ 4/0	$v_{4,7}$ 2/0	$G[i][2]$
target	0's	0's	0's	0's	0's	
5	2	1	×	×	1	0001
6	3	×	×	1	×	0001
8	1	1	1	×	1	0001
⇓						
5	01	0	×	×	0	−
6	001	×	×	0	×	−
8	1	0	0	×	0	−
	19/3	3/0	1/0	5/0	4/0	

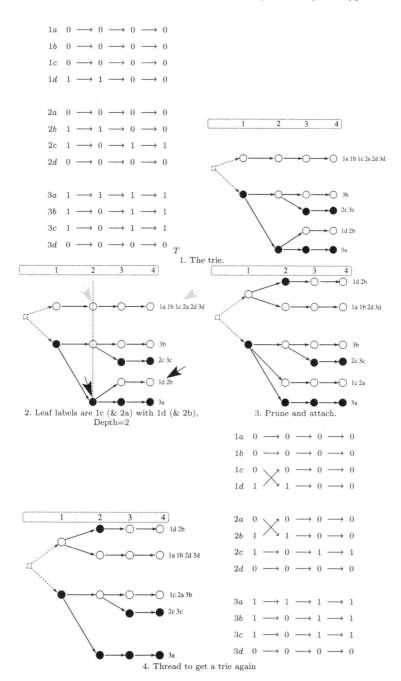

Fig. 3. Trie-shake algorithm. An illustrative example. Only those branches can be exchanged if the leaf labels are from the same individual (for instance 1d with 1c and 2b with 2a). Secondly, the marked internal nodes should be at the same depth and of opposite labels, i.e. one hollow and the other solid.

another run Phase 4. Het II Ad II nodes and labels based on global optimization

	(a)			(b)						
i	$v_{2,4}$	$v_{2,5}$	$v_{2,9}$	$v_{2,1}$	$v_{2,2}$	$v_{2,3}$	$v_{2,6}$	$v_{2,7}$	$v_{2,8}$	$G[i][1]$
	0/0	0/0	0/0	14/0	1/0	1/0	1/0	1/0	2/0	
target	*	*	*	0's	0's	0's	0's	0's	0's	
2	1	2	×	×	×	×	×	1	×	0001
3	×	×	×	2	×	×	×	2	×	0011
4	×	×	1	2	×	×	×	1	×	0001
5	×	×	×	1	1	1	×	×	1	0001
8	×	1	×	×	1	1	×	×	1	0001

⇓ **Step 1**: Initialize all to 0's

i	$v_{2,4}$	$v_{2,5}$	$v_{2,9}$	$v_{2,1}$	$v_{2,2}$	$v_{2,3}$	$v_{2,6}$	$v_{2,7}$	$v_{2,8}$	$G[i][1]$
2	0	00	×	×	×	×	×	0	×	0001
3	×	×	×	00	×	×	×	00	×	0011
4	×	×	0	00	×	×	×	0	×	0001
5	×	×	×	0	0	0	×	×	0	0001
8	×	0	×	×	0	0	×	×	0	0001

⇓ **Step 2**: a. For all cols with target 1's; flip 0's to 1's, $G[i][j]$ permitting
    b. Update the genotype col; If homozygous in 0's; remove the row
    −no such col−

Step 3: Repeat a-d until no change:
    a. Solve *Min Set Cover* i.e., min number of cols covering all rows
    One *Min Set Cover* solution: $v_{2,3}$, $v_{2,7}$
⇓   b. Assign the 1's in cols $v_{2,3}$, $v_{2,7}$, $G[i][j]$ permitting
    c. Update the genotype col; If homozygous in 0's; remove the row
    d. If col's target is fulfilled remove the col

i	$v_{2,4}$	$v_{2,5}$	$v_{2,9}$	$v_{2,1}$	$v_{2,2}$	$v_{2,3}$	$v_{2,6}$	$v_{2,7}$	$v_{2,8}$	$G[i][1]$
2	0	00	×	×	×	×	×	1	×	0001
3	×	×	×	00	×	×	×	11	×	0011
4	×	×	0	00	×	×	×	1	×	0001
5	×	×	×	0	0	1	×	×	0	0001
8	×	0	×	×	0	1	×	×	0	0001
	1/0	3/0	1/0	5/0	2/0	0/2	0/0	0/4	2/0	
				19/0	3/0	1/2	1/0	1/4	4/0	
	$v_{1,5}$	$v_{1,6}$	$v_{1,11}$			$v_{1,3}$		$v_{1,8}$		

6.2   On the Algebraic Method

Notation and Basic Definitions. Each element of the matrix is a genotype, say X.

Definition 1 (coded genotype X and $x_1, x_0, x_l, x_L, X_p, \{X\}$ of X). *Each genotype is equivalent to a 3-tuple (triple)*

$$X \equiv (x_1, x_0, x_l), \text{ where } \begin{cases} x_1 & number\ of\ 1's \\ x_0 & number\ of\ 0's \\ x_l & number\ of\ variables \\ x_L & is\ the\ set\ of\ variables\ and \\ & x_l = |x_L| \\ X_p & is\ the\ ploidy\ and \\ & X_p = x_1 + x_0 + x_l \\ \{X\} & set\ coded\ by\ X \end{cases}$$

The implementation tracks the states of the variables as v or $\bar{v}$ where $v \in x_L$. For brevity, we skip the details. Some concrete illustrative examples of genotypes:

X	3-tuple	$\{X\}$	$|X|$	X_p	x_L
11100	$(3,2,0)$	{ 11100 }	1	5	$\emptyset$
1110q	$(3,1,1)$	{ 11100 , 11110 }	2	5	$\{q\}$
11qr	$(2,0,2)$	{ 1100 , 1110 , 1111 }	3	4	$\{q,r\}$

Definition 2 (VOID, empty genotype). *A coded genotype X is empty when* $\{X\} = \emptyset$. *X is VOID when $x_1 < 0$ or $x_0 < 0$ holds.*

Definition 3 ($X \leq Y$). *Let X and Y be genotypes. $X \leq Y \Leftrightarrow x_1 \leq y_1, x_0 \leq y_0, x_l \leq y_l$.*

Lemma 3. *For a genotype $X \equiv (x_1, x_0, x_l)$ the following hold:*

1. $|X| = x_l + 1$.
2. $X \subseteq Y \Leftrightarrow (X_p = Y_p)$ *AND* $y_1 \leq x_1 \leq x_1 + x_l \leq y_1 + y_l$.

Sketch of Proof: 2. Since $X_p = Y_p$, it is adequate to base the arguments only on the number of 1's in X and Y. The possible number of 1's in X is in the interval $[x_1, x_1 + x_l]$ and similarly in Y. So if $X \subseteq Y$, then $[x_1, x_1 + x_l]$ is contained in $[y_1, y_1 + y_l]$ and vice-versa, leading to the above.   □

Definition 4 ($\langle X \rangle$, ploidy $\langle X \rangle_p$). *$\langle X \rangle$ is defined to an ordered finite list of coded genotypes $X^1, X^2, \ldots, X^j \ldots$ with the same ploidy k. Then k is defined to be $\langle X \rangle_p$, the ploidy of the $\langle X \rangle$.*

Algebra of Genotypes

Resolving Variables. Two *randomized* procedures variable-to-constant (v2c) and variable-to-variable (v2v) are defined below. Also, a composition of these two primitive operations in resVar() on two coded genotypes.

input conds	operation details	comments
	$X' = \mathrm{res}(X)$	
$x_L \neq \emptyset$	For each pair $v, \bar{v} \in x_L$ $x_1' = x_l + 1, x_0' = x_0 + 1,$ $x_L' \leftarrow x_L \setminus \{v, \bar{v}\}, x_l' = x_l - 2$	(clean up of var set x_L; can be periodically invoked) $X_p' = X_p$
	$X' = \mathrm{v2c}(k, d, X)$	
$0 \leq k \leq x_l$ $d = 0, 1$	Pick some k vars, $v_1, \ldots, v_k$ in x_L $x_d' = x_d + k$ $x_L' = x_L \setminus \{v_1, \ldots, v_k\}; x_l' = x_l - k$	(some k vars in x_L are assigned d) $\{X'\} \subset \{X\}$ $X_p' = X_p$
	$y_L' = \mathrm{v2v}(k, x_L, y_L)$	
$0 \leq k \leq |x_L|, |y_L|$	some k vars in y_L are assigned to some k vars in x_L VOID, if no such mapping exists (note a var v can be assigned to itself but not to its complement)	x_L does not change; y_L is modified, i.e., k vars in y_L are re-labeled, & so are all the upstream instances (Note if Y' is the modified Y then $\{Y'\} = \{Y\}; Y_p' = Y_p$)

$(X_{\mathrm{new}}, Y_{\mathrm{new}}) = \mathrm{resVar}(t_{x1}, t_{x0}, t_{y1}, t_{y0}, t_v, X, Y)$	
Partition x_L and y_L as follows $x_L = S_{x1} \uplus S_{x0} \uplus S_{xv} \uplus S_{xr}$ $y_L = S_{y1} \uplus S_{y0} \uplus S_{yv} \uplus S_{yr},$ with $|S_{xv}| = |S_{yv}| = t_v,$ for $d = 0, 1 : |S_{xd}| = t_{xd}; |S_{yd}| = t_{yd},$ and $(S_{x1} \cup S_{x0} \cup S_{xv}) \cap (S_{y1} \cup S_{y0} \cup S_{yv}) = \emptyset$. IF the above partitioning is not possible THEN resVar() is VOID t_v vars in S_{yv} are randomly assigned to vars in S_{xv} For $d = 0, 1$ the vars of S_{xd} are assigned d For $d = 0, 1$ the vars of S_{yd} are assigned d $X_{\mathrm{new}} = (x_1 + t_{x1}, x_0 + t_{x0}, S_{xv} \cup S_{xr})$ $Y_{\mathrm{new}} = (y_1 + t_{y1}, y_0 + t_{y0}, S_{yv} \cup S_{yr})$	$\uplus$ denotes disjoint union

Primitive Genotype Operations. When X and Y are two given coded genotypes, Z is produced based on the operations as follows:

Z	z_1	z_0	z_L	comments
	symmetric & transitive binary operations			
	$\min\{x_1,y_1\}$	$\min\{x_0,y_0\}$	$x_L \cup \{q_1,...,q_k\}$	if cond I $\left\{\begin{array}{l} x_l = y_l \\ y_L' = \text{v2v}(x_l, x_L, y_L) \\ \|x_1 - y_1\| = k \\ \text{New variables } q_1,...,q_k \end{array}\right.$
$X \cup Y$ when $X_p = Y_p$ (randomized)	y_1	y_0	y_L	if cond II $\left\{\begin{array}{l} x_l < y_l\ \&\\ X \subseteq Y \end{array}\right.$ $x_L' = \text{v2v}(y_l, y_L, x_L)$
	x_1	x_0	x_L	if cond III $\left\{\begin{array}{l} y_l < x_l\ \&\\ Y \subseteq X \end{array}\right.$ $y_L' = \text{v2v}(x_l, x_L, y_L)$
	VOID (no unique triple notation)			otherwise
$X \cap Y$	$\min_1$	$\min_0$	$x_L \cap y_L$	$\left\{\begin{array}{l} \min_1 = \min\{x_1,y_1\} \\ \min_0 = \min\{x_0,y_0\} \end{array}\right.$
	EMPTY when $Z \equiv (0,0,\emptyset)$			
$X \cap_k Y$ (randomized)	$\bigcup_{W_p=k} (W \leq (X \cap Y))$			$k \leq (X \cap Y)_p$ Note $Z_p = k$
	unary operations			
$X_{v \leftarrow 1}$ $X_{v \leftarrow 0}$	$x_1 + 1$ x_1	x_0 $x_0 + 1$	$x_L \setminus \{v\}$	$Z = \text{v2c}(1, 1/0, X)$ with $v \in x_L$
	asymmetric binary operations			
$X \setminus Y$ when $X_p \geq Y_p$	$x_1 - y_1 - y_l'$	$x_0 - y_0 - y_l'$	$x_L' \cup \bar{y}_L'$	$\left\{\begin{array}{l} x_L' = x_L \setminus (x_L \cap y_L) \\ y_L' = y_L \setminus (x_L \cap y_L) \\ \bar{y}_L' = \{\bar{v} \mid v \in y_L'\} \\ \text{Note } Z_p = X_p - Y_p \end{array}\right.$
	VOID if $z_1 < 0$ or $z_0 < 0$			

VOID/Empty Genotypes. When a primitive operation fails, i.e., either results in an empty genotype $Z \equiv (0, 0, \emptyset)$ or at least one of z_0, z_1 is negative, then we resolve some of the variables, either by assigning explicit 1 or 0 (v2c) or assigning it to other variables (v2v). Note that if z_L is empty, then there is no variable to resolve and this failure cannot be rescued (unless the model admits possible errors in the input). However, when z_L is non-empty, there is a possibility that it can be rescued and in the following operations we minimize the number of resolved variables to do so:

$Z = X \bullet_f Y$	randomization component in $\bullet_f$	comments/resolve vars
	For $d = 0, 1$ $\text{gap}_d = y_d - x_d$; $\text{buff}_d = x_d - y_d$; $\text{nN}(\text{gap}_d)$; $\text{nN}(\text{buff}_d)$	(force a negative val v to 0) $\text{nN}(v) \equiv$ IF $v < 0$ THEN $v \leftarrow 0$
$X \cap_f^t Y$ when $X \cap Y$ is EMPTY t is the desired increase in ploidy of Z	t is randomly split into 3 non-negative integers, i.e. a random point in shaded region of Fig 4, as $t = (t_{1x} \oplus t_{1y}) + (t_{0x} \oplus t_{0y}) + t_v$ where all the following hold: $t_v \leq \min(x_l, y_l)$ $(t_{1x} \oplus t_{0x}) + t_v \leq x_l$ $(t_{1y} \oplus t_{0y}) + t_v \leq y_l$ and For $d = 0, 1$ IF $y_d - x_d \geq 0$, $t_{dx} \leq \min(\text{gap}_d, x_l)$; $t_{dy} = 0$ ELSE $t_{dx} = 0$; $t_{dy} \leq \min(\text{buff}_d, y_l)$	$t_{1x} \oplus t_{1y}$ denotes one of t_{1x}, t_{1y} is positive, but not both Similarly $t_{1y} \oplus t_{0y}$ $(X_{\text{new}}, Y_{\text{new}}) =$ resVar$(t_{1x}, t_{0x}, t_{1y}, t_{0y}, t_v, X, Y)$ $Z = X_{\text{new}} \cap Y_{\text{new}}$ $(Z_p = t)$
$X \setminus_f Y$ when $X \setminus Y$ is VOID	(at least $y_l' + \text{gap}_1 + \text{gap}_0$ vars needs to be resolved) $\text{gap}_1 + \text{gap}_0 \leq x_l'$ must hold ELSE VOID y_l' is randomly split into 3 non-negative integers, i.e. a random point in shaded region of Fig 4, as $y_l' = t_1 + t_0 + t_v$ where the following hold: $t_1 \leq \text{buff}_1$; $t_0 \leq \text{buff}_0$ $t_v \leq x_l' - (\text{gap}_1 + \text{gap}_0)$	x_l', x_L', y_l', y_L' from $X \setminus Y$ defn $(X_{\text{new}}, Y_{\text{new}}) =$ resVar$(\text{gap}_1, \text{gap}_0, t_1, t_0, t_v, X, Y)$ $Z = X_{\text{new}} \setminus Y_{\text{new}}$

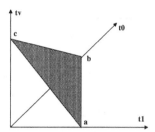

Fig. 4. The bounded plane of plausible solutions of variable resolution in the t_1, t_0, t_v space. In $\cap_f$ operation: a = gap$_1$ or buff$_1$; b = gap$_0$ or buff$_0$; c = $\min(x_l, y_l)$. In $\setminus_f$ operation: a = buff$_1$; b = buff$_0$; c = x_l' - (gap$_1$ + gap$_0$).

Lemma 4. *1. If $Z = X \cap_k Y$, then $Z_p = k$.*
2. If $Z = X \setminus Y$, then $Z_p = X_p - Y_p$.

Sketch of Proof: 1. Note that the union is over genotypes that each have a ploidy of k. Since the union operation maintains the ploidy, the result must hold.
2. If the operation is not a failure, then

$$\begin{aligned}
Z_p &= z_1 + z_0 + z_l \\
&= (x_1 - y_1 - y_l') + (x_0 - y_0 - y_l') + (x_l' + y_l') \\
&= (x_1 + x_0 + x_l') - (y_1 + y_0 + y_l') \\
&= (x_1 + x_0 + x_l' + |x_L \cap y_L|) - (y_1 + y_0 + y_l' + |x_L \cap y_L|) \\
&= X_p - Y_p.
\end{aligned}$$

$\square$

Primitive Operation Illustrative Examples

Z	X	Y	Z	X	Y	Z
$X \cup Y$	110vq	100qr	10wqv	$(2,1,\{v,q\})$	$(1,2\{q,r\})$	$(1,1,\{w,q,r=v\})$
$X \cup Y$	110vq	1100r	110wvq	$(2,1,\{v,q\})$	$(2,2,\{r\})$	$(2,1,\{q=r,v,w\})$
$X \cap Y$ $X \cap_2 Y$	1100qr	1000q	100q bq	$(2,2,\{q,r\})$	$(1,3,\{q\})$	$(1,2,\{q\})$ $(0,0,\{b,q\})$
$X \cap Y$ $X \cap_3 Y$	1000qr	100rq	100rq 0ar	$(1,3,\{q,r\})$	$(1,2,\{q,r\})$	$(1,2,\{r,q\})$ $(0,1,\{a,r\})$
$X \setminus Y$	110q	rq	$1\bar{r}$	$(2,1,\{q\})$	$(0,0,\{r,q\})$	$(1,0,\{\bar{r}\})$
$X \setminus Y$	1100qa	10rq	$a\bar{r}$	$(2,2,\{q,a\})$	$(1,1,\{r,q\})$	$(0,0,\{a,\bar{r}\})$
$X \setminus Y$	110qr	0abc	FAILURE	$(2,1,\{q\})$	$(0,1,\{a,b,c\})$	FAILURE

Any negative value of the tuple is flagged as VOID. The relaxed intersection $X \cap_2 Y$ is carried out as follows.

$$X \cap Y = (1,2,1) = (1,2,\{q\}) \hspace{3cm} \text{note var } r \text{ is lost}$$

$$X \cap_2 Y = \bigcup \left(\underbrace{(1,0,\{q\}), (0,1,\{q\})}, \underbrace{((1,1,\emptyset), (0,2,\emptyset))} \right) \hspace{1cm} \text{by defn of } \cap_k; \text{ transitivity}$$

$$= (0,0,\{a,q\}) \bigcup (0,1,\{b\}) \hspace{3cm} \text{cond I; new vars } a \; b$$

$$= (0,0,\{a=b,q\}) \hspace{2cm} \text{cond III; internal var } a \text{ can be dropped}$$

The next example:

$$X \cap Y = (1,2,2) = (1,2,\{r,q\}).$$

$$X \cap_3 Y = \bigcup \left((1,1,\{q\}), (0,2,\{q\}), (1,1,\{r\}), (0,2,\{r\}), (0,1,\{r,q\}), (1,0,\{r,q\}), (1,2,\emptyset) \right)$$

$$= \underbrace{((0,1,\{a,q\}) \bigcup (0,1,\{b,r\}))} \bigcup \left(\underbrace{(0,1,\{r,q\}) \bigcup (1,0,\{r,q\})} \right) \bigcup (1,2,\emptyset) \hspace{1cm} \text{cond I in both}$$

$$= (0,1,\{a=b,q=r\}) \bigcup \left(\underbrace{(0,1,\{r,q\}) \bigcup (1,0,\{r,q\})} \right) \bigcup (1,2,\emptyset) \hspace{1cm} \text{cond II}$$

$$= (0,1,\{a=b,q=r\}) \bigcup (1,2,\emptyset) \hspace{1cm} \text{discard genotypes whose ploidy<3 due to var identity } r=q$$

$$= (0,1,\{a,r=q\}) \hspace{2cm} \text{cond III; internal new vars b, c, dropped}$$

Operations on Row $\langle X \rangle$

Definition 5 ($\langle X \rangle \cap \langle Y \rangle, \langle X \rangle \setminus \langle Y \rangle$). *If $\langle X \rangle$ and $\langle Y \rangle$ are two rows then The intersection and difference operations on $\langle X \rangle$ and $\langle Y \rangle$ are defined as:*

$$\langle X \rangle \cap \langle Y \rangle = \langle X \cap_k Y \rangle, \text{ where } k = \min_j \left\{ (X^j \cap Y^j)_p \right\}, \hspace{2cm} (5)$$

$$\langle X \rangle \setminus \langle Y \rangle = \langle X \setminus Y \rangle. \hspace{4cm} (6)$$

Executing the Row-Row Operation. Let S_x be the sample haplotypes associated with ith row say $\langle X \rangle$ and S_y be the sample haplotypes associated with i'th row say $\langle Y \rangle$. Note that the set S tracks multiplicities as well, i.e., multiple haplotypes of the same sample. In other words if $S = \{a(2), b\}$, this is interpreted as two haplotypes of sample a and 1 haplotype of sample b.

The row-row operation on $\langle X \rangle$ and $\langle Y \rangle$ is defined as follows.

- **Case I** $X_p > 1$, $Y_p > 1$: The intersection or overlap operation between the two row results in the following three new rows (that replace the ith and i'th rows):
  1. $\langle Z \rangle \leftarrow \langle X \rangle \cap \langle Y \rangle$ with $S_z = S_x \cup S_y$ and $\langle Z \rangle_p = k$, where k is defined in Eq. 5.
  2. $\langle V \rangle \leftarrow \langle X \rangle \setminus \langle Z \rangle$ with $S_v = S_x$ and $\langle V \rangle_p = \langle X \rangle_p - k$.
  3. $\langle W \rangle \leftarrow \langle Y \rangle \setminus \langle Z \rangle$ with $S_w = S_y$ and $\langle W \rangle_p = \langle Y \rangle_p - k$.
- **Case II** $X_p > 1$, $Y_p = 1$: The intersection or overlap operation between the two row results in the following new row (that replace the ith row):
  1. $\langle V \rangle \leftarrow \langle X \rangle \setminus \langle Y \rangle$ with $S_v = S_x$ and $\langle V \rangle_p = \langle X \rangle_p - 1$.
  2. $S_y \leftarrow S_y \cup S_x$.

Row-Row Operation FAILURE. Let X and Y be two genotypes. Then $X \cap Y$ is successful, if and only if the following hold.

- **Case I** $X_p > 1$, $Y_p > 1$: None of the following result in EMPTY/VOID: (1) $Z = X \cap Y$ (2) $X \setminus Z$ and (3) $Y \setminus Z$.
- **Case II** $X_p > 1$, $Y_p = 1$: $X \setminus Y$ is not VOID.

Use "$\cap_f$" instead of "$\cap$" and "$\setminus_f$" instead of "$\setminus$" for the genotype pair when there is EMPTY or VOID result.

Empirical Lemmas. Let n be the number of samples and m the number of SNPs.

Lemma 5. *Accuracy of the algorithm improves with increase in n and m.*

Lemma 6. *For a given fraction of heterozygous alleles and m, the value of n can be estimated where the accuracy of reconstruction saturates.*

References

1. Aguiar, D., Istrail, S.: Haplotype assembly in polyploid genomes and identical by descent shared tracts. Bioinformatics **29**(13), i352–i360 (2013). https://doi.org/10.1093/bioinformatics/btt213. http://dx.doi.org/10.1093/bioinformatics/btt213
2. Browning, S., Browning, B.: Haplotype phasing: existing methods and new developments. Nat. Rev. Genet. **12**, 703 (2011)
3. Chaisson, M.J., Mukherjee, S., Kannan, S., Eichler, E.E.: Resolving multicopy duplications *de novo* using polyploid phasing. In: Sahinalp, S.C. (ed.) RECOMB 2017. LNCS, vol. 10229, pp. 117–133. Springer, Cham (2017). https://doi.org/10.1007/978-3-319-56970-3_8
4. Halldórsson, B.V., Bafna, V., Edwards, N., Lippert, R., Yooseph, S., Istrail, S.: A survey of computational methods for determining haplotypes. In: Istrail, S., Waterman, M., Clark, A. (eds.) RSNPsH 2002. LNCS, vol. 2983, pp. 26–47. Springer, Heidelberg (2004). https://doi.org/10.1007/978-3-540-24719-7_3

5. He, D., Saha, S., Finkers, R., Parida, L.: Efficient algorithms for polyploid haplotype phasing. BMC Genom. **19**(2), 110 (2018)
6. Motazedi, E., Finkers, R., Maliepaard, C., de Ridder, D.: Exploiting next-generation sequencing to solve the haplotyping puzzle in polyploids: a simulation study. Brief. Bioinform. **19**(3), 387–403 (2018)
7. Siragusa, E., Haiminen, N., Utro, F., Parida, L.: Linear time algorithms to construct populations fitting multiple constraint distributions at genomic scales. IEEE/ACM Trans. Comput. Biol. Bioinform. **16**, 1132–1142 (2018)
8. Siragusa, E., Haiminen, N., Finkers, R., Visser, R., Parida, L.: Haplotype assembly of autotetraploid potato using integer linear programing. Bioinformatics (2019). https://doi.org/10.1093/bioinformatics/btz060
9. Utro, F., et al.: iXora: exact haplotype inferencing and trait association. BMC Genet. **14**(1), 48 (2013)

Classification of Epileptic Activity Through Temporal and Spatial Characterization of Intracranial Recordings

Vanessa D'Amario[1(✉)], Gabriele Arnulfo[1], Lino Nobili[2,3], and Annalisa Barla[1]

[1] DIBRIS, Università degli Studi di Genova, Genova, Italy
[2] Ospedale Niguarda Ca' Granda, Milano, Italy
vanessa.damario@dibris.unige.it, gabriele.arnulfo@edu.unige.it,
annalisa.barla@unige.it
[3] DINOGMI, Università degli Studi di Genova, Genova, Italy
lino.nobili@unige.it

Abstract. Focal epilepsy is a chronic condition characterized by hyperactivity and abnormal synchronization of a specific brain region. For pharmacoresistant patients, the surgical resection of the critical area is considered a valid clinical solution, therefore, an accurate localization is crucial to minimize neurological damage. In current clinical routine the characterization of the Epileptogenic Zone (EZ) is performed using invasive methods, such as Stereo-ElectroEncephaloGraphy (SEEG). Medical experts perform the tag of neural electrophysiological recordings by visually inspecting the acquired data, a highly time consuming and subjective procedure. Here we show the results of an automatic multi-modal classification method for the evaluation of critical areas in focal epileptic patients. The proposed method represents an attempt in the characterization of brain areas which integrates the anatomical information on neural tissue, inferred using Magnetic Resonance Imaging (MRI) in combination with spectral features extracted from SEEG recordings.

Keywords: Focal epilepsy · Machine learning · Spectral analysis · Signal processing · Multi-modal data analysis

1 Introduction

Epilepsy is a neurological disorder characterized by abnormal neural activity that leads to abrupt onset of seizures. Among world population, about 50 million people suffer from generalized epilepsy. For the majority of the affected patients, symptoms can be pharmacologically controlled. Unfortunately, 30% of patients are refractory to medication and, when diagnosed with focal onset, brain surgery can be considered as treatment. In these cases, complex and multimodal investigations are mandatory to accurately localize the Epileptogenic

© Springer Nature Switzerland AG 2020
M. Raposo et al. (Eds.): CIBB 2018, LNBI 11925, pp. 69–79, 2020.
https://doi.org/10.1007/978-3-030-34585-3_6

Zone (EZ), defined as the minimum amount of cortex that should be removed to produce seizure-free subjects. While the patient is hospitalized, multiple Magnetic Resonance Imaging tests (MRIs) as well as scalp electroencephalography data are acquired to define putative EZ. Nevertheless, this protocol shows clear evidence of malformations (e.g., tumors or dysplasia) only in a small percentage of patients candidate for surgery. Moreover, even in presence of positive results the border of the EZ or the localization of the onset zone might be elusive. In these cases, neurophysiologists require the acquisition of invasive intracerebral recordings such as Stereo-Electroencephalography (SEEG) [1]. It consists in the implantation in the brain tissue of depth filiform electrodes, whose number depends on the severity of the case, each endowed with several acquisition channels, that record local field potential at high sampling frequency.

Clinicians then perform a very time-consuming and highly subjective visual inspection on the signal acquired from each channel, looking for epileptic biomarkers, such as spike or spike-and-wave patterns and characterizing the relationship between brain regions by co- or lagged-occurrence of these pathological patterns [2]. This tagging procedure is a time- and resource-consuming task, with medical experts spending, on average, about two hours for the analysis of a 10 min neural activity recording. Moreover, even if SEEG is a highly precise acquisition method, surgical resection, does not lead to positive outcomes in a relevant portion of patients [3]. Among the possible reasons of this unsatisfactory success rate there is the highly subjectivity due to the EZ identification procedure.

Therefore, it is clear that the definition of an automatic tool for the detection of the pathological tissue may prove as a great advancement in this context.

1.1 Scientific Background

Most of the available works are not only restricted to the localization of the epileptogenic areas but also to seizure prediction.

Since the advent of new medical devices, developed to monitor the neural activity and to forecast critical events, high interest concerns the transition from *interictal* to *ictal* states, that is the change from a normal state to a seizure. Motivated by this, a great deal of recent literature focuses on automatic seizure detection tools. We mention, for instance the Kaggle challenge on the Melbourne dataset[1].

State of the art methods both for classification epileptogenic areas and seizure forecasting perform a feature extraction stage. The scientific community agrees in the importance of spectral quantities both in temporal and frequency domain as descriptors of the neurological signal. Relative amount of signal power in frequency bands of interests and temporal events characterized by critical amplitudes are indeed considered pathological biomarkers [4–7]. In Truong [4] et al. the authors analyze the Melbourne dataset using standard spectral analysis

---

[1] https://www.kaggle.com/c/melbourne-university-seizure-prediction.

methods as Fast Fourier Transform (FFT) to measure correlations across channels, in order to infer which are the mostly involved in the seizure generation. In the work of Vila et al. [5] some criteria for the localization of Seizure Onset Zone (SOZ) are established using spectral measures. The study is performed on seven patients, and characterizes the transition between interictal and ictal states. Mean activation measure, defined as the average of the instantaneous activity during the seizure epochs and relative time average power of every channel is shown to be significantly higher in SOZ across all patients in the α (8–13 Hz) and β (13–30 Hz) rhythms. Oscillations in the β and γ (30–70 Hz) range, rapid discharges, spectral and temporal aspects are taken into account to discriminate SOZ from physiological areas. The presence of rapid discharges in a given brain area immediately before the ictal state is also shown to be a good measure of epileptogenicity of the zone [6]. In this regard significant changes of the activity are evaluated using thresholds. The authors demonstrate a statistical correlation between duration of high energy phenomena and epileptogenicity of the area.

In this work we define a feature extraction pipeline which leverages on spectral features in line with the ones defined above. In particular we are interested in (i) implementing the feature extraction in such a way to obtain a characterization of the neural signal independent from the specific patient, so to realize an automatic classification method across patients, (ii) including anatomical knowledge derived from imaging test (MRI) and (iii) measuring the classification performances of our pipeline in the analysis of interictal signal.

The paper is organized as follows: Sect. 2 provides a description of the SEEG and MRI dataset, the feature extraction pipeline and the machine learning methods used for the analysis. Section 3 regards the statistics and the obtained results, in Sect. 4 we conclude by describing the ongoing work on epileptic signal classification through network analysis.

2   Material and Methods

We acquired the dataset at the Hospital Niguarda Neurology Unit (Milan, Italy). Patients provided written consent for the analysis of the data. The dataset consists of SEEG and MRI data for forty patients and SEEG data only for another set of twenty patients.

We registered local field potential with common reference in white matter, using platinum-iridium, multi-lead electrodes. The number of contacts for each electrodes varies from 8 to 15, each is 2 mm long, 0.8 mm of thickness and have distance of 1.5 mm from its neighbours (DIXI medical, Besancon, France). We acquired 10 min of spontaneous resting state activity, at a sampling frequency of 1 kHz, with eyes closed, using a 192-channels SEEG amplifier system (NIHON-KOHDEN NEUROFAX-110). We automatically ascertained the position of each recording contact using a dedicated segmentation software [8].

We fused MRI-pre with CT-post (Computed Tomography) using affine rigid-body coregistration [1]. After the coregistration phase, the algorithm automatically segments each contact contained in the multi-lead electrodes by searching its center of mass.

The total amount of channels for this set of patients is 5315, only 1342 have been marked as pathological by an equipe of medical experts. On average, the total number of channels per patient is 140 ± 20 (mean $\pm$ std), of which 34 ± 21 (mean $\pm$ std) are epileptogenic or characterized by critical activity.

For what concerns the SEEG acquisition, the extraction of relevant spectral features is first preceded by a preprocessing stage which consists of two steps: (i) local reference of potential, (ii) removal of power line effects. We proceed in the former case to the computation of the potential difference of neighbour channels on the same electrode. This local reference of potential is shown to decrease the correlation of spurious electrical activity which propagates through fibers in the white tissue [9] and for this reason it is preferable than average reference of the potential or other settings. We remove the power line effects through notch filters peaked at 50 Hz and harmonics (Butterworth, 2nd order).

2.1   Time-Frequency Features

In order to get effective descriptors of the average activity in the interictal stage, we extracted spectral features on temporal windows of 300 s. First, we measured basic features as the first moments: *variance, skewness* and *kurtosis*.

Then, to capture the variability across patients, we measure the *mean energy* values [7] for slow rhythms ($f < 1$ Hz, B0), the frequency bands of clinical interest δ 1–4 Hz (B1), θ 4–8 Hz (B2), α 8–13 Hz (B3), β 13–30 Hz (B4), γ 30–70 Hz (B5), high-γ 70–90 Hz (B6). Recent results [10–12] show a relevant contribution of high frequency patterns in determining the pathological state. To this aim, we include in the analysis mean energy values at frequencies higher than high-γ band. Band pass filters of width 50 Hz has been used, spanning the frequency space from high-γ up to Nyquist frequencies. We define the higher frequency bands as B7 90–140 Hz, B8 140–190 Hz, B9 190–240 Hz, B10 240–290 Hz, B11 290–340 Hz, B12 340–390 Hz, B13 390–440 Hz, B14 440–490 Hz.

We use an orthogonal discrete mother wavelet (Daubechies, 2nd order) to perform the decomposition of the signal in approximation and detail coefficients, with cD a list of coefficients of different length, dependent from the scale considered $[cA, cD] = DWT(X)$. Similarly to the relative energy, we compute the relative amplitude at each scale by summing the square of wavelet coefficients in the temporal dimensions $E_s = \sum_{j \in \text{time}} (cD_s)^2$. For each scale we divide this quantity by the sum at all scales $E_s / \sum_{k \in \text{scale}} E_k$. We also evaluate the *wavelet entropy* measure, which has been shown to be a discriminative quantity in the evaluation of signal coherence in neurophysiology [13], especially in pathological activity detection [14].

As in Bartolomei et al. [6] we also measure the *hyperactivity* of each channel, defined as abnormal signal amplitudes, with respect to the baseline activity. To distinguish between baseline and hyperactivity, we set different threshold values on the filtered signal at different bands. We use a 2nd order Butterworth filter for each of the frequencies defined above to estimate the length of hyperactivity periods. The thresholds were learned on the 20 subjects with only the SEEG data available. We excluded this subset of patients from the rest of the classification pipeline, in order to prevent from potential overfit issues. From these recordings, we consider the 1968 physiological channels to estimate an adaptive signal baseline. We first filter these recordings in the frequency bands defined above, and consider the standard deviation of the activity at each band. We denote through $\bar{\sigma}_f$ the mean standard deviation for the band f for this subset of patients. The values relative to each band are shown in Table 1. Then we measure the time spent over a variable threshold $a \cdot \sigma_f$, where a assumes discrete values in the range $[2, 7]$, by first filtering the time series in the band f.

Table 1. Mean values of $\bar{\sigma}_f$ for different energy bands f evaluated on 1968 physiological channels for which the MR images where not available. These values can be considered as a physiological standard activity across patients

Band	B0	B1	B2	B3	B4	B5	B6	B7
$\bar{\sigma}_f$ [μV]	17.24	13.19	15.54	12.45	11.09	4.58	1.19	1.07
Band	B8	B9	B10	B11	B12	B13	B14	
$\bar{\sigma}_f$ [μV]	0.578	0.391	0.298	0.244	0.203	0.183	0.190	

2.2 Spatial Feature

To the best of our knowledge there has been no attempt to integrate anatomical quantities together with spectral features in automatic learning pipelines for detection of critical areas.

With this regard, the recent work of Mercier [9] gives a broad insight in the role played by white matter in the signal propagation through the brain. He points out the improvement in the analysis of brain potential obtained by computing local difference of potential for neighboring channels, in order to decouple spurious activity. His work also shows the relevance of quantifying the anatomical nature of the brain tissue in the area of acquisition of the signal.

The characterization of brain regions is based on the differentiation of gray and white matter evaluated through `FreeSurfer` [15], a software tool which parcellates cortical and subcortical regions from MRI acquisition. By defining each voxel imaging as 1 mm^3, the Partial Tissue Density (PTD) index is defined as follows

$$\text{PTD} = \frac{\text{Vox Gray} - \text{Vox White}}{\text{Vox Gray} + \text{Vox White}} \tag{1}$$

where Vox Gray and Vox White correspond respectively to the number of gray and white voxels contained in a volume of $3 \times 3 \times 3$ mm^3 centered around the electrode position [9]. Indeed in the same work, the electrode position is proved to be crucial for clinical evaluations, as there is a high correlation between signal power and PTD index, and signal amplitude is greater in the gray matter than in the white matter regions. It is well known in the clinical routine that the amplitude of the signal at low frequencies is a pathological biomarker of the epileptic activity, which is originated by the excitation of neural population, localized in the gray matter. The PTD quantifies the proximity between the gray matter and white matter and assumes continuous values between $[-1, 1]$.

2.3   Data Representation

Each sample collects the features computed for a fixed channel position for a window of 5 min of activity. By applying the feature extraction pipeline we get the data matrix used for classification, which contains 10630 samples, each described by 156 time-spatial features.

2.4   Machine Learning Methods

We consider several machine learning techniques for classification, both linear and non-linear. In particular, we use sparse Logistic Regression (LR) [16], Support Vector Machines with linear kernel (SVMs) [17], Random Forest (RF) [18] and Gradient Boosting (GB) [19]. All the learning methods require the tuning of hyperparameters, which is performed through cross validation.

We used several metrics for the evaluation of results, performed on the test set, which take into account the unbalance of the classes, such as Precision (P), Recall (R), Balanced Accuracy (BA) [20] and F1 score (F1).

3   Experimental Results

Classification of Channels Activity. The proportion of epileptic and non epileptic channels is unbalanced in favor of non epileptic channels, with random guess corresponding to 0.74%. For this reason, we computed the performance of our methods using metric scores that take into account the unbalance of the dataset. We split the dataset in 85% samples for learning and 15% for test, using automatic `scikit-learn` [21] procedures that split the dataset with respect to the unbalance of the original problem. The choice of the optimal hyperparameters for all the algorithms was performed by using three-fold cross-validation in the learning procedure.

In LR we imposed sparsity through the L_1 norm on the regularization term, with the regularization constant C varying in a logarithmically spaced range of twenty values between $(10^{-2}, 10^2)$. For SVM, we fixed a linear kernel and let the cross validation choose the best values of C, in the same range of LR. For what concerns RF we fixed the number of estimators to 10^3 where the tunable parameters were the percentage of maximum features with respect to the total, in the range $(0.1, 0.2, 0.3, 0.4, 0.5, 0.6)$. In GB we fixed the learning rate to 10^{-3}, the tunable parameters were the max depth of trees, free to vary linearly in

Table 2. We evaluate several scores for each algorithm considered. We report the mean ± std for each metric, obtained from 50 repetitions of the experiment. Random forest classifier gives the best performances for all the score metrics considered.

Classifier	P	R	BA	F1
LR	0.68 ± 0.03	0.34 ± 0.03	0.64 ± 0.01	0.45 ± 0.03
SVM	0.57 ± 0.04	0.15 ± 0.03	0.55 ± 0.01	0.23 ± 0.05
RF	**0.88 ± 0.02**	**0.57 ± 0.02**	**0.77 ± 0.01**	**0.69 ± 0.01**
GB	0.80 ± 0.07	0.35 ± 0.06	0.66 ± 0.02	0.48 ± 0.05

a range between (3, 31) and the number of estimators, chosen between three linearly spaced values in (100, 500).

The learning and testing procedures were repeated 50 times in order to get a statistically reliable outcome for the four classifiers.

The results obtained on the test set across the 50 repetitions are shown in Table 2. Random Forest performs best for all the considered metrics. The values of balanced accuracy and F1 score are highlighted. Both metrics show a performance which is highly above chance level and is promising in the discrimination of epileptic areas. The precision value for Random Forest indicates that the number of false positives is relatively low.

Localization of the Critical Area. As Random Forest gave the best classification performance, we report here the results in the area localization task for this case. We performed this step by considering the channels positions given as output from the segmentation of the MRI obtained through FreeSurfer, see [8] for more details. The position of each channel is specified through (x, y, z) spatial coordinates. These features were not used in the learning procedure. For each patient we evaluate the *mean position* of the focus, defined as the average of the (x, y, z) coordinates of all channels which were tagged as epileptic by medical experts. We also define as *spread of the focus* the standard deviation of the epileptic channels positions from the mean position of the focus. On each patient we evaluate the average position of the montage and its spatial extension, by measuring the standard deviation of the (x, y, z)-positions for all channels, that we define as montage spread.

In Fig. 1 we report the distribution across all patients for the montage spread and the one relative to the focus, based on the clinical experts' characterization of the neural activity.

We proceed by measuring the performance in the area localization task. For this scope we considered the misclassified channels at each repetition and we give a measure of the error made by the learning algorithm in this case. We evaluate separately the contributes of false positives (FP) and false negatives (FN), by considering the classification errors made by our model. In both cases we compute for each patient the distance from the mean position of the focus, divided by the spread of the focus, in order to obtain a descriptive result over

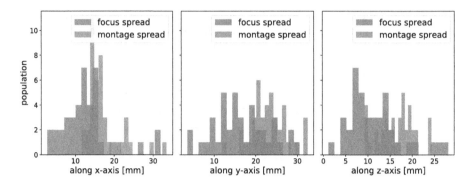

Fig. 1. Histograms relative to the montage spread (orange) and focal areas spread (blue) for all patients, based on the characterization given by medical experts. The three plots correspond to the measures of spread along the (x, y, z)-coordinates. (Color figure online)

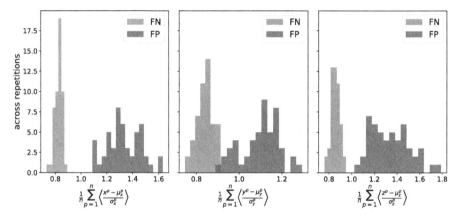

Fig. 2. Distribution of the discrepancy values for channels classified as false positive (red) and false negative (gray) across 50 repetition of the experiments. As expected the false positive channels fall out of the focal spread. (Color figure online)

all patients. We measure the discrepancy in the three directions (x, y, z) which we denote as d_x, d_y, d_z.

$$(d_x)_{FP} = \frac{1}{n} \sum_{p=1}^{n} \left\langle \frac{|x_{FP}^p - \mu_x^p|}{\sigma_x^p} \right\rangle, \quad (d_x)_{FN} = \frac{1}{n} \sum_{p=1}^{n} \left\langle \frac{|x_{FN}^p - \mu_x^p|}{\sigma_x^p} \right\rangle \quad (2)$$

$$(d_y)_{FP} = \frac{1}{n} \sum_{p=1}^{n} \left\langle \frac{|y_{FP}^p - \mu_y^p|}{\sigma_y^p} \right\rangle, \quad (d_y)_{FN} = \frac{1}{n} \sum_{p=1}^{n} \left\langle \frac{|y_{FN}^p - \mu_y^p|}{\sigma_y^p} \right\rangle \quad (3)$$

$$(d_z)_{FP} = \frac{1}{n} \sum_{p=1}^{n} \left\langle \frac{|z_{FP}^p - \mu_z^p|}{\sigma_z^p} \right\rangle, \quad (d_z)_{FN} = \frac{1}{n} \sum_{p=1}^{n} \left\langle \frac{|z_{FN}^p - \mu_z^p|}{\sigma_z^p} \right\rangle \quad (4)$$

In these formulas μ_i^p and σ_i^p denote respectively the i-th component of the mean position and spread of the epileptic focus for the p-th patient. For each patient we evaluated the standardized mean discrepancy, denoted through the average, respectively for false positive and false negative samples. The index p runs over all patients. In Fig. 2 we report the discrepancies d_x, d_y, d_z obtained from 50 repetition of the experiment.

The distribution of the false negative samples is centered at $(d_x)_{FN} = 0.82 \pm 0.02$, $(d_y)_{FN} = 0.84 \pm 0.04$, $(d_z)_{FN} = 0.86 \pm 0.04$ with respect to the mean position of the focus. The distribution of the false positive samples is centered at $(d_x)_{FP} = 1.3 \pm 0.1$, $(d_y)_{FP} = 1.10 \pm 0.09$, $(d_z)_{FP} = 1.3 \pm 0.2$ with respect to the mean position of the focus.

4 Conclusions

In this work we defined a pipeline for the analysis of brain activity in focal epileptic patients during the interictal period, with the aim of localizing critical areas, involved in seizure generation and propagation.

By considering relative measures based on single patient, several machine learning methods have been trained across patients, all of them with results highly above chance. In particular Random Forest classifiers has been shown to achieve the best performances (balanced accuracy $= 0.77 \pm 0.01$, F_1 score $= 0.69 \pm 0.01$). To the best of our knowledge, the integration of spectral features with anatomical characteristics of the recorded areas for posterior localization based on MRI test represents a first attempt to merge multiple clinical tests, fixing a set of features which are both functional and structural descriptors of the epileptic brain.

We measure the discrepancy in the area localization for the misclassified channels, and we found that the distribution for the false negative is centered in regions relatively far from the mean position of the focus. Indeed the discrepancy in the three directions is higher than 0.8 the spread of the focus for all the repetitions of the experiment.

In the analysis we have considered long interictal period, segments of 5 min at high sampling frequency. The features extracted represents an average behavior of each area. Even if the results are encouraging, we plan to focus on shorter windows of time to get a deeper insight on the role of specific temporal patterns which could represent the biomarkers of pathological neural activity. The pipeline should be integrated with analysis of the preictal and ictal stages of the pathology, which are considered highly informative for the focus localization. It must be observed that the learning has been performed by taking into account the activity recorded from all the subjects under study. In the future we aim to consider separately subsets of patients, in such a way to avoid possible correlation effects.

This preliminary work goes in the direction of the definition of relevant features for the discriminative task and the study of the epileptic activity. The ground-truth on which we based the classification relies on the visual tagging of the neural recordings performed by clinical experts. To the best of our knowledge there is no available free code which performs this feature extraction and classification task. We provide for the analysis an open source Python code https://github.com/vanessadamario/multichannelAnalysis.

Acknowledgements. This work was generously supported by the grant "Advancing of non-invasive procedures for the support of early diagnosis of partial epilepsies", funded by Compagnia di San Paolo protocol 2017.AAI4513.U5101/SD/pv.

References

1. Cardinale, F., et al.: Stereoelectroencephalography: surgical methodology, safety, and stereotactic application accuracy in 500 procedures. Neurosurgery **72**(3), 353–366 (2012)
2. de la Prida, L.M., Staba, R.J., Dian, J.A.: Conundrums of high-frequency oscillations (80–800 Hz) in the epileptic brain. J. Clin. Neurophysiol. Off. Publ. Am. Electroencephalogr. Soc. **32**(3), 207 (2015)
3. Spencer, S., Huh, L.: Outcomes of epilepsy surgery in adults and children. Lancet Neurol. **7**(6), 525–537 (2008)
4. Truong, N.D., Kuhlmann, L., Bonyadi, M.R., Yang, J., Faulks, A., Kavehei, O.: Supervised learning in automatic channel selection for epileptic seizure detection. Expert Syst. Appl. **86**, 199–207 (2017)
5. Vila-Vidal, M., Principe, A., Ley, M., Deco, G., Campo, A.T., Rocamora, R.: Detection of recurrent activation patterns across focal seizures: application to seizure onset zone identification. Clin. Neurophysiol. **128**(6), 977–985 (2017)
6. Bartolomei, F., Chauvel, P., Wendling, F.: Epileptogenicity of brain structures in human temporal lobe epilepsy: a quantified study from intracerebral EEG. Brain **131**(7), 1818–1830 (2008)
7. Omerhodzic, I., Avdakovic, S., Nuhanovic, A., Dizdarevic, K.: Energy distribution of EEG signals: EEG signal wavelet-neural network classifier. arXiv preprint arXiv:1307.7897 (2013)
8. Narizzano, M., et al.: SEEG assistant: a 3DSlicer extension to support epilepsy surgery. BMC Bioinform. **18**(1), 124 (2017)
9. Mercier, M.R., et al.: Evaluation of cortical local field potential diffusion in stereotactic electro-encephalography recordings: a glimpse on white matter signal. NeuroImage **147**, 219–232 (2017)
10. Lachaux, J.-P., Axmacher, N., Mormann, F., Halgren, E., Crone, N.E.: High-frequency neural activity and human cognition: past, present and possible future of intracranial EEG research. Progress Neurobiol. **98**(3), 279–301 (2012)
11. Crépon, B., et al.: Mapping interictal oscillations greater than 200 Hz recorded with intracranial macroelectrodes in human epilepsy. Brain **133**(1), 33–45 (2009)
12. Dümpelmann, M., Jacobs, J., Kerber, K., Schulze-Bonhage, A.: Automatic 80–250 Hz "ripple" high frequency oscillation detection in invasive subdural grid and strip recordings in epilepsy by a radial basis function neural network. Clin. Neurophysiol. **123**(9), 1721–1731 (2012)
13. Rosso, O.A., et al.: Wavelet entropy: a new tool for analysis of short duration brain electrical signals. J. Neurosci. Methods **105**(1), 65–75 (2001)
14. Mooij, A.H., Frauscher, B., Amiri, M., Otte, W.M., Gotman, J.: Differentiating epileptic from non-epileptic high frequency intracerebral EEG signals with measures of wavelet entropy. Clin. Neurophysiol. **127**(12), 3529–3536 (2016)

15. Fischl, B.: Freesurfer. Neuroimage **62**(2), 774–781 (2012)
16. Hastie, T., Tibshirani, R., Wainwright, M.: Statistical Learning with Sparsity: the Lasso and Generalizations. CRC Press, Boca Raton (2015)
17. Cortes, C., Vapnik, V.: Support-vector networks. Mach. Learn. **20**(3), 273–297 (1995)
18. Breiman, L.: Random forests. Mach. Learn. **45**(1), 5–32 (2001)
19. Friedman, J.H.: Greedy function approximation: a gradient boosting machine. Ann. Stat. **29**, 1189–1232 (2001)
20. Brodersen, K.H., Ong, C.S., Stephan, K.E., Buhmann, J.M.: The balanced accuracy and its posterior distribution. In: 2010 20th International Conference on Pattern Recognition, pp. 3121–3124. IEEE (2010)
21. Pedregosa, F., et al.: Scikit-learn: machine learning in Python. J. Mach. Learn. Res. **12**, 2825–2830 (2011)

Committee-Based Active Learning to Select Negative Examples for Predicting Protein Functions

Marco Frasca, Maryam Sepehri, Alessandro Petrini, Giuliano Grossi, and Giorgio Valentini$^{(\boxtimes)}$

Dipartimento di Informatica, Università degli Studi di Milano,
Via Celoria 18, 20133 Milan, Italy
{frasca,grossi,valentini}@di.unimi.it,
{maryam.sepehri,alessandro.petrini}@unimi.it

Abstract. The Automated Functional Prediction (AFP) of proteins became a challenging problem in bioinformatics and biomedicine aiming at handling and interpreting the extremely large-sized proteomes of several eukaryotic organisms. A central issue in AFP is the absence in public repositories for protein functions, e.g. the Gene Ontology (GO), of well defined sets of negative examples to learn accurate classifiers for AFP. In this paper we investigate the Query by Committee paradigm of active learning to select the negatives most informative for the classifier and the protein function to be inferred. We validated our approach in predicting the Gene Ontology function for the *S.cerevisiae* proteins.

Keywords: Query By Committee · Active learning · Protein function prediction

1 Scientific Background

The Automated Function Prediction (AFP) of proteins involves sophisticated computational techniques to accurately predict the annotations of proteins and proteomes. AFP is characterized by several issues, including the selection of negative examples to train high quality predictors. The Gene Ontology (GO) [1], the reference repository of protein functions, usually stores positive associations (annotations) between GO terms and proteins, whereas unannotated proteins are rarely marked as negative for a given term—a protein not currently annotated with a GO term, might be a positive example which has not been detected yet due to insufficient investigations. Surprisingly, a few studies investigated this problem, mainly leveraging the GO structure (a directed acyclic graph) to choose negative examples [2–4].

We present here a preliminary work which addresses the negative selection problem by leveraging Query By Committee (QBC) active learning [5] to appropriately select the negative proteins. Unlike usual approaches to active learning,

© Springer Nature Switzerland AG 2020
M. Raposo et al. (Eds.): CIBB 2018, LNBI 11925, pp. 80–87, 2020.
https://doi.org/10.1007/978-3-030-34585-3_7

which typically aim to obtain the true labels of some selected data points, our approach undertakes the selection of negative examples (whose labels are obviously known). The rationale behind this approach is that the capability of active learning to focus on the most informative examples can be leveraged to filter out from the training set unhelpful non-positive proteins—or even harmful. Pool-based QBC considers most informative the examples from a pool of unlabeled examples on which the committee members (classifiers) most disagree. Hence, in our setting QBC is used to select as negative examples a subset of proteins from the pool represented by non-positive proteins. We experimentally validated our approach using two well-known classifiers to predict, in a genome-wide fashion, the GO functions of yeast proteins.

2   Materials and Methods

2.1   Preliminaries and Notations

Vectors and matrices are denoted using standard, lower bold and upper bold symbols as x and X. Protein pairwise similarities are represented by an undirected weighted graph $G\langle V, W\rangle$, where $V = \{1, \ldots, n\}$ is the set of nodes/proteins and W is the $n \times n$ matrix of intra-protein functional similarity: $W_{ij} \in [0, 1]$ is the similarity between proteins $i, j \in V$, with $W_{ij} = 0$ when i and j are not connected. Given a protein function, the labels are described by the binary vector $y = (y_1, y_2, \ldots, y_n)$, where $y_i = 1$ if protein i is annotated with that function (positive instance), -1 otherwise. Here the GO terms are adopted as protein functions. Let $V_+ := \{i \in V | y_i = 1\}$ and $V_- := \{i \in V | y_i = 0\}$ be the subsets of positive and non positive proteins, respectively. A relevant issue in AFP is the labeling imbalance: most GO functions posses a highly unbalanced labeling, that is $\frac{|V_+|}{|V_-|} \ll 1$. Furthermore, the labeling is known only for a subset $S \subset V$ of proteins, where it is unknown for its complement set $U := V \backslash S$.

The *Automated protein Function Prediction* (AFP) problem consists in inferring the labeling for proteins U using the known labels and the connection matrix W.

The complexity of AFP is increased by the fact that the GO rarely stores *negative* annotations between proteins and functions, and only positive annotations are usually available. Thus, non positive proteins (proteins in $S \cap V_-$) typically do not correspond to *negative* annotations, and some of them might be redundant for the current task. Moreover, some non positive proteins might become positive in future, in case further studies would annotate them. This makes central the need to select informative negatives among non positive proteins to be used as negative examples during the learning of automated models for solving AFP—indirectly, it would also cope with the label imbalance, since the disproportion between positive and negative would be reduced.

Instance Representation. Following [6], the input proteins are represented through a two-dimensional feature vector, obtained by operating a projection of nodes S onto the space $\mathbb{R}^2$, so that the node $i \in S$ is associated with the point

$\boldsymbol{x}_i \equiv (x_{i1}; x_{i2})$, where $x_{i1} = \sum_{j \in S_+} W_{ij}$ and $x_{i2} = \sum_{j \in S_-} W_{ij}$. This embedding casts into the position of point $\boldsymbol{x}_i$ the imbalance in the neighborhood of protein i in G, and sensibly reduces the input space dimension, thus speeding up the computation. Moreover, recent studies have confirmed that this two features are informative for inferring GO functions [7]. The obtained training set is $L = \{(\boldsymbol{x}_i, y_i) | i \in S\}$.

2.2   Data

We retrieved the protein network for *S. cerevisiae* (yeast) from the STRING database, version 10.5 [8], which merges several sources of information about proteins, including databases collecting experimental data, such as BIND, DIP, GRID, HPRD, IntAct, MINT, and databases collecting curated data, such as Biocarta, BioCyc, KEGG, and Reactome. The matrix $\boldsymbol{W}$ of Sect. 2.1 is obtained from the STRING connections $\widehat{\boldsymbol{W}}$ after the normalization $\boldsymbol{W} = \boldsymbol{D}^{-1/2}\widehat{\boldsymbol{W}}\boldsymbol{D}^{-1/2}$, which preserves the connection symmetry. $\boldsymbol{D}$ is the diagonal matrix with non-null elements $d_{ii} = \sum_j \widehat{W}_{ij}$. As suggested by STRING curators, we set the threshold for connection weights to 700. The final network contains 6391 proteins. Annotations for the three GO branches, namely Biological Process (BP), Molecular Function (MF), and Cellular Component (CC), have been downloaded from the UniProt GOA, release 69 (9 May 2017), by retaining solely experimentally validated annotations. To discard too generic terms and having a minimum of information to learn, we chose functions with 10–100 annotations, obtaining 162, 227 and 660 terms for CC, MF, and BP branches, respectively.

2.3   Algorithm

We propose a novel approach to address the negative selection in AFP, which leverages a variant of Query-by-Committee (QBC) active learning (AL) to appropriately select the most informative negative examples. In particular, our technique focuses on the selection of most informative negatives for the specific classification model, rather than selecting those negative examples "most informative" in general. We empirically validated our proposal on two well-known supervised classifiers.

2.3.1   QBC Active Learning for Negative Selection

Denoted by $S_+ = S \cap V_+$ and $S_- = S \cap V_-$ respectively the training sets of positive and non positive proteins for a given GO term, a budget $0 < B < |S_-|$ is given, representing the cardinality of a subset of negative examples $\widehat{S}_- \subset S_-$ to be selected in order to maximize the performance of the classifier trained using the examples $S_+ \cup \widehat{S}_-$.

This problem is tackled through pool-based QBC active learning, which typically examines a pool of unlabeled examples and selects only those that are most informative according to the committee models, and asks for their labels. This avoids to save annotation cost by discarding redundant labeling examples that contribute little new information [5]. Common approaches for pool-based QBC is to ask the label of those points on which committee members most disagree [9].

We adopt a variant of active learning, since we want the QBC algorithm to select instances whose label is known already (equal to 0). Nevertheless, we may exploit AL to pick out the "most informative" negative points for training our model. Our AL algorithm is defined as follows.

QBC Active Learning Procedure (Template)

1. A seed training set $I(0) = S_+ \cup S_-(0)$ is selected, where $S_-(0) \subset S_-$ is randomly drawn and balanced (i.e., $|S_-(0)| = |S_+|$).[1] Due to the rarity of positives, $I(0)$ contains all available positives.
2. At iteration $t \geq 1$, learn m committee models $f_k : I(t-1) \rightarrow \{-1, 1\}$, $k \in \{1, 2, \ldots, m\}$.
3. Build $I(t)$ by adding to $I(t-1)$ the s instances in $S_- \setminus I(t-1)$ with highest degree of disagreement among the committee members.
4. Update the committee classifiers using $I(t)$.
5. Iterate steps 2–4 until time $\bar{t}$, with $|I(\bar{t})| = |S_+| + B$ (budget is exhausted).

The rationale is that examples on which the classifiers most disagree have a higher 'utility' for the committee. Further, it is beneficial in QBC ensuring diversity among committee classifiers [10]: a common approach is to use bagging for learning the m committee classifiers [11], in which m random subsets $I_1, I_2, \ldots, I_m$ of $I(\bar{t})$ are randomly drawn (in our setting each I_i contains all available positives and a randomly drawn subset of the non positive examples in $I(\bar{t})$), and the member f_k is trained using the set I_k.

The Vote Entropy has been employed as measure of disagreement, a natural measure for quantifying the uniformity of classes assigned to an example by the different committee member [9]. Given an instance $x \in \mathbb{R}^q$, its Vote Entropy disagreement is $V(x) = -\nu_x \log \nu_x - (1 - \nu_x) \log(1 - \nu_x)$, where $\nu_x = \frac{\sum_{k=1}^m \mathbb{I}\{f_k(x)=1\}}{m}$, and $\mathbb{I}$ is the indicator function. Thus ν_x is the proportion of members that predicted x as positive. Accordingly, the closer ν_x to 0.5, the higher the Vote Entropy disagreement.

We validated the QBC algorithm to select negatives in AFP by adopting two popular feature-based models at Step 2 of the procedure, briefly described below.

Support Vector Machines. Given a training set $L = \{(x_i, l_i)\} \in \mathbb{R}^q \times \{-1, 1\}$, the Support Vector Machine (SVM) [12,13] learns the hyperplane $\hat{\omega} \in \mathbb{R}^q$ unique solution of the following optimization problem:

$$\min_{\omega \in H_\mathcal{K}} \frac{1}{2} \|\omega\|_\mathcal{K}^2 + C \sum_i^{|L|} e_i(\omega) \tag{1}$$

where $e_i(\omega) = 1 - l_i \langle \omega, \phi_\mathcal{K}(x_i) \rangle$, if $l_i \langle \omega, \phi_\mathcal{K}(x_i) \rangle < 1$ (margin constraint violation), 0 otherwise, and $\mathcal{K}$ is a kernel implementing the inner product

---

[1] A balanced seed training set counterbalances the predominance of 0 labels.

$\mathcal{K}(\boldsymbol{x}, \boldsymbol{z}) = \langle \phi_{\mathcal{K}}(\boldsymbol{x}), \phi_{\mathcal{K}}(\boldsymbol{z}) \rangle$ of two vectors $\boldsymbol{x}, \boldsymbol{z} \in \mathbb{R}^d$ according to a feature map $\phi_{\mathcal{K}} : \mathbb{R}^d \to H_{\mathcal{K}}$. $H_{\mathcal{K}}$ is a higher dimensional space. The margin of an instance $\boldsymbol{x}_i$ is $\left| \sum_{j=1}^{|L|} \alpha_j l_j \mathcal{K}(\boldsymbol{x}_j, \boldsymbol{x}_i) \right|$, where $\alpha_j \geq 0$ are the Lagrange multipliers (see for instance [13]).

Two popular choices of $\mathcal{K}$ are adopted in this work, namely the *linear* kernel $\mathcal{K}_1(\boldsymbol{x}, \boldsymbol{z}) = \boldsymbol{x}^T \boldsymbol{z}$, and the *Gaussian* kernel $\mathcal{K}_2(\boldsymbol{x}, \boldsymbol{z}) = \exp(-\frac{\|\boldsymbol{x} - \boldsymbol{z}\|^2}{2\sigma^2})$.

Decision Trees. Let $X_1, \ldots, X_q$ be q predictor variables (discrete or continuous), $L = \{(\boldsymbol{x}_i, l_i)\} \in \mathbb{R}^q \times \{0, 1\}$ be a set of labeled observations on a class variable Y (binary in our case) that takes values $-1, 1$. Briefly, the decision tree (DT) algorithm [14] learns a model $T : \mathbb{R}^q \to \{-1, 1\}$ for predicting the values of Y from observations $\boldsymbol{x}_i$ by simply a partitioning the space $\mathbb{R}^q$ into 2 disjoint sets A_-, A_+, such that the predicted value of Y is 0 if $T(\boldsymbol{x}_i) \in A_-$, 1 if $T(\boldsymbol{x}_i) \in A_+$.

Classification tree methods grows from an initial (root) node by recursively partitioning the data set one predictor variable at a time. Each node is assigned a label (0 or 1), and accordingly it is associated with a classification error (based on labels l_i), used to measure the node impurity. At each step, the node to be split is determined by exhaustively searching the split, e.g. $X_j > t$, over all nodes and predictors X_j which minimizes the total impurity of its two child nodes. Then an instance $\boldsymbol{x}_i$ at the split node is assigned to one of the two children according to the value of its j-th component x_{ij}. The process iterates till a stopping criterion is met (e.g. maximum depth reached). To predict an instance $\boldsymbol{z}$, the algorithm follows the path from the root to a leaf node, and classify $\boldsymbol{z}$ with the label of that leaf node. As impurity measure a common choice is the Gini index [15], which has been adopted also in this work.

3   Results

We name SVM QBC (resp. DT QBC) the method using SVMs (resp. DTs) both in step 2 of the QBC procedure and to learn the final model over the set $I(\bar{t})$. Firstly, to evaluate the usefulness of QBC, we implemented an active learning negative selection using only one member (*baseline AL, m = 1*), where the most informative instances are those with smaller margin for SVMs, and those belonging to the leaves with higher impurity for DTs. Generalization capabilities have been evaluated using a 3-fold cross validation (CV) procedure, and measured in terms of F_1 measure (F in short), which is a measure suitable for unbalanced labelings. The model parameters, C for linear SVM, C and σ for Gaussian SVM, and the tree maximum depth for DT, have been learned through inner 3-fold CV.

We first investigated the impact of parameters s and B on the model performance, by tuning them on the CC GO terms. Furthermore, the variant of AL in which all the negatives ($B - |S(0)|$) are selected at the first iteration of step 3 in the QBC procedure has been implemented, and named *One shot* strategy. Figure 1 depicts the overall results. There is at least one QBC configuration which outperforms the baseline AL in all the settings, and the improvements are

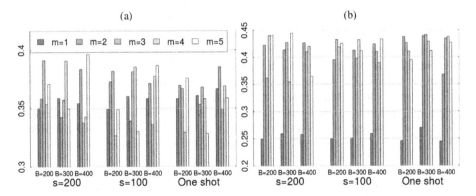

Fig. 1. F values averaged across CC terms for DT (left) and linear SVM (right) algorithms. B is the negative budget, s is the active learning parameter, m the number of committee members.

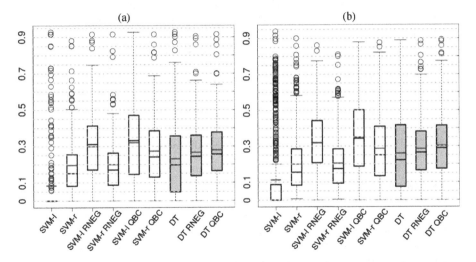

Fig. 2. F values averaged across (a) MF and (b) BP terms on yeast data. White boxes correspond to SVM methods, gray boxes to DT methods. SVM-l and SVM-r denotes respectively the SVM using linear and Gaussian kernels. Red horizontal segments correspond to mean values. (Color figure online)

remarkable when using SVM. Interestingly, with just 200 negatives, SVM QBC already achieves its top performance, and adding further examples ($B = 300$ or $B = 400$) rarely helps. This behaviour is verified also for DT QBC, with lower magnitude.

To assess the effectiveness in AFP for models using QBC, we also tested the vanilla SVM and DT, learned on all available training data (no negative selection applied), and the SVM and DT where the B negative examples are uniformly extracted (named SVM RNEG and DT RNEG). According to the insights provided by results in Fig. 1, we set $s = 100, B = 400$ for DTs, and $s = 200, B = 300$ for SVMs.

As shown in Fig. 2 for MF anc BP classes (CC shown analogous trends), QBC negative selection always allows to outperform both the vanilla variant and the variant using random negative selection, which in turn improves the performance of the vanilla methods. Noticeable is the case of SVM-l, which is the worst method in the vanilla fashion, but with QBC is the top performing method. The improvements of QBC over RNEG negative selection are always statistically significant according to the Wilcoxon signed rank test ($p\text{-}value < 0.05$) [16].

4   Conclusion

Preliminary results have shown that Query By Committee active learning might be employed as effective tool to address the negative selection problem in AFP. Despite the promising results, further studies must be carried out to investigate the impact that several features of the method have on the classification abilities, like the adoption of different measures of committee disagreement among the numerous measures proposed in the literature, and of different stopping criterion than fixing a budget of negatives, along with experimentations on other organisms/datasets.

Acknowledgments. This work was supported by the grant title *Machine learning algorithms to handle label imbalance in biomedical taxonomies*, code PSR2017_DIP_010_MFRAS, Università degli Studi di Milano.

References

1. Ashburner, M., et al.: Gene ontology: tool for the unification of biology. The gene ontology consortium. Nat. Genet. **25**, 25–29 (2000)
2. Eisner, R., Poulin, B., Szafron, D., Lu, P.: Improving protein prediction using the hierarchical structure of the gene ontology. In: IEEE Symposium on Computational Intelligence in Bioinformatics and Computational Biology (2005)
3. Mostafavi, S., Morris, Q.: Using the gene ontology hierarchy when predicting gene function. In: Proceedings of the Twenty-Fifth Annual Conference on Uncertainty in Artificial Intelligence (UAI-09), (Corvallis, Oregon), pp. 419–427. AUAI Press (2009)
4. Youngs, N., Penfold-Brown, D., Bonneau, R., Shasha, D.: Negative example selection for protein function prediction: the NoGO database. PLoS Comput. Biol. **10**, 1–12 (2014)
5. Freund, Y., Seung, H.S., Shamir, E., Tishby, N.: Selective sampling using the query by committee algorithm. Mach. Learn. **28**, 133–168 (1997)
6. Bertoni, A., Frasca, M., Valentini, G.: *COSNet*: a cost sensitive neural network for semi-supervised learning in graphs. In: Gunopulos, D., Hofmann, T., Malerba, D., Vazirgiannis, M. (eds.) ECML PKDD 2011. LNCS (LNAI), vol. 6911, pp. 219–234. Springer, Heidelberg (2011). https://doi.org/10.1007/978-3-642-23780-5_24

7. Frasca, M., Lipreri, F., Malchiodi, D.: Analysis of informative features for negative selection in protein function prediction. In: Rojas, I., Ortuño, F. (eds.) IWBBIO 2017, Part II. LNCS, vol. 10209, pp. 267–276. Springer, Cham (2017). https://doi.org/10.1007/978-3-319-56154-7_25

8. Szklarczyk, D., et al.: String v10: protein-protein interaction networks, integrated over the tree of life. Nucleic Acids Res. **43**(D1), D447–D452 (2015)

9. Dagan, I., Engelson, S.P.: Committee-based sampling for training probabilistic classifiers. In: Proceedings of the Twelfth International Conference on Machine Learning, pp. 150–157. Morgan Kaufmann (1995)

10. Melville, P., Mooney, R.J.: Diverse ensembles for active learning. In: Proceedings of the Twenty-first International Conference on Machine Learning, ICML 2004, p. 74. ACM, New York (2004)

11. Abe, N., Mamitsuka, H.: Query learning strategies using boosting and bagging. In: Proceedings of the Fifteenth International Conference on Machine Learning, ICML 1998, San Francisco, CA, USA, pp. 1–9 (1998)

12. Vapnik, V.N.: The Nature of Statistical Learning Theory. Springer, New York (1995)

13. Cristianini, N., Shawe-Taylor, J.: An Introduction to Support Vector Machines: and Other Kernel-based Learning Methods. Cambridge University Press, New York (2000)

14. Breiman, L., Friedman, G., Olshen, R., Stone, C.: Classification and Regression Trees. Wadsworth, Belmont (1984)

15. Gini, C.: Variabilità e Mutuabilità. Contributo allo Studio delle Distribuzioni e delle Relazioni Statistiche, C. Cuppini, Bologna (1912)

16. Wilcoxon, F.: Individual comparisons by ranking methods. Biometrics **1**, 80–83 (1945)

A Graphical Tool for the Exploration and Visual Analysis of Biomolecular Networks

Cheick Tidiane Ba[1], Elena Casiraghi[1], Marco Frasca[1],
Jessica Gliozzo[1,2], Giuliano Grossi[1], Marco Mesiti[1(✉)], Marco Notaro[1],
Paolo Perlasca[1], Alessandro Petrini[1], Matteo Re[1],
and Giorgio Valentini[1]

[1] Department of Computer Science, Università degli Studi di Milano, Via Celoria 18,
20133 Milan, Italy
[2] Department of Dermatology, Fondazione IRCCS Ca' Granda - Ospedale Maggiore
Policlinico, 20122 Milan, Italy

Abstract. Many interactions among bio-molecular entities, e.g. genes, proteins, metabolites, can be easily represented by means of property graphs, i.e. graphs that are annotated both on the vertices (e.g. entity identifier, Gene Ontology or Human Phenotype Ontology terms) and on the edges (the strength of the relationship, the evidence of the source from which the weight has been taken, etc.). These graphs contain a relevant information that can be exploited for conducting different kinds of analysis, such as automatic function prediction, disease gene prioritization, drug repositioning. However, the number and size of the networks are becoming quite large and there is the need of tools that allow the biologists to manage the networks, graphically explore their structures, and organize the visualization and analysis of the graph according to different perspectives. In this paper we introduce the web service that we have developed for the visual analysis of biomolecular networks. Specifically we will show the different functionalities for exploring big networks (that do not fit in the current canvas) starting from a specific vertex, for changing the view perspective of the network, and for navigating the network and thus identifying new relationships. The proposed system extends the functionalities of off-the-shelf graphical visualization tools (e.g. GraphViz and GeneMania) by limiting the production of big cloud of points and allowing further customized visualizations of the network and introducing their vertex-centric exploration.

Keywords: Biological network · Protein function prediction · Information visualization · Graph visualization

1 Scientific Background

Biological and biomedical data require advanced integration and visualization tools and methods for representing and modeling the intrinsic complexity of such domain. Several visualization tools and methods have been proposed for covering

© Springer Nature Switzerland AG 2020
M. Raposo et al. (Eds.): CIBB 2018, LNBI 11925, pp. 88–98, 2020.
https://doi.org/10.1007/978-3-030-34585-3_8

a range of different features (Kuznetsova et al. 2018, Pavlopoulos et al. 2008) (Napolitano et al. 2008). Following the approach adopted by the cited systems, we have adopted the JavaScript library Cytoscape.js for the development of the interactive network visualization part of the application. This library indeed offers the flexibility required for developing a web application that need to handle graphs of big dimensions and provide a professional visualization. Different web-tools for protein function prediction are available, such as N-Browse (Kao and Gunsalus 2002), SIFTER (Sahraeian et al. 2015), MouseNet v2 (Kim et al. 2016), the IMP tool (Wong et al. 2015), and the GeneMANIA server (Warde-Farley et al. 2010). N-Browse provides a graphical user interface (GUI), leveraging inter-action in the network display on node and edge information, allowing the user to select the networks involved in the analysis; however, solely three organisms are supported, and N-Browse runs as a Java web start, which might be not imme-diate for a generic user. SIFTER is a sequence-based web interface exploring a protein family's phylogenetic tree as a statistical graphical model of function evo-lution. The search is limited to one protein at a time, or must include the whole proteome, and the user cannot specify a subset of query proteins. MouseNet v2 extends MouseNET (Guan et al. 2008), a previous prediction server for labora-tory mouse, by including new microarray data derived from diverse biological contexts and embedding other 8 model vertebrates to exploit the orthology-based projection of their genes on MouseNet. However the search is limited to one organism. The IMP system provides an easy to use interface to query one or more proteins at the same time, even from different organism, by exploiting gene homology information. SIFTER, MouseNet and IMP hide the data inte-gration phase to the user, which consequently cannot evaluate the impact of specific connection types on the final integrated network. Moreover, they do not provide the user with the possibility to interact with the resulting integrated protein network. Finally, the GeneMANIA prediction server allows the user to specify customized queries and to interact in the visualization process, depict-ing even a graphical view of the obtained consensus network. Nevertheless, it assigns weights using a Gaussian random field framework that cannot cope with label-imbalance characterizing the GO terms.

All these systems provide features for graphically representing the networks, but they are not usually able to represent and highlight specific local character-istics of the biomolecular graph under study. Moreover, when the networks are large, a dark cloud of points is shown from which it is quite hard to understand the structure of the network and the visual analysis of the graph becomes unfea-sible. For this reason, we developed a web tool that supports a "vertex centric" visualization of the subgraph connected to the node under study. In this way the user can focus on a specific biomolecular entity (e.g. a specific gene or protein), and explore the topologically close subnetwork that include its neighborohood nodes and their annotations. Moreover the user can interactively include new vertices and navigate on the network structure in order to identify useful pat-terns. The presence of different perspectives, i.e. mode of visualization of the subnetwork, gives the user the possibility to directly obtain biological insights and infer novel characteristics of the biomolecular entities under study.

2 Materials and Methods

The web service we have developed (available at http://unipred.di.unimi.it) is flexible and extensible enough to take into account new constraints and requirements. The internal representation of information is separated from its rendering to the end-users according to the model-view-control paradigm. For this purpose we have used Angular.js for maintaining the interaction with the database and Cytoscape.js for the graphical representation of networks. Biological networks are stored into a mysql database and we are able to compute at run time new integrated networks and to extract the sub networks to be displayed according to the user requests. Angular.js data-binding and suitable mechanisms to handle asyncronous data access have been exploited to manage the server-side computation required for preparing the network and for extracting the sub network to be displayed, while Cytoscape.js has been used for the graph visualization and analysis. In order to test the functionalities of the application, we have considered biological networks downloaded from STRING (Szklarczyk et al. 2015) and from the GeneMANIA website.

3 Results

A Web Application has been realized for the integration and visualization of biomolecular networks that have been collected in a web server. Once the preferred network is loaded in the server, different functionalities for the vertex-centric exploration of the network, for the customized visualization from different perspective, and for the navigation can be selected. As an example, Figs. 1 and 2 show how an initial visualization of a loaded biomolecular network (represented as a point cloud) can be improved by applying a sequence of customized rendering options specifically developed in the Web application: by increasing the node repulsion layout option of the graph displayed in Fig. 1(a) a new positioning of the nodes is obtained, as shown in Fig. 1(b); hiding labels the graph becomes uncluttered (see Fig. 2(a)); filtering the weights allows to show only edges of interest (see Fig. 2(b)). Beside the specification of the graph to be visualized, the interface allows to identify the vertex from which the graph exploration should be started. This vertex represents a biomolecule that the user wishes to analyze and the "radius" of the subgraph to be displayed (where the radius corresponds to the number of hops from the center). As an example, Fig. 3 shows the interface for preparing the rendering of the network view that is centered on the protein Fbgn0267347 of the *Drosophila melanogaster organism* with GO:0000001 class with depth 2. The user can drag each vertex of the subnetwork and obtain a

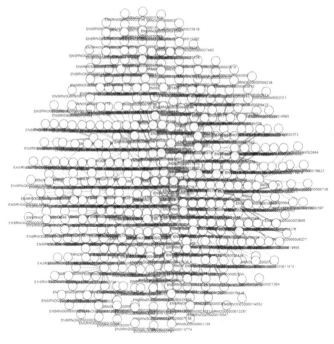

(a) Initial point cloud

(b) Point cloud after increasing node repulsion

Fig. 1. Options for visualization optimization of a biomolecular network

(a) Point cloud after hiding labels

(b) Point cloud after weights filtering

Fig. 2. Optimized visualization of the biomolecular network in Fig. 1

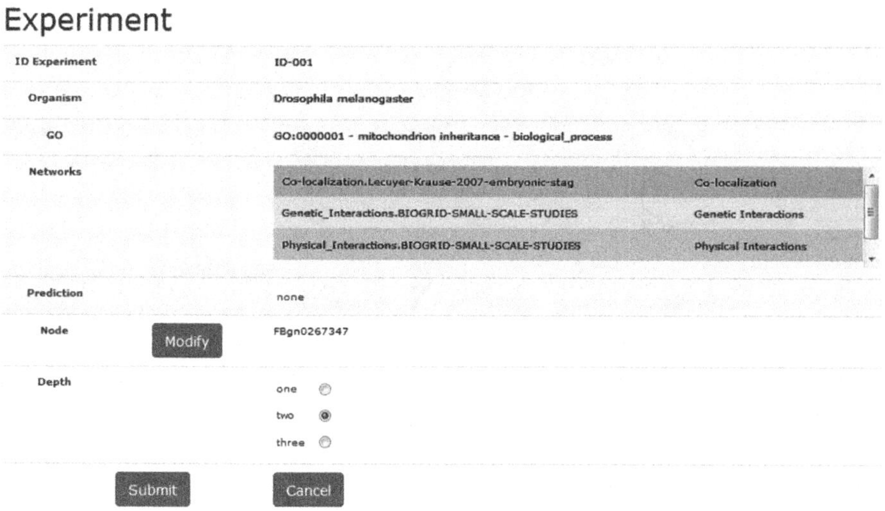

Experiment

ID Experiment	ID-001
Organism	Drosophila melanogaster
GO	GO:0000001 - mitochondrion inheritance - biological_process

Networks		
	Co-localization.Lecuyer-Krause-2007-embryonic-stag	Co-localization
	Genetic_Interactions.BIOGRID-SMALL-SCALE-STUDIES	Genetic Interactions
	Physical_Interactions.BIOGRID-SMALL-SCALE-STUDIES	Physical Interactions

Prediction	none
Node Modify	FBgn0267347
Depth	one ⊙ two ⊙ three ⊙

Submit   Cancel

Fig. 3. Web interface for preparing the rendering of the network view

personalized visualization; by clicking on a node or an edge, he/she can obtain the corresponding information as shown in Fig. 4(a).

The web tool is also equipped with different visualization options for making the visual analysis of the generated network more user-friendly (note that each layout might be customized to obtain an optimized visualization, as previously shown in Figs. 1 and 2). *Cose, grid, concentric, circle* and *breadthfirst* are the available visualization layouts provided also by Cytoscape.js. Figures 4 and 5 show the application of different layouts to the experiment described above. The *cose* (Dogrusoz et al. 2009) visualization option uses a physics simulation to layout graphs and is based on the traditional force-directed layout algorithm with extensions to handle multi-level nesting (compound nodes), edges between nodes of arbitrary nesting levels and varying node sizes (see Fig. 4(a)). With the *grid* visualization option, the proteins in the subnetwork are placed in a grid and their connections are shown in the canvas. With the *concentric* visualization option, the target protein is positioned at the center of the canvas and vertices at distance one, two or three, according to the chosen experiment depth, are drawn in different concentric circles. This rendering allows one to better understand the connectivity of the target with its neighborhood as shown in Fig. 4(b). With the *circle* visualization option, all vertices are posed in a circle and shown their connections with the other vertices. Vertices with an higher in-out-edge-degree are positioned closer in the circle. This visualization allows one to better appreciate the nodes for which there is a high interconnection strength from those whose connections are minimal. This feature helps to graphically detect *hub proteins*,

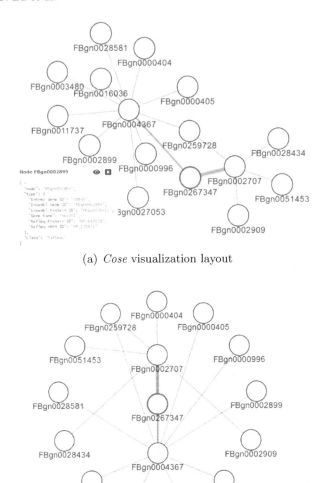

(a) *Cose* visualization layout

(b) *Concentric* visualization layout

Fig. 4. Different layout visualizations

i.e. those possessing higher centrality indexes, such as node betweeness, and global clustering coefficient, as shown in Fig. 5(a). Finally, the *breadthfirst* visualization option puts nodes in a hierarchy, based on a breadthfirst traversal of the graph (see Fig. 5(b)).

By opening a panel in the right-top corner of the web interface, the network visualization can be personalized from different perspectives (beside the visualization option described above). The user can decide to prune connections

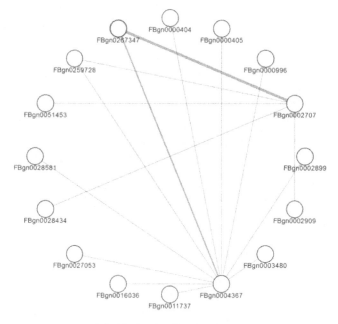

(a) *Circle* visualization layout

(b) *Breadthfirst* visualization layout

Fig. 5. Different layout visualizations

in the visualized network relying on their weights. This feature is quite useful for keeping in the canvas only the edges with the higher connectivity relevance. Moreover, the user can change the color of the network nodes depending of the biological functions of the networks used for the integration. This is an important feature in order to make the user immediately aware of a specific property

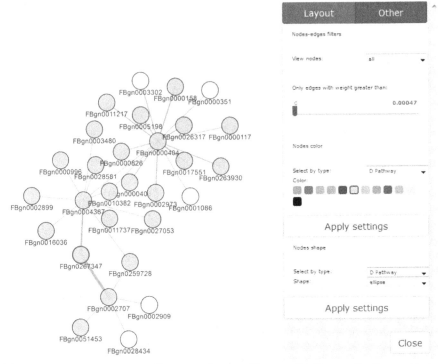

(a) Nodes belonging to a specific network are yellow colored

(b) Personalization options

Fig. 6. Layout personalization example (Color figure online)

such as belonging to a specific network as shown in the Fig. 6(a). In the same spirit, it is possible to change the shape of the network nodes. Figure 6(b) shows the different options so far discussed.

Figure 7 shows an important feature, named *one-step navigation*, of our application. By clicking on a node of the subnetwork currently visualized in the canvas, further nodes can be included that stay a step-forward from the clicked one. In this way the user can enrich the subgraph with the neighbor nodes that the user considers to be interesting.

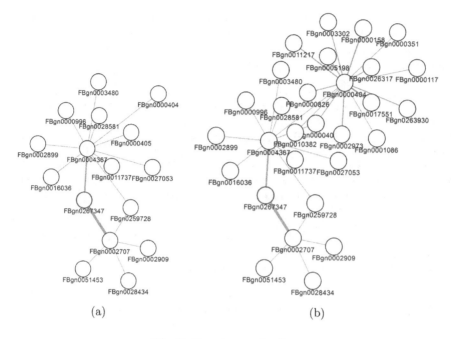

(a)   (b)

Fig. 7. One step navigation

4   Conclusion

In this paper we have presented the features of a Web application for the visualization, analysis and navigation of biological networks of large size. By adopting a vertex-centric rendering of a subnetwork, the system offers different customized visualizations that can be exploited for identifying useful patterns in the analyzed network. Moreover, we have discussed the characteristics for extending the visualized subnetwork by clicking on the nodes of the network and provide a larger visualization of the subnetwork. As future work we are planning to introduce machine learning algorithms for suggesting to the user the best visualization by considering user feedbacks in the proposed visualizations. Moreover, we wish to consider visualizations at different granularities in order to reduce the amount of vertices to be included in the current canvas.

References

Kuznetsova, I., Lugmayr, A., Holzinger, A.: Visualisation methods of hierarchical biological data: a survey and review. Int. Ser. Inf. Syst. Manag. Creative eMedia (CreMedia) (2017/2), 32–39 (2018)

Pavlopoulos, G.A., Wegener, A.-L., Schneider, R.: A survey of visualization tools for biological network analysis. BioData Min. **1**(1), 12 (2008)

Napolitano, F., Raiconi, G., Tagliaferri, R., Ciaramella, A., Staiano, A., Miele, G.: Clustering and visualization approaches for human cell cycle gene expression data analysis. Int. J. Approx. Reasoning **47**(1), 70–84 (2008)

Kao, H.-L., Gunsalus, K.C.: Browsing Multidimensional Molecular Networks with the Generic Network Browser (N-Browse). Wiley, Hoboken (2002)

Sahraeian, S.M., Luo, K.R., Brenner, S.E.: Sifter search: a web server for accurate phylogeny-based protein function prediction. Nucleic Acids Res. **43**(W1), W141–W147 (2015)

Kim, E., et al.: MouseNet v2: a database of gene networks for studying the laboratory mouse and eight other model vertebrates. Nucleic Acids Res. **44**(D1), D848–D854 (2016)

Wong, A.K., Krishnan, A., Yao, V., Tadych, A., Troyanskaya, O.G.: Imp 2.0: a multi-species functional genomics portal for integration, visualization and prediction of protein functions and networks. Nucleic Acids Res. **43**(W1), W128–W133 (2015)

Warde-Farley, D., et al.: The genemania prediction server: biological network integration for gene prioritization and predicting gene function. Nucleic Acids Res. **38**(suppl 2), W214–W220 (2010)

Guan, Y., Myers, C.L., Lu, R., Lemischka, I.R., Bult, C.J., Troyanskaya, O.G.: A genomewide functional network for the laboratory mouse. PLoS Comput. Biol. **4**(9), 1–15 (2008)

Szklarczyk, D., et al.: String v10: protein-protein interaction networks, integrated over the tree of life. Nucleic Acids Res. **43**(D1), D447–D452 (2015)

Dogrusoz, U., Giral, E., Cetintas, A., Civril, A., Demir, E.: A layout algorithm for undirected compound graphs. Inf. Sci. **179**(7), 980–994 (2009)

Improved Predictor-Corrector Algorithm

Hassan Pazira$^{(\boxtimes)}$ (iD)

Johann Bernoulli Institute, University of Groningen, Groningen, The Netherlands
h.pazira@amsterdamumc.nl

Abstract. The differential geometric least angle regression method consists essentially in computing the solution path. In Augugliaro et al. [4], this problem is satisfactorily solved by using a predictor-corrector (PC) algorithm, that however has the drawback of becoming intractable when working with thousands of predictors. Using the PC algorithm leads to an increase in the run times needed for computing the solution curve. In this paper we explain an improved version of the PC algorithm (IPC), proposed in Pazira et al. [9], to decrease the effects stemming from this problem for computing the solution curve. The IPC algorithm allows the dgLARS method to be implemented by using less number of arithmetic operations that leads to potential computational saving.

Keywords: Differential geometry · dgLARS · Predictor-corrector algorithm · Sparsity · High-dimensional inference

1 Introduction

Modern statistical methods developed to study high-dimensional data sets are usually based on the idea to use a penalty function to estimate a solution curve embedded in the parameter space and then to find the point that represents the best compromise between sparsity and predictive behaviour of the model. Recent statistical literature has a great number of contributions devoted to this problem, such as the ℓ_1-penalty function [11] and the Smoothly Clipped Absolute Deviation method (SCAD) [7]. Differently from the methods cited here, Augugliaro et al. [4] proposed a new approach based on the differential geometrical representation of a Generalized Linear Model (GLM). The derived method, that does not require an explicit penalty function, has been called differential geometric LARS (dgLARS) method because it is defined generalizing the geometrical ideas on which the least angle regression (LARS), proposed in [6], is based. Moreover, Pazira et al. [9] extended the dgLARS method to the high-dimensional GLMs based on the exponential dispersion models with arbitrary link functions. In the same paper the authors proposed a new estimation method of the dispersion parameter based on high-dimensional feature space showed that is more accurate than the classic estimator. Furthermore, the authors, in [12], proposed an extension of the dgLARS method for sparse inference in relative risk regression models.

© Springer Nature Switzerland AG 2020
M. Raposo et al. (Eds.): CIBB 2018, LNBI 11925, pp. 99–106, 2020.
https://doi.org/10.1007/978-3-030-34585-3_9

The dgLARS method uses a predictor-corrector (PC) algorithm, proposed by Augugliaro et al. [4], to compute the implicitly defined solution curve, that however has the drawback of becoming intractable when working with thousands of predictors. From a computational point of view, using the PC algorithm leads to an increase in the run times needed for computing the solution curve. In this paper we explain an improved version of the PC algorithm (IPC), proposed in [9], to decrease the effects stemming from this problem for computing the solution curve. Since the IPC algorithm allows the dgLARS method to be implemented using less number of arithmetic operations needed to compute the Euler predictor scales as the cube of the number of predictors, it greatly reduces the computational burden. In addition these two algorithms, Augugliaro et al. [3] proposed a much more efficient cyclic coordinate descend (CCD) algorithm to fit the dgLARS solution curve when we work with a high-dimensional data set. Although this algorithm is computationally fast, the solution curve (parameter estimation) is not accurate. Our focus was only on the PC and IPC algorithms, although all three algorithms are available in the new version of the R-package `dglars` [2]. The package is available on the Comprehensive R Archive Network (CRAN) at http://CRAN.R-project.org/package=dglars.

The paper is organized as follows; In Sect. 2, we introduce the dgLARS method by giving some essential clues to the theory underlying a GLM from a differential geometric point of view and present the general case of equations based on the class of the exponential family. In Sect. 3, we propose our improved predictor-corrector algorithm. In Sects. 4, the comparisons in terms of performance between the PC and IPC algorithms are done by using the simulation and application studies, respectively. Finally, in Sect. 5 we draw some conclusions.

2   Scientific Background

Let $\mathbf{Y} = (Y_1, Y_2, \cdots, Y_n)^\top$ be an n-dimensional random vector with independent components. In what follows we shall assume that Y_i is a random variable with probability density function belonging to an exponential dispersion family, i.e.,

$$p_{Y_i}(y_i; \theta_i, \phi) = \exp\{y_i\theta_i - b(\theta_i)/a(\phi) + c(y_i, \phi)\}, \quad y_i \in \mathcal{Y}_i \subseteq \mathbb{R}, \quad (1)$$

where $\theta_i \in \Theta_i \subseteq \mathbb{R}$ is the canonical parameter, $\phi \in \Phi \subseteq \mathbb{R}^+$ is the dispersion parameter, and $a(\cdot)$, $b(\cdot)$ and $c(\cdot, \cdot)$ are given functions. In the following, we assume that each Θ_i is an open set and $a(\phi) = \phi$ is considered as a known parameter. The expected value of $\mathbf{Y}$ is related to the canonical parameter by $\boldsymbol{\mu} = \{\mu(\theta_1), \cdots, \mu(\theta_n)\}^\top$, where $\mu(\theta_i) = \frac{\partial b(\theta_i)}{\partial \theta_i}$ is called mean value mapping, and $\mathrm{Var}(\mathbf{Y}) = \phi\mathbf{V}(\boldsymbol{\mu})$, where $\mathbf{V}(\boldsymbol{\mu}) = diag\{V(\mu_1), \ldots, V(\mu_n)\}$ so that $V(\mu_i) = \frac{\partial^2 b(\theta_i)}{\partial \theta_i^2}$ is called the variance function. Since μ_i is a reparameterization, model (1) can be also denoted as $p_{Y_i}(y_i; \mu_i, \phi)$.

A GLM is defined by means of a known function $g(\cdot)$, called link function, relating the expected value of each Y_i to the vector of covariates $\mathbf{x}_i =$

$(1, x_{i1}, \ldots, x_{ip})^\top$ by the identity $g\{E(Y_i)\} = \eta_i = \mathbf{x}_i^\top \boldsymbol{\beta}$ where η_i is called the ith linear predictor and $\boldsymbol{\beta} = (\beta_0, \beta_1, \ldots, \beta_p)^\top$ is the vector of regression coefficients. In order to simplify our notation we let $\boldsymbol{\mu}(\boldsymbol{\beta}) = \{\mu_1(\boldsymbol{\beta}), \ldots, \mu_n(\boldsymbol{\beta})\}^\top$ where $\mu_i(\boldsymbol{\beta}) = g^{-1}(\mathbf{x}_i^\top \boldsymbol{\beta})$, $p_{\mathbf{Y}}(\mathbf{y}; \boldsymbol{\mu}(\boldsymbol{\beta})) = \prod_{i=1}^n p_{Y_i}(y_i; \mu_i(\boldsymbol{\beta}))$, and $\ell(\boldsymbol{\beta}; \mathbf{y}) = \log p_{\mathbf{Y}}(\mathbf{y}; \boldsymbol{\mu}(\boldsymbol{\beta}))$.

Augugliaro et al. [4] showed that the dgLARS estimator follows naturally from a differential geometric interpretation of a GLM, generalizing the LARS method [6], using the angle between scores $\partial_m \ell(\boldsymbol{\beta}; \mathbf{Y})$ and tangent residual vector $\mathbf{r}(\boldsymbol{\beta}, \mathbf{y}; \mathbf{Y}) = \sum_{i=1}^n (y_i - \mu_i) \frac{\partial \ell(\boldsymbol{\beta}; \mathbf{y})}{\partial \mu_i}$, defined as $\rho_m(\boldsymbol{\beta}) = \arccos[r_m(\boldsymbol{\beta})/\Upsilon]$ where $\Upsilon = \|\mathbf{r}(\boldsymbol{\beta}, \mathbf{y}; \mathbf{Y})\|_{p\{\boldsymbol{\mu}(\boldsymbol{\beta})\}}$ and $r_m(\boldsymbol{\beta}) = \partial_m \ell(\boldsymbol{\beta}; \mathbf{y}) \mathcal{I}_m(\boldsymbol{\beta})^{-1/2}$ is the Rao's score test statistic. LARS and dgLARS algorithms define a coefficient solution curve by identifying the most important variables step by step and including them into the model at specific points of the path. The original algorithms took as starting point of the path the model with the intercept only. This is a sensible choice as it makes the model invariant under affine transformations of the response or the covariates. However, the choice of the starting point of the least angle approach can be used to incorporate prior information about which variables are expected to be part of the final model and which ones one does not want to make subject to selection. The extended dgLARS method allows for a set of covariates, possibly including the intercept, that are always part of the model. However, in this paper, we consider as starting point of the path the model with the intercept only.

The extended dgLARS solution curve, which is denoted by $\hat{\boldsymbol{\beta}}_{\mathcal{A}}(\gamma) \subset \mathbb{R}^p$ where $\gamma \in [0, \gamma^{(1)}]$ and $0 \leqslant \gamma^{(p)} \leqslant \cdots \leqslant \gamma^{(2)} \leqslant \gamma^{(1)}$, is defined in the following way: for any $\gamma \in (\gamma^{(k+1)}, \gamma^{(k)}]$, if $\mathcal{A}(\gamma) = \{a_1, a_2, \cdots, a_k\}$ and $\mathcal{N}(\gamma) = \{a_1^c, a_2^c, \cdots, a_h^c\}$, then the extended dgLARS estimator satisfies the following conditions called the *generalized equiangularity condition*:

$$|r_{a_i}(\hat{\boldsymbol{\beta}}(\gamma))| = |r_{a_j}(\hat{\boldsymbol{\beta}}(\gamma))| = \gamma, \qquad \forall a_i, a_j \in \mathcal{A}(\gamma), \qquad (2)$$

$$r_{a_i}(\hat{\boldsymbol{\beta}}(\gamma)) = s_{a_i} \cdot \gamma, \qquad \forall a_i \in \mathcal{A}(\gamma),$$

$$|r_{a_l^c}(\hat{\boldsymbol{\beta}}(\gamma))| < |r_{a_i}(\hat{\boldsymbol{\beta}}(\gamma))| = \gamma, \qquad \forall a_l^c \in \mathcal{N}(\gamma) \text{ and } \forall a_i \in \mathcal{A}(\gamma),$$

where $s_{a_i} = \text{sign}\{r_{a_i}(\hat{\boldsymbol{\beta}}(\gamma))\}$, $k = |\mathcal{A}(\gamma)| = \#\{m : \hat{\beta}_m(\gamma) \neq 0\}$ and $h = |\mathcal{N}(\gamma)| = \#\{m : \hat{\beta}_m(\gamma) = 0\}$ are the number of covariates in the active and non-active sets, respectively, at location γ. The new covariate is included in the active set at $\gamma = \gamma^{(j)}$, with $j = 2, \ldots, p$, when the following condition is satisfied:

$$\exists a_l^c \in \mathcal{N}(\gamma): \quad |r_{a_l^c}(\hat{\boldsymbol{\beta}}(\gamma^{(j)}))| = |r_{a_i}(\hat{\boldsymbol{\beta}}(\gamma^{(j)}))|, \quad \forall a_i \in \mathcal{A}(\gamma). \qquad (3)$$

3   Materials and Methods

To compute the solution curve we can use the Predictor-Corrector (PC) algorithm [1], which explicitly finds a series of solutions by using the initial conditions (solutions at one extreme value of the parameter) and continuing to find the adjacent solutions on the basis of the current solutions. From a computational point of view, using the standard PC algorithm leads to an increase in the run times needed for computing the solution curve. In this section we propose an improved version of the PC algorithm to decrease the effects stemming from this problem for computing the solution curve. Using the improved PC algorithm leads to potential computational saving.

The PC method computes the exact coefficients at the values of γ at which the set of non-zero coefficients changes. This strategy yields a more accurate path in an efficient way than alternative methods and provides the exact order of the active set changes. Let us suppose that k predictors are included in the active set $\mathcal{A} = \{a_1, \cdots, a_k\}$ at location γ, such that $\gamma \in (\gamma^{(k+1)}, \gamma^{(k)}]$ is a fixed value of the tuning parameter. The corresponding point of the solution curve will be denoted by $\hat{\boldsymbol{\beta}}_{\mathcal{A}}(\gamma) = (\hat{\beta}_{a_0}(\gamma), \hat{\beta}_{a_1}(\gamma), \ldots, \hat{\beta}_{a_k}(\gamma))^\top$ where $\hat{\beta}_{a_0}$ is the intercept. Using (2), the dgLARS solution curve $\hat{\boldsymbol{\beta}}_{\mathcal{A}}(\gamma)$ satisfies the relationship

$$|r_{a_1}(\hat{\boldsymbol{\beta}}_{\mathcal{A}}(\gamma))| = |r_{a_2}(\hat{\boldsymbol{\beta}}_{\mathcal{A}}(\gamma))| = \cdots = |r_{a_k}(\hat{\boldsymbol{\beta}}_{\mathcal{A}}(\gamma))|, \qquad (4)$$

and is implicitly defined by the following system of non-linear equations:

$$\begin{cases} r_{a_0}(\hat{\boldsymbol{\beta}}_{\mathcal{A}}(\gamma)) = 0 \,, \\ r_{a_1}(\hat{\boldsymbol{\beta}}_{\mathcal{A}}(\gamma)) = s_{a_1}\gamma \,, \\ \quad\vdots \qquad\qquad \vdots \\ r_{a_k}(\hat{\boldsymbol{\beta}}_{\mathcal{A}}(\gamma)) = s_{a_k}\gamma \,. \end{cases} \qquad (5)$$

where $s_{a_i} = \text{sign}\{r_{a_i}(\hat{\boldsymbol{\beta}}_{\mathcal{A}}(\gamma))\}$.

We define $\tilde{\boldsymbol{\varphi}}_{\mathcal{A}}(\gamma) = \boldsymbol{\varphi}_{\mathcal{A}}(\gamma) - \mathbf{s}_{\mathcal{A}}\gamma$, where $\boldsymbol{\varphi}_{\mathcal{A}}(\gamma) = (r_{a_0}(\hat{\boldsymbol{\beta}}_{\mathcal{A}}(\gamma)), r_{a_1}(\hat{\boldsymbol{\beta}}_{\mathcal{A}}(\gamma)),$ $\ldots, r_{a_k}(\hat{\boldsymbol{\beta}}_{\mathcal{A}}(\gamma)))^\top$ and $\mathbf{s}_{\mathcal{A}} = (0, s_{a_1}, \ldots, s_{a_k})^\top$. We can locally approximate the solution curve at $\gamma - \Delta\gamma$ by the following expression

$$\hat{\boldsymbol{\beta}}_{\mathcal{A}}(\gamma - \Delta\gamma) \approx \tilde{\boldsymbol{\beta}}_{\mathcal{A}}(\gamma - \Delta\gamma) = \hat{\boldsymbol{\beta}}_{\mathcal{A}}(\gamma) - \Delta\gamma \left(\partial\boldsymbol{\varphi}_{\mathcal{A}}(\gamma)/\partial\hat{\boldsymbol{\beta}}_{\mathcal{A}}(\gamma)\right)^{-1} \mathbf{s}_{\mathcal{A}} \,, \qquad (6)$$

where $\Delta\gamma \in [0; \gamma - \gamma^{(k+1)}]$ and $\frac{\partial\boldsymbol{\varphi}_{\mathcal{A}}(\gamma)}{\partial\hat{\boldsymbol{\beta}}_{\mathcal{A}}(\gamma)}$ is the Jacobian matrix of the vector function $\boldsymbol{\varphi}_{\mathcal{A}}(\gamma)$ evaluated at the point $\hat{\boldsymbol{\beta}}_{\mathcal{A}}(\gamma)$. Equation (6) with the step size given in (7) are used for the predictor step of the PC algorithm. In the corrector step, $\tilde{\boldsymbol{\beta}}_{\mathcal{A}}(\gamma - \Delta\gamma)$ is used as starting point for the Newton-Raphson algorithm that is used to solve (5).

An efficient implementation of the PC method requires a suitable method to compute the smallest step size $\Delta\gamma$ that changes the active set of the non-zero coefficients. For each $a_j^c \in \mathcal{N}(\gamma)$ we have a value for $\Delta\gamma^{a_j^c}$ as follows

$$\Delta\gamma^{a_j^c} = \begin{cases} (\gamma - r_{a_j^c}(\hat{\boldsymbol{\beta}}_{\mathcal{A}}(\gamma)))/(1 - \dfrac{dr_{a_j^c}(\hat{\boldsymbol{\beta}}_{\mathcal{A}}(\gamma))}{d\gamma}) & \text{if } 0 \leq \Delta\gamma_1 \leq \gamma; \\ (\gamma + r_{a_j^c}(\hat{\boldsymbol{\beta}}_{\mathcal{A}}(\gamma)))/(1 + \dfrac{dr_{a_j^c}(\hat{\boldsymbol{\beta}}_{\mathcal{A}}(\gamma))}{d\gamma}) & \text{if } otherwise, \end{cases}$$

so that we consider the smallest value of the set of $\Delta\gamma^{a_j^c}$s as an optimal value

$$\Delta\gamma^{opt} = \min\left\{\Delta\gamma^{a_j} \mid a_j^c \in \mathcal{N}(\gamma)\right\}. \tag{7}$$

The main problem of the original PC algorithm is related to the number of arithmetic operations needed to compute the Euler predictor, which requires the inversion of the Jacobian matrix. From a computational point of view, using the PC algorithm leads to an increase in the run times needed to compute the solution curve. To improve the PC algorithm we propose a method to reduce the number of steps, thereby greatly reduce the computational burden because of decreasing the number of points of the solution curve.

Since the optimal step size is based on a local approximation, we also include an exclusion step for removing incorrectly included variables in the model. When an incorrect variable is included in the model after the corrector step, we have that there is a non-active variable such that the absolute value of the corresponding Rao score test statistic is greater than γ. To adjust the step size in the case of incorrectly including certain variables in the active set, Augugliaro et al. [4] reduced the optimal step size from the previous step, $\Delta\gamma^{opt}$, by using a small positive constant ε and then the inclusion step is repeated until the correct variable is joined to the model. They proposed a half of $\Delta\gamma^{opt}$ for ε as a possible choice. Augugliaro et al. [2,4,5] used a contractor factor, cf, which is a fixed value (i.e., $\gamma_{cf} = \gamma_{old} - \Delta\gamma$, where $\gamma_{old} = \gamma_{new} + \Delta\gamma^{opt}$ and $\Delta\gamma = \Delta\gamma^{opt} \cdot cf$), where $cf = 0.5$ as a default. In this case, this method acts like a *Bisection* method. However, the predicted root, γ_{cf}, may be closer to γ_{new} or γ_{old} than the mid-point between them. The poor convergence of the Bisection method as well as its poor adaptability to higher dimensions (i.e., systems of two or more non-linear equations) motivate the use of better techniques. In this case, we apply the method of Regula-Falsi (sometimes called False-Position) which always converges. The Regula-Falsi method uses the information about the function, $h(\cdot)$, to arrive at γ_{rf}, while in the case of the Bisection method finding γ is a *static* procedure since for a given γ_{new} and γ_{old}, it gives *identical* γ_{cf}, no matter what the function we wish to solve.

The Regula-Falsi method draws a secant from $h(\gamma_{new})$ to $h(\gamma_{old})$, and estimates the root as where it crosses the γ-axis, so that in our case $h(\gamma) = r_{a_j^c}(\hat{\boldsymbol{\beta}}_{\mathcal{A}}(\gamma)) - s_{a_j^c} \cdot \gamma$ where $s_{a_j^c} = \text{sign}\{r_{a_j^c}(\hat{\boldsymbol{\beta}}_{\mathcal{A}}(\gamma_{new}))\}$ and $a_j^c \in \mathcal{N}(\gamma)$. From (2), we have $h(\gamma) = r_{a_i}(\hat{\boldsymbol{\beta}}_{\mathcal{A}}(\gamma)) - s_{a_i}\gamma = 0$ for all $a_i \in \mathcal{A}(\gamma)$. Indeed, after the corrector step, when there is a non-active variable such that the absolute value of the corresponding Rao score test statistic is greater than γ, we want to find an exact point (γ_{rf}), which is very close or even equal to the true point, called the transition point $(\gamma^{(k+1)})$, that changes the active set, so that at the end, we have lower number of the arithmetic operations.

For applying the Regula-Falsi method to find the root of the equation $h(\gamma_{rf}) = 0$, let us suppose that k predictors are included in the active set, such that $\gamma_{new} < \gamma^{(k)}$. After the corrector step, when $\exists a_j^c \in \mathcal{N}(\gamma_{new})$ such that $|r_{a_j^c}(\hat{\boldsymbol{\beta}}_{\mathcal{A}}(\gamma_{new}))| > \gamma_{new}$, we find an γ_{rf} in the interval $[\gamma_{new}, \gamma_{old}]$,

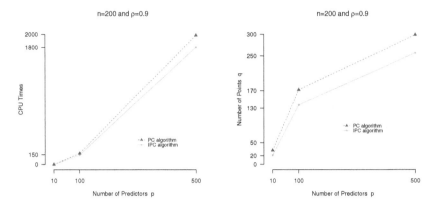

Fig. 1. (Left) CPU times, (right) mean number of the points of the solution curve, q, for the IPC and PC algorithms from the simulation study based on the Gamma regression model with $n = 200$ and $\rho = 0.9$, which are showed as a function of p.

where $\gamma_{old} = \gamma_{new} + \triangle\gamma^{opt}$, which is given by the intersection of the γ-axis and the straight line passing through $(\gamma_{new}, r_{a_j^c}(\hat{\boldsymbol{\beta}}_{\mathcal{A}}(\gamma_{new})) - s_{a_j^c} \cdot \gamma_{new})$ and $(\gamma_{old}, r_{a_j^c}(\hat{\boldsymbol{\beta}}_{\mathcal{A}}(\gamma_{old})) - s_{a_j^c} \cdot \gamma_{old})$. It is easy to verify that the root γ_{rf} is given by

$$\gamma_{rf} = \frac{\gamma_{new}\, r_{a_j^c}(\hat{\boldsymbol{\beta}}_{\mathcal{A}}(\gamma_{old})) - \gamma_{old}\, r_{a_j^c}(\hat{\boldsymbol{\beta}}_{\mathcal{A}}(\gamma_{new}))}{r_{a_j^c}(\hat{\boldsymbol{\beta}}_{\mathcal{A}}(\gamma_{old})) - r_{a_j^c}(\hat{\boldsymbol{\beta}}_{\mathcal{A}}(\gamma_{new})) + s_{a_j^c} \cdot (\gamma_{new} - \gamma_{old})}, \tag{8}$$

where $\forall a_j^c \in \mathcal{N}(\gamma_{new})$. Afterwards, we first set $\triangle\gamma = \triangle\gamma^{opt} - (\gamma_{rf} - \gamma_{new})$ and then set $\gamma = \gamma_{rf}$, to be able to go to the predictor step.

4   Experimental Results

In this section, we compare the improved PC algorithm with the original PC algorithm proposed in [9] and [4], respectively, using the simulation and application studies. Although, as mentioned before, these two algorithms compute the same active set, they have different number of arithmetic operations composing the active set. The main problem of the PC algorithm is related to the number of the points of the solution path or the number of arithmetic operations needed to compute the solution path (q, which is always greater than or equal to k), so that increasing the amount of q means increasing the run times.

In order to better understand the effects of the number of the points of the solution path (q) on the run times of these two algorithms, we use a simple simulation study based on a Gamma model with a non-canonical link function (log) with sample size equal to $n = (50, 100, 200)$ and $p = (10, 100, 500)$. The study is based on three different configurations of the covariance structure of the p predictors, such that $X_1, X_2, \cdots, X_n$ sampled from an $N(\mathbf{0}, \Sigma)$ distribution, where the diagonal elements of Σ are 1 and the off-diagonal elements follow $corr(X_i; X_j) = \rho^{|i-j|}$, where $i \neq j$ and $\rho = (0, 0.9)$. To simulate the response vector we use a model with intercept and choose $\beta_0 = 1$, $\beta_1 = \beta_2 = \beta_3 = 2$ and $\beta_4 = \beta_5 = \cdots = \beta_p = 0$.

The results show that the IPC algorithm has a lower average CPU time than the PC algorithm. Moreover, the mean number of the points of the solution curve yielded by the IPC algorithm is always lower than those yielded by the standard PC.

In Fig. 1, both timing and q are showed as a function of the number of predictors $p = (10, 100, 500)$. The differences between the two algorithms can be clearly seen in these figures although these differences are more tangible when $\rho = 0$ (no correlation among the predictors). For brevity, more results are not reported.

In the following, we demonstrate the use of the proposed algorithm in the dgLARS method applied to a *diabetes* data set obtained by [6]. The response y is a quantitative measure of disease progression for patients with diabetes one year later. The data has $n = 442$ observations on $p = 64$ variables $(n > p)$. The aim of the study is to identify which of the covariates are important factors in disease progression. A dgLARS *Gamma* regression model was fitted with the *inverse* canonical link function. The number of the points of the solution curve (q) for this low-dimensional data set by using the PC and IPC algorithms are 302 and 111, respectively, which shows that the IPC algorithm works faster than the PC. For brevity, more results can be found in [8] and [10].

5 Conclusion

In this paper, we briefly reviewed the differential geometrical theory underlying the dgLARS method, and described the new algorithm implemented in the new version of the `dglars` package, and we also compared run times between this algorithm and the previous one implemented in the old version. In simulations and the actual datasets we have shown that the improved PC algorithm is faster than the original PC algorithm. A new version of `dglars` with new algorithm and functions is available on CRAN.

References

1. Allgower, E., Georg, K.: Introduction to Numerical Continuation Methods. Society for Industrial and Applied Mathematics, New York (2003)
2. Augugliaro, L., Mineo, A., Wit, E.C., Pazira, H.: dglars: Differential Geometric LARS (dgLARS) Method. R package version 2.0.5 (2019). http://CRAN.R-project.org/package=dglars
3. Augugliaro, L., Mineo, A.M., Wit, E.C.: Differential geometric lars via cyclic coordinate descent method. International Conference on Computational Statistics (COMPSTAT 2012), pp. 67–79 (2012). limassol, Cyprus

4. Augugliaro, L., Mineo, A.M., Wit, E.C.: Differential geometric least angle regression: a differential geometric approach to sparse generalized linear models. J. Royal.Stat. Soc.: Ser. B **75**(3), 471–498 (2013)
5. Augugliaro, L., Mineo, A.M., Wit, E.C.: dglars: an R package to estimate sparse generalized linear models. J. Stat. Softw. **59**(8), 1–40 (2014a)
6. Efron, B., Hastie, T., Johnstone, I., Tibshirani, R.: Least angle regression. Ann. Stat. **32**(2), 407–499 (2004)
7. Fan, J., Li, R.: Variable selection via nonconcave penalized likelihood and its oracle properties. J. Am. Stat. Assoc. **96**(456), 1348–1360 (2001)
8. Pazira, H.: High-Dimensional Variable Selection for GLMs and Survival Models. University of Groningen, The Netherlands (2017)
9. Pazira, H., Augugliaro, L., Wit, E.C.: Extended differential geometric LARS for high-dimensional GLMs with general dispersion parameter. Stat. Comput. **28**(4), 753–774 (2018). https://doi.org/10.1007/s11222-017-9761-7
10. Pazira, H., Augugliaro, L., Wit, E.C.: A software tool for sparese estimation of a general class of high-dimensional GLMs. J. Stat. Softw. (2019)
11. Tibshirani, R.: Regression shrinkage and selection via the lasso. J. Roy. Stat. Soc. Ser. B **58**(1), 267–288 (1996)
12. Wit, E.C., Augugliaro, L., Pazira, H., Gonzalez, J., Abegaz, F.: Sparse relative risk regression models. Biostatistics **28**(4), 1–17 (2018). https://doi.org/10.1093/biostatistics/kxy060

Identification of Key miRNAs
in Regulation of PPI Networks

Antonino Fiannaca[1]([envelope]) [iD], Laura La Paglia[1] [iD], Massimo La Rosa[1] [iD],
Giosué Lo Bosco[2] [iD], Riccardo Rizzo[1] [iD], and Alfonso Urso[1] [iD]

[1] ICAR-CNR, National Research Council of Italy,
via Ugo La Malfa 153, 90146 Palermo, Italy
{antonino.fiannaca,laura.lapaglia,massimo.larosa,
riccardo.rizzo,alfonso.urso}@icar.cnr.it
[2] Dipartimento di Matematica e Informatica, UNIPA,
Universitá degli Studi di Palermo, Palermo, Italy
giosue.lobosco@unipa.it

Abstract. In this paper, we explore the interaction between miRNA
and deregulated proteins in some pathologies. Assuming that miRNA can
influence mRNA and consequently the proteins regulation, we explore
this connection by using an interaction matrix derived from miRNA-
target data and PPI network interactions. From this interaction matrix
and the set of deregulated proteins, we search for the miRNA subset
that influences the deregulated proteins with a minimum impact on the
not deregulated ones. This regulation problem can be formulated as a
complex optimization problem. In this paper, we have tried to solve it by
using the Genetic Algorithm Heuristic. As the main result, we have found
a set of miRNA that is known to be involved in disease development.

Keywords: miRNA expression profiles · Protein-protein interaction
networks · Genetic algorithms

1 Introduction

Protein-protein interaction networks (PPIN) represent conventionally a static
lattice of protein connections. Through interactome analysis, it is possible to
investigate biological processes both in physiological and pathological conditions. More recently many researchers began to expand the network analysis
through the inclusion of microRNA (miRNA) molecules [1]. miRNAs are post-
translational regulator of gene expression [2,3] found in plants, animals, some
viruses, and bacteria [4] that acts through mRNA repression or protein translation inhibition [5]. Recent studies has suggested that a deep understanding
of the complexity of gene regulation is due to miRNAs and chromatin features
[6–8]. The action of miRNAs would imply to dynamic modelling of PPIN, leading the activation or inhibition of some important pathways linked to specific
cellular processes. The understanding of these mechanisms is even more relevant

M. Raposo et al. (Eds.): CIBB 2018, LNBI 11925, pp. 107–117, 2020.
https://doi.org/10.1007/978-3-030-34585-3_10

when considering pathways related to diseases such as cancer [9]. Of course, to deeply investigate on the dynamic processes linked to network modelling, it is necessary to overcome simple miRNA-target interactions, but it is necessary to study also further connections. Indeed, miRNAs can act both directly or indirectly on target molecules regulating their expression. Moreover, PPIN analysis could lead to the identification of new potential biomarkers for cancer prediction, prognosis and treatment. Liang and Li [1] evidenced a strong correlation between microRNA repression and PPI. Indeed, they showed a positive correlation between the number of 3'UTR miRNA-target sites of interacting proteins and PPI, and they also showed a central regulative role of miRNAs in proteins that act as hubs of the analysed network.

Hsu [10], expanded Liang and Li's work introducing in miRNA-PPIN analysis the evaluation of L1 connections.

More recently, Chia-Hsien Lee et al. [11], analysed miRNA-regulated PPIN in breast cancer patients, by using miRNA target prediction data and mRNA/miRNA expression data, for the construction of network interactions. The network analysis evidenced some miRNAs functionally linked to breast cancer pathology. The study of PPIN together with miRNA interactions could expand the network analysis, better defining the functional modules of the different interacting partners.

In this work, we aim to discover relationships among some deregulated proteins and miRNAs playing a key role in their deregulation. Considering that interactions among miRNAs and proteins are mainly referable to mRNA degradation or translation inhibition, we chose to integrate PPIN with miRNA-target interactions, where the target is the gene coding for that protein in the network.

2   Materials and Methods

2.1   miRNA-Protein Interactions Network Model

In order to build a miRNA-protein interactions network, we retrieved protein-protein interactions from the publicly available online resource STRING-db (http://string-db.org/) (release 10.5), that contains, for different species, a set of known and predicted protein-protein interactions. STRING-db assigns to each pair of proteins an association score representing the probability of their interaction. As regards miRNA to protein interactions, we consider validated miRNA-target interactions because they are given from both mRNA degradation and translation inhibition. We retrieved these human miRNA-target interactions from miRTarBase database (http://mirtarbase.mbc.nctu.edu.tw) (release 7.0). miRTarBase provides a list of connection between mature miRNA expressed as miRNA ID, whereas genes are reported with Entrez gene ID (or gene symbol ID). Without loss of generality, the proposed approach can also be applied eventually considering predicted miRNA-target interactions, provided by several in-silico predictors [12–15], or taking into account tissue-specific miRNAs, that can be found using services such as [16] and [17].

In order to combine protein-protein interactions with miRNA-target interactions, we created a miRNA-regulated protein-protein interaction network, where a target gene is considered as the corresponding protein. Then, we built a graph data structure where nodes are miRNAs or proteins and edges are weighted accordingly to STRING-db score. Notice that edges between miRNAs and target proteins have the weight equal to 1 because we assign to a validated interaction the highest score. Since in specific biological functions the protein expression could be influenced by other miRNA-regulated proteins, the proposed model will take into account both protein-miRNA connections and protein-protein-miRNA connections. In other words, for each protein, we not only take into account the connected miRNA but also miRNA connected to their interacting partner proteins. According to the nomenclature adopted by [10], for each protein, we denoted as L0 the sub-network composed of itself and its closest miRNAs, and L1 the sub-network composed by itself, its closest miRNAs and its closest L0 sub-networks. Figure 1 shows an example of miRNA-regulated protein-protein interaction network with both L0 and L1 sub-networks: in this case, with respect to Prot_1 protein, if we consider subnetwork L0 only miR_1 and miR_2 miRNAs must be taken into account, whereas if we consider sub-network L1 also miR_3 and miR_4 miRNAs are reached with a weight of 0.8 and 0.5, respectively.

2.2   Validation Technique

We adopted a semi-supervised evaluation criterion, based on scientific literature since we cannot perform in vitro experiments. We used different database collecting information on miRNA expression profiles detected by high-throughput methods (dbDEMC) (www.picb.ac.cn/dbDEMC/), miRNA expression profiles in various human cancers, automatically extracted from published literature in PubMed (miRCancer) (http://mircancer.ecu.edu/), and custom literature research through Pubmed (www.ncbi.nlm.nih.gov/pubmed/).

2.3   Analysis of miRNA-Regulated PPI Network

From the miRNA-protein network obtained as described in the previous Section, it is possible to extract two interaction matrices $\mathbf{M}_0$ and $\mathbf{M}_1$, both of size $N_{prot} \times N_{miRNA}$, corresponding to L0 and L1 sub-networks, respectively. N_{prot} is the total number of proteins and N_{miRNA} is the total number of miRNAs. For each couple miRNA-protein, the related value of the interaction matrix is equal to the weight between those couple in the L0 or L1 subnetworks. In case of L0 subnetwork, $\mathbf{M}_0$ coincides with an adjacent matrix. The elements in $\mathbf{M}_1$ matrix is calculated considering the path that links the protein i with the miRNA j. This path is made by 2 connections (see Fig. 1) and the element m_{ij} of $\mathbf{M}_1$ is obtained by multiplying the weights associated to the two links (one of them is always one). For example, the value corresponding to the miR_3 and Prot_1 will be 0.8.

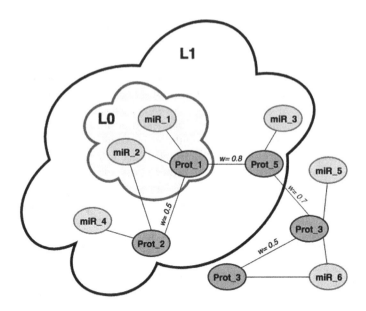

Fig. 1. An example of miRNA-regulated protein interaction network with 5 proteins and 6 miRNAs. With respect to the protein Prot_1, the cloud L0 includes 2 miRNAs, whereas the cloud L1 extends the range of influence to 4 miRNAs.

Given those matrices, we are looking for the miRNA that influences the deregulated proteins and do not "impact" on the proteins that are not deregulated.

At L0 level, we can consider the generic element of the matrix $\mathbf{M}_0$ as:

$$m_{ij} \in \{0, 1\} \quad i = 1, 2, \ldots, N_{prot} \quad j = 1, 2, \ldots, N_{miRNA} \tag{1}$$

because there is a direct interaction between the miRNA and the protein.

A representation of this matrix is in the left part of Fig. 2 where black dots indicate that the corresponding miRNA in the column influences the protein in the row. The horizontal line separates the K proteins deregulated form the other ones.

Assuming that there are K deregulated proteins we can arrange the matrix rows so that $i = 1, 2, \ldots, K$ are the deregulated proteins and $i = K + 1, K + 2, \ldots, N_{prot}$ are the not deregulated ones.

The hypothesis is that the miRNAs that influence the network are among the one that influences the deregulated proteins, but have a small impact on the not deregulated ones.

In a first attempt we can attach to each miRNA j two values:

$$v_j = \sum_{i=1}^{K} m_{ij} \qquad p_j = \sum_{i=K+1}^{N_{prot}} m_{ij} \tag{2}$$

where v_j represent the total impact of the miRNA j on the set of the deregulated proteins, while p_j represents the total impact on the not deregulated ones.

According to this framework we can try to obtain the set of the n_1 miRNA (with indexes l_p, $p = 1, 2, \ldots, n_1$) that maximize the total impact on the K deregulated proteins $\sum_{p=1}^{n_1} v_{l_p}$ and minimize the total impact on the $N_{prot} - K$ not deregulated proteins $\sum_{p=1}^{n_1} p_{l_p}$.

If a maximum capacity P is chosen, the problem we are going to solve can be formulated as a *maximum knapsack*. We know that it is an NP complete problem that can be solved using various heuristics. However, to be influenced by more than one miRNA, and do not guarantee that all the deregulated proteins are "targeted" by, at least, one miRNA. These characteristics are really important for the solution of our problem.

Another approach to further refine this research can be obtained by applying some other constrains to this optimization problem. We should select a set of miRNA that influence the maximum number of deregulated proteins with a minimum overlap, and also have a minimum impact on the not deregulated ones. We can observe that the columns of the matrix corresponding to the miRNA j, $\mathbf{w}_j$, can be separated in two components, $\mathbf{w}_j^1$ and $\mathbf{w}_j^2$:

$$\mathbf{w}_j^1 = [m_{1j}, m_{2j}, \ldots, m_{Kj}]^T \tag{3}$$
$$\mathbf{w}_j^2 = [m_{K+1j}, m_{K+2j}, \ldots, m_{N_{prot}j}]^T \tag{4}$$

Now we want to select a set of n_2 miRNA (with indexes l_p, $p = 1, 2, \ldots, n_2$) that minimize the overlap among miRNA that impact on the set of deregulated proteins. So that summing the set of weights we have:

$$\mathbf{W}^1 = \sum_{p=1}^{n_2} \mathbf{w}_{l_p}^1 = \left[\sum_{p=1}^{n_2} m_{1l_p}, \sum_{p=1}^{m_2} m_{2l_p}, \ldots, \sum_{p=1}^{n_2} m_{Kl_p}, \right]^T \tag{5}$$

In order to have all the proteins covered by just one miRNA we should minimize:

$$\left| \sum_{i=1}^{K} \sum_{p=1}^{n_2} m_{il_p} - K \right| \quad \text{and} \quad \sum_{p=1}^{n_2} p_{l_p} \tag{6}$$

with the constrain:

$$\sum_{p=1}^{n_2} m_{il_p} \geq 1 \tag{7}$$

The constraint in Eq. 7 means that the protein i should be targeted by at least one miRNA (ideally should be just one miRNA and $\sum_{p=1}^{n_2} m_{il_p} = 1$). This formulation makes our problem configurable as a *weighted maximum coverage problem* because we have to "cover" the whole deregulated protein set with the $\mathbf{w}_j^1$ miRNA vectors that have a weight given by p_j. Also, in this case, the optimization problem belongs to the class of NP.

In case of L1 sub-network, the values of the matrix $\mathbf{M}_1$ for the K up-regulated proteins are threshold by using the following rule

$$m_{ij}' = \begin{cases} 0, & \text{if } m_{ij} = 0 \quad i = 1, 2, \ldots, K \\ 1, & \text{if } m_{ij} > 0 \quad i = 1, 2, \ldots, K \end{cases} \tag{8}$$

and considering m_{ij}' throughout the rest of the algorithm. A visible example of the matrix M_1 is depicted in Fig. 2, where it is shown as a binary image. The values associated with the $i = K + 1, K + 2, \ldots N_{prot}$ are not subjected to a threshold. The calculation of the weights p_j was also modified and the Eq. 2 is now:

$$p_j = \frac{\sum_{i=K+1}^{N_{prot}} m_{ij}}{\frac{\sum_{i=1}^{K} m_{ij}}{\sum_{i=1}^{K} m_{ij}'}} \tag{9}$$

Equation 9 could be also used for the matrix $\mathbf{M}_0$, where $m_{ij}' = m_{ij}$.

In this work, we have used a genetic algorithm (GA) to find the set of miRNA that solves the optimization problem related to our purpose. It uses as fitness function the one introduced in Eq. 6 and as the set of constraints the one shown in Eq. 7. The used genetic algorithm is the simple canonical one, the representation of the solution is a string of bits, one bit for each miRNA that can or can not be part of the set that impact to the deregulated proteins. The problem is a multi-objective optimization one, because it is necessary to minimize both terms in Eq. 6. The optimization program was developed in python language using the Deap package (https://github.com/DEAP/deap). The probability of mating was set to 0.5 and the single point cross over was used, the mutation probability was set to 0.001.

3   Results

We used a breast cancer cohort of patients presenting ER-negative (ER-) histopathological markers and the corresponding dataset is given by [18]. It contains a set of 2319 ER-breast cancer tissue proteins. Among these proteins, there are 95 selectively up-regulated in ER- breast tumours. We decided to select just up-regulated proteins from ER- samples, in order to test the specificity of our approach in a specific case study having a subclass of IHC markers and a subclass of deregulated proteins. As described in the previous section, starting from a set of proteins, we got protein-protein interactions from STRING-db and validated miRNA-target interactions from miRTarBase. According to the available interactions in STRING-db and miRTarBase, we filtered the number of analysed proteins from 2319 to 2043, and therefore the subset of the up-regulated ones is reduced from 95 to 87 proteins. So that the number of deregulated protein was $K = 87$ and the number of not deregulated was $N_{prot} - K = 2043 - 87 = 1956$.

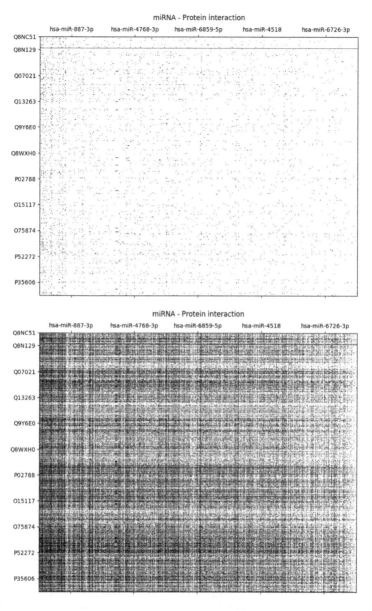

Fig. 2. A visualization of the interaction matrices $\mathbf{M}_0$ (up) and $\mathbf{M}_1$ (down) for the analysed case study. Proteins and miRNAs are in rows and columns, respectively. For each protein-miRNA interaction, the score is represented as a grey-scale pixel. A horizontal line separates the de-regulated proteins (over the line) from the not de-regulated ones (below the line). Notice as the $\mathbf{M}_0$ matrix is sparser than the $\mathbf{M}_1$ one.

Table 1. Set of miRNAs resulting from L0 and L1 network analysis grouped for deregulation and immunohistochemical types. miRNAs in italic font are discovered exclusively in L1; underlined miRNAs are given only in L0. Finally, bold type stands for
well-known interesting miRNAs in breast cancer.

Set of miRNAs from network analysis		DE in BC	IHC Type
L0	L1		
hsa-miR-1-3p; hsa-miR-133b; hsa-miR-223-3p; hsa-miR-140-5p; hsa-miR-125b-5p; hsa-miR-30a-3p; hsa-miR-192-5p; hsa-miR-124-3p; hsa-miR-335-5p; **hsa-let-7a-5p**; hsa-miR-222-3p; hsa-miR-200b-3p; hsa-miR-1296-5p; hsa-miR-16-5p	hsa-miR-1-3p; hsa-miR-133b; hsa-miR-223-3p; hsa-miR-140-5p; hsa-miR-125b-5p; hsa-miR-30a-3p; hsa-miR-192-5p; hsa-miR-124-3p; hsa-miR-335-5p; **hsa-let-7a-5p**; hsa-miR-222-3p; hsa-miR-200b-3p; hsa-miR-1296-5p; hsa-miR-16-5p; *hsa-miR-146a-5p; hsa-miR-320a; hsa-miR-26a-5p;* *hsa-miR-29c-3p;* ***hsa-miR-92a-3p;*** *hsa-miR-31-5p;* *hsa-miR-7-5p; hsa-miR-122-5p; hsa-miR-143-3p;* *hsa-miR-193a-3p; hsa-miR-328-3p;* *hsa-miR-519b-3p; hsa-miR-145-5p; hsa-miR-451a;* ***hsa-miR-206;*** *hsa-miR-199a-3p; hsa-miR-130a-3p;* *hsa-miR-138-5p; hsa-miR-9-5p; hsa-miR-584-5p;* *hsa-miR-101-3p; hsa-miR-133a-3p;* *hsa-miR-205-5p; hsa-miR-224-5p;* *hsa-miR-200c-3p; hsa-miR-15a-5p;* *hsa-miR-200a-3p; hsa-miR-559*	down	ER-
hsa-miR-424-3p; hsa-miR-424-5p	hsa-miR-424-3p; hsa-miR-424-5p; *hsa-miR-27a-3p; hsa-miR-25-3p; hsa-miR-29a-3p;* *hsa-miR-519c-3p; hsa-miR-126-3p;* *hsa-miR-377-3p;* ***hsa-miR-30e-5p;*** *hsa-miR-433-3p;* *hsa-miR-375; hsa-miR-29b-3p; hsa-miR-210-3p*	down	ER+
	hsa-miR-199b-5p; hsa-miR-218-5p	down	her2+/PR+
hsa-miR-92b-5p; hsa-miR-155-5p	hsa-miR-155-5p; *hsa-miR-449a; hsa-miR-106a-5p;* *hsa-miR-17-5p; hsa-miR-141-3p; hsa-miR-17-3p;* *hsa-miR-24-3p; hsa-miR-519a-3p; hsa-miR-20a-5p;* *hsa-miR-520c-3p; hsa-miR-373-3p;* *hsa-miR-199a-5p; hsa-miR-661; hsa-miR-545-3p*	up	ER-
hsa-miR-20b-5p; hsa-miR-196a-5p; hsa-miR-30d-5p; hsa-miR-197-3p; hsa-miR-96-5p	hsa-miR-20b-5p; hsa-miR-196a-5p; hsa-miR-30d-5p; hsa-miR-197-3p; *hsa-miR-376a-5p; hsa-miR-32-5p*	up	ER+
hsa-miR-605-5p	hsa-miR-605-5p	DE in other cancers	-

Exploring the miRNA regulatory PPI network, we extracted two matrices M_0
and M_1 (see Fig. 2). We investigated two different depth of the network to test if
there were differences in miRNAs identified in the two matrices and consequently
to evidence potential miRNAs relevant for the regulation of some of the analysed
up-regulated protein that could indirectly act on the target regulation. Both
miRNA lists identified in L0 and L1 depth were validated in literature through
the use of different web tools as described in the previous section, investigating
(1) if identified miRNAs were associated with breast cancer pathology, (2) if they
were down-regulated, (3) if they were deregulated in ER- IHC subclass. These
three criteria allow reinforcing the potential relation evidenced with the proposed
method, as miRNAs are post-transcriptional regulators. Thus we expected to
find for the same sub-class of patients (ER-) those down-regulated miRNAs that
are related to the up-regulated protein set.

The analysis of $\mathbf{M}_0$ with the genetic algorithm produced a total of 24 miRNAs interacting with up-regulated proteins. Among these, 14 were down-regulated in ER-, 2 were down-regulated in ER+ patients, 2 were up-regulated in ER-patients, 5 were up-regulated in ER+ patients and 1 miRNA (hsa-miR-605-5p) was deregulated in other cancer types different from breast cancer. The genetic algorithm applied to $\mathbf{M}_1$ matrix produced a total of 77 miRNAs interacting with up-regulated proteins. Among these, 42 were down-regulated in ER-, 13 were down-regulated in ER+ patients, 14 were up-regulated in ER- patients, 6 were up-regulated in ER+ patients, and 2 were down-regulated in other IHC types. Just 2 miRNA (i.e., hsa-miR-92b-5p and hsa-miR-96-5p) obtained from $\mathbf{M}_0$ matrix were not found in $\mathbf{M}_1$. Finally, we can deduce that L1 depth can expand the miRNA-PPI regulative network, evidencing also known relevant miRNAs for breast cancer pathology. Both L0 and L1 depth evidence miRNAs strongly related to breast cancer. As an example, hsa-let-7 regulates the expression of multiple genes related to the metastasis and stem cell phenotype, as shown by Thammaiah et al [19]. Other miRNAs have been identified by the genetic algorithm exclusively from $\mathbf{M}_1$ matrix, i.e. hsa-miR-92a, hsa-miR-206 and hsa-miR-30e. The first one is a key oncogenic component of the miR-17-92 cluster [20]; it is known to directly target the anti-apoptotic molecule BCL-2-interacting mediator of cell death (BIM), leading to an alteration in the apoptotic signalling cascade in different tissues including breast cancer [20]. The second one is another down-regulated miRNAs evidenced in ER- samples, involved in breast cancer progression [21]. Finally, hsa-miR-30e is involved in breast cancer initiation by inhibiting its downstream targets as ITGB3 and UBC9 [22].

Table 1 reports the results obtained in 100 iterations of the genetic algorithm.

4   Conclusion

In this preliminary study, we introduce a methodology with the aim of obtaining a list of interesting miRNA, starting from a set of deregulated proteins in the presence of a cancer disease. The methodology makes use of a protein-protein interaction network of a specific disease, where the information about the miRNA-target interactions is integrated. This allows obtaining a matrix that represents the weighted connection between miRNA and proteins. Different matrices can be obtained by considering different levels of interactions between the proteins in the network. On this matrix, it is possible to highlight the deregulated proteins and obtain the corresponding miRNA by assuming a direct correlation between them. Applying this methodology to the case of breast cancer, we have obtained a list of miRNA that contains relevant miRNA for the disease, confirming the effectiveness of the method. We plan to investigate more on the way the protein-protein interaction network is set. Moreover, other diseases could be studied.

References

1. Liang, H., Li, W.H.: MicroRNA regulation of human protein-protein interaction network. RNA **13**(9), 1402–1408 (2007)
2. Giancarlo, R., Lo Bosco, G., Pinello, L., Utro, F.: The three steps of clustering in the post-genomic era: a synopsis. In: Rizzo, R., Lisboa, P.J.G. (eds.) CIBB 2010. LNCS, vol. 6685, pp. 13–30. Springer, Heidelberg (2011). https://doi.org/10.1007/978-3-642-21946-7_2
3. Ciaramella, A., et al.: Interactive data analysis and clustering of genomic data. Neural Netw. **21**(2–3), 368–378 (2008)
4. Fiannaca, A., et al.: Deep learning models for bacteria taxonomic classification of metagenomic data. BMC Bioinform. **19**, 198 (2018)
5. Kwak, P.B., Iwasaki, S., Tomari, Y.: The microRNA pathway and cancer. Cancer Sci. **101**(11), 2309–2315 (2010)
6. Tao, B.B., et al.: Evidence for the association of chromatin and microRNA regulation in the human genome. Oncotarget **8**(41), 70958–70966 (2017)
7. Pinello, L., Lo Bosco, G., Yuan, G.-C.: Applications of alignment-free methods in epigenomics. Briefings Bioinform. **15**(3), 419–430 (2014)
8. Di Gangi, M., Lo Bosco, G., Rizzo, R.: Deep learning architectures for prediction of nucleosome positioning from sequences data. BMC Bioinform. **19**, 418 (2018)
9. Camastra, F., Di Taranto, M.D., Staiano, A.: Statistical and computational methods for genetic diseases: an overview. Comput. Math. Methods Med. **954598**, 2015 (2015)
10. Hsu, C.W., Juan, H.F., Huang, H.C.: Characterization of microRNA-regulated protein-protein interaction network. Proteomics **8**(10), 1975–1979 (2008)
11. Lee, C.H., Kuo, W.H., Lin, C.C., Oyang, Y.J., Huang, H.C., Juan, H.F.: MicroRNA-regulated protein-protein interaction networks and their functions in breast cancer. Int. J. Mol. Sci. **14**(6), 11560–11606 (2013)
12. Paraskevopoulou, M.D., Georgakilas, G., Kostoulas, N., Vlachos, I.S., et al.: DIANA-microT web server v5.0: service integration into miRNA functional analysis workflows. Nucleic Acids Res. **41**(W1), W169–W173 (2013)
13. Friedman, R., Farh, K., Burge, C., Bartel, D.P.: Most mammalian mRNAs are conserved targets of microRNAs. Genome Res. **19**(1), 92–105 (2009)
14. Fiannaca, A., La Rosa, M., La Paglia, L., Rizzo, R., Urso, A.: MiRNATIP: a SOM-based miRNA-target interactions predictor. BMC Bioinform. **17**, 321 (2016)
15. Bino, J., Enright, A.J., Aravin, A., Tuschl, T., et al.: Human microRNA targets. PLoS Biol. **2**(11), e363 (2004)
16. Sooyoung, C., Insu, J., Yukyung, J., Suhyeon, Y., et al.: MiRGator v3.0: a microRNA portal for deep sequencing, expression profiling and mRNA targeting. Nucleic Acids Res. **41**(D1), D252–D257 (2012)
17. Fiannaca, A., La Rosa, M., La Paglia, L., Urso, A.: miRTissue: a web application for the analysis of miRNA-target interactions in human tissues. BMC Bioinform. **9**(S15), 434 (2018)
18. Rezaul, K., Thumar, J.K., Lundgren, D.H., Eng, J.K., Claffey, K.P., Wilson, L., Han, D.K.: Differential protein expression profiles in estrogen receptor-positive and -negative breast cancer tissues using label-free quantitative proteomics. Genes Cancer **1**(3), 251–271 (2010)
19. Thammaiah, C.K., Jayaram, S.: Role of let-7 family microRNA in breast cancer. Non-coding RNA Res. **1**(1), 77–82 (2016)

20. Smith, L., Baxter, E.W., Chambers, P.A., et al.: Down-regulation of miR-92 in breast epithelial cells and in normal but not tumour fibroblasts contributes to breast carcinogenesis. PLoS One **10**(10), e0139698 (2015)
21. Fu, Y., Shao, Z.M., He, Q.Z., Jiang, B.Q., Wu, Y., Zhuang, Z.G.: Hsa-MiR-206 represses the proliferation and invasion of breast cancer cells by targeting Cx43. Eur. Rev. Med. Pharmacol. Sci. **19**(11), 2091–2104 (2015)
22. Yang, S.J., Yang, S.Y., Wang, D.D., et al.: The MiR-30 family: versatile players in breast cancer. Tumor Biol. **39**(3), 1–13 (2017)

Recurrent Deep Neural Networks for Nucleosome Classification

Domenico Amato[1]⦿, Mattia Antonino Di Gangi[2]⦿, Giosuè Lo Bosco[1](✉)⦿,
and Riccardo Rizzo[3]⦿

[1] Dipartmento di Matematica e Informatica, Università degli studi di Palermo,
Palermo, Italy
{domenico.amato01,giosue.lobosco}@unipa.it
[2] Fondazione Bruno Kessler, Trento, Italy
digangi@fbk.eu
[3] ICAR-CNR - National Research Council of Italy, Palermo, Italy
riccardo.rizzo@icar.cnr.it

Abstract. Nucleosomes are the fundamental repeating unit of chromatin. A nucleosome is an 8 histone proteins complex, in which approximately 147–150 pairs of DNA bases bind. Several biological studies have clearly stated that the regulation of cell type-specific gene activities are influenced by nucleosome positioning. Bioinformatic studies have improved those results showing proof of sequence specificity in nucleosomes' DNA fragment. In this work, we present a recurrent neural network that uses nucleosome sequence features representation for their classification. In particular, we implement an architecture which stacks convolutional and long short-term memory layers, with the main purpose to avoid the features extraction and selection steps. We have computed classifications using eight datasets of three different organisms with a growing genome complexity, from yeast to human. We have also studied the capability of the model trained on the highest complex species in recognizing nucleosomes of the other organisms.

Keywords: Nucleosome classification · Epigenetic · Deep learning networks · Recurrent Neural Networks

1 Scientific Background

Chromatin is a particular DNA-protein complex that package eukaryote DNA in a more compact form, by means of histone octamers called nucleosomes [1]. More in details, the genome is contained into the nucleus of the cell, wrapped in 1.7 turns (around 150 bp) around each nucleosome. The DNA sequences that separate the nucleosomes are called *linkers*. Chromatin structure controls gene expression, so dynamic modification of nucleosomes' positions is of great importance in understanding changes in cells' behaviour. *ATP-dependent chromatin remodelling complexes*, for example, is one of the factors able to modify the structure and the position of nucleosomes [2,3]. Aberrations in chromatin remodelling

M. Raposo et al. (Eds.): CIBB 2018, LNBI 11925, pp. 118–127, 2020.
https://doi.org/10.1007/978-3-030-34585-3_11

proteins are found to be associated with human diseases, including cancer [4]. For this reason, the class of computational methods for studying gene expression and genetic diseases [5–10] has received new contributions with the main purpose of investigating the property of nucleosomes by dealing with their automatic recognition [11,12]. The specific nucleotide content of the DNA fragments wrapped into nucleosomes shows specific dinucleotides patterns [13,14]. This consideration has led to the development of specific computational models that uses only the DNA string information as input [11,15–17]. These methodologies use a numerical vectorial representation of DNA strings, extracted as feature vectors during a preliminary step with the supervision of an expert. An example is the *k-mers* representation used by several machine learning methodologies for sequence classification [16,18–20]. In the last years, the adoption of *deep learning neural networks* (DLNN) [21] has shown very significant progress in several problematic artificial intelligence tasks [22]. DLNN models have the ability to automatically extract features from the input pattern without any *a priori* knowledge about the problem. For classification problems involving sequential input data, the *Convolutional Neural Networks* (CNNs) and the *Recurrent Neural Networks* (RNNs) have been successfully adopted. CNNs have the first layer of convolutional filters whose output is further processed by a non-linear activation function. The final classification is computed by a fully connected layer [23]. Instead, RNNs use a hidden layer state as internal memory, implemented by feedback connections directed through the hidden state to perform update taking into consideration the past steps. Thanks to this hidden state, the RNNs are excellent models for the processing of sequences with temporal information. Both for generic sequences [24–26] and for the case of nucleosome related ones [27–29], several DLNN have been proposed so far. In particular, it has been already proposed a basic CNN architecture capable of classifying sequences represented as k-mers [27]. A more recent version [28] performs automatically the sequence features extraction by using a convolutional layer, while a subsequent recurrent layer can deal with the longer-range positional dependency between part of the sequence. This latter model has shown that the specific ability of a recurrent network to catch specific orders of the dinucleotides in the DNA string, is really effective. In this work, we compute new experiments for nucleosome classification, extending this last contribution. Apart from a comparison with one state of the art method, using different datasets of several organisms, we have also investigated the information content behind the nucleosome pattern related to the most complex organism. In particular, we have used the nucleosomes human sequence data to train a model, performing the classification test on the simplest organisms.

Table 1. The distribution of samples in the dataset. The labels HM,DM,YS indicate the Human, Drosophila and Yeast sets of sequences respectively; WG indicates the whole genome, LC indicates the largest chromosome, PM indicates the Promoter sequences; 5U indicates the sequences from the 5UTR exon region.

	HM			DM			YS	
	LC	PM	5U	LC	PM	5U	WG	PM
N	97209	56404	11769	46054	48251	4669	39661	27373
L	65563	44639	4880	30458	28763	2704	4824	4463
T	162772	101043	16649	76512	77014	7373	44485	31836

2 Materials and Methods

2.1 Dataset

The dataset used for the experiments [15] is composed of nucleosome sequences belonging to three organisms. Their differences are in terms of species complexity, ranging from yeast to human. Data related to each organism comes from different areas of the genome, for a total number of eight distinct sets. Besides, different kinds of sequences are provided in each of the species. In detail, three species are available: Yeast (YS), Drosophila (DM) and Human (HM). We found entire sequences of the genome (WG) and promoter (PM) of YS, and the largest sequences of the chromosome (LC), promoter (PM) and exon 5'UTR region (5U) of DM and HM. The PM and 5'UTR sequences are important in our study because they have a specific role in transcription and translation respectively. WC and LC are sequences without any specific function. Since Liu et al. [15] provided only bed files containing genome coordinates positions of nucleosomes and linkers, we had to retrieve the related sequences processing the entire genome. In particular, we first extracted the coordinates of the midpoints of each sequence from the bedfiles and obtained the sequences by looking at an interval of 75 bp to the left and 75 bp to the right of the midpoints. The sequences have been fetched from the genome files of the organisms, downloaded from UCSC Table Browser. Using this extraction methodology, we have set 8 different datasets for three species of nucleosomal and linker sequences, each of length 151 bp.

The distribution of the classes for this group of datasets is shown in Table 1.

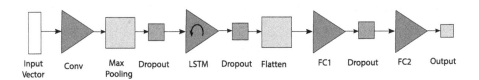

Fig. 1. The deep neural network architecture of the classifier.

2.2 The Adopted Network

A Network architecture is defined by the type, number and sequence of layers, along with parameter values. A deep learning network uses various types of layers that can be applicable to different types of data and can be chosen based on the type of features we expect will be relevant to the task.

In Fig. 1 we show the network layers composition which characterizes the deep learning network proposed in this work: a convolutional layer, a max pooling layer, a dropout layer, an LSTM layer, and two fully connected layers. The input X to the model is a matrix of $4 \times R$ binary values, where R is the length of the sequence. Each column of X is a binary vector of length 4, having all zero entries except for a single one in position i. The unique 1 denotes the presence of a nucleotide in the alphabet $\{A, C, G, T\}$. This representation is known as *character-level one-hot encoding.*

The first layer of the network (from left to right) is a 1-D convolutional layer whose primary role is the extraction of features from input data X. In general, a 1-D convolutional layer gives as output a matrix h of size $L \times R$, where L is the number of output feature maps, processing input X in the form of a numerical matrix of size $4 \times R$. h is the result of the convolution of the data and a kernel w^l of length k. In the proposed network, this layer is composed by a bank of $n = 50$ $1D$ convolutions between the kernel vectors w^l $l = 1, 2, \ldots n$ and the input sequence X. The vector h is padded by filling with zeroes the positions in order to make it of the same dimension of the input vector X. The transfer function of the convolutional layer is a Rectified Linear Unit (ReLU) activation function.

The next *Max Pooling* layer operates as a non-linear down-sampling. in particular, it divides the input vector into a set of non-overlapping regions and, for each sub-region, their maximum value is chosen as the output. The width and stride parameters of the network have been set to 2. This layer introduces the possibility to capture the most significant features of convolution and reduces the output size of the input vectors. Overfitting during the training steps is avoided using a dropout layer [31] with probability $p = 0.5$.

The next LSTM layer [32] is used to find long-range relations between the time steps along all the sequence. The adoption if this layer, which belongs to the category of recurrent networks, is motivated in the specific case of nucleosome classification since it is known that there exist specific sequences of nucleotide strongly related to the presence of nucleosomes. LSTM layers are explicitly designed to alleviate the *vanishing gradient problem* [33] and to select the inputs that are relevant for updating the hidden state of the recurrence. The LSTM layer proposed in our network is composed of 50 hidden memory units. The convolutional and LSTM layers have a regularizer L_2 with $\lambda = 0.001$. The outputs from all the LSTM time steps are then concatenated in a single vector.

Finally, 2 subsequent fully-connected layers reduce vector length to 150 and then to 1. The output of the network is a real value in the interval $[0, 1]$ calculated by using a sigmoid function. In order to properly adjust the weights of the network, we use Adam [34], without learning decay, as an optimization algorithm for the training phase.

Table 2. Area under the ROC curve for each dataset. The first column shows the AUC values of the method by Kaplan et al. [35] reported in the paper by Liu et al. [15], sometimes with approximate values (interval range or close to symbol '∼'). The last column regards our proposed method. Best values are in bold.

	Kaplan et al. [35]	DLNN
HM-LC	∼0.65	**0.79**
DM-LC	**(0.70,0.75)**	0.71
YS-WG	∼0.7	**0.79**
HM-PM	∼0.6	**0.77**
DM-PM	**(0.70,0.75)**	0.72
YS-PM	**(0.70,0.75)**	0.73
HM-5U	∼0.65	**0.67**
DM-5U	**(0.65,0.70)**	0.66

3 Results

In this work, we performed two kinds of experiments. The first one regard the comparison of the proposed deep learning network with a state of the art method for nucleosome classification. The second one is a study of the informative content of the nucleosome pattern with regard to the complexity of the organism.

3.1 Comparison with a State of the Art Method

In the first one, we have computed the receiver operating characteristic (ROC) curves and the Area under the ROC curves (AUC) using the evaluation protocol proposed by Liu et al. [15]. This is dictated by the consequence of having the possibility to compare the results computed by our model with the ones reported by other models presented in the manuscript. The protocol consists of computing an average roc curve on 100 test samples (with replacement) of 100 sequences each. We have noticed that using a PC desktop with Xeon CPU, 24 GB RAM and a GTX980 nVidia graphic card, networks required hours for the training, due to the concatenation of CNN and LSTM layers. For this reason, the Liu et al. method is preferable instead of k-fold cross validation that would surely increase the time needed to perform the experiments. We have split each dataset *a priori* into training and test so that there is sufficient data to form a strong model while having a large pool for the selection of test data with the above mentioned experimental protocol. After that, we selected the validation set from the training set as a number of samples equivalent to the 10% of the dataset size, clipped to 1000 if the size exceeded this number. In our experiments, we have used the Adam optimizer in the training phase and we have set the epochs to 100 with a learning rate of $3 * 10^{-4}$. For each dataset, we have trained only one

Table 3. Mean Area under the ROC curve for each dataset. The man is computed on 10 fold. The first column reports the AUC results when using Human as training set and the other species as tests. The last column shows results using Yeast as training and the other species as tests. Best values are in bold.

	DLNN-HM	DLNN-YS
HM-LC	-	0.67
DM-LC	**0.65**	0.53
YS-WG	0.67	-
HM-PM	-	0.64
DM-PM	**0.68**	0.54
YS-PM	0.71	-
Average total	0.67	0.59

model and then we have computed the ROC curve and the AUC over the 100 tests. Specificity and sensitivity were computed by setting a score threshold, i.e. a sequence was classified as a nucleosome if the respective networks' output was greater than the threshold. The convolution kernels of the first layer have a size of 3. Table 2 compares our results with the ones relative to the method by Kaplan et al. [35], that, according to the paper by Liu et al. [15], has demonstrated the best performances in the experiments. Note that in most of the cases the paper by Liu et al. does not report precise values of the AUC, so we have reported the closest value or the minimum and the maximum values of the interval in brackets. Our method works exclusively on the annotated data set and is able to learn only on the basis of a simple character-level representation with one-hot encoding, unlike the Kaplan et al. method [35] which uses a probabilistic model using an ad hoc in vitro nucleosomal map, calculated by expert biologists during a preliminary phase of extraction of the features. The bold values in Table 2 indicate the best AUC performances. In the case of DLNN, its AUC values are better than the competitor on HM-LC, YS-WG, HM-PM, HM-5U datasets, and very close to the competitor on the rest of the dataset. Results presented here have been extended in a recent work [30].

3.2   Informative Content of Nucleosomes with Regard to Organism

We have also started to investigate on the informative content of the nucleosome, in relation to organisms' complexity. The question is about the possibility to observe how the performance of the model changes in relation to the dataset used for the training. In particular, we have considered two scenarios. One that considers the model trained on the highest complex organism (HM), and another trained on the lowest complex one (YS). For each one of them, we have choosen the best trained model of the previous experiment and we have computed the test performances on the complementary species. Therefore, we have used 10 random partitions for each complementary species and we have computed the

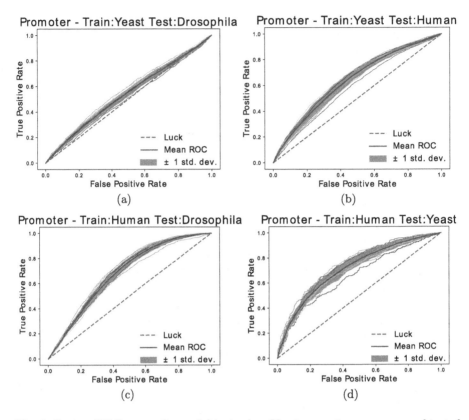

Fig. 2. On top, ROC curves for model trained on Yeast promoter sequences and tested on Drosophila (a) and Human (b) promoter sequences. On Bottom, ROC curves for model trained on Human promoter sequences and tested on Drosophila (c) and Yeast (d) promoter sequences. The blue plot shows the average Roc curve computed on 10 fold, the dashed red plot shows the random case, the other colored plots show the ROC relative to each fold. (Color figure online)

ROC curves and the AUC. In Fig. 2, mean ROC and mean AUC are shown together with the ROC computed at each fold. Table 3 shows the specific values. The first column (DLNN-HM) shows the test results when using HM as training. The second column (DLNN-YS) shows the test results when YS is using as training. We can observe that the model trained on Yeast has a very weak capability on recognizing DM nucleosome sequences (see also Fig. 2(a, b)). Conversely, the model trained on Human indicates AUC values greater than 65% (see also Fig. 2(c, d)). In total, the model trained on HM is able to recognize other nucleosome species with an average AUC of 67%, while the YS is below the 60% The thesis we can suppose is that the sequence features complexity of the DNA wrapped into nucleosomes increases with the complexity of the species. This could be due to the presence of a set of dinucleotides C for the case of the highest complex species, which includes the set of dinulceotides B of the lowest complex one. Of course, this hypothesis should be verified with more experiments which is one of the goals of our future research.

4   Conclusion

The study of the role of DNA sequences in the chromatin dynamics is an actual studied challenging problem. In this study, a deep neural network model based on a mix of convolutional and recurrent layers has been presented. It has been used for the automatic classification of nucleosome forming and not forming sequences. The efficacy of the method is proven in terms of AUC values. Results confirm that this deep neural network reaches better performances with respect to a state of the art method. Its interesting property is that it does not make any use of supervised knowledge, such as the extraction and selection of DNA sequence features, in fact, it processes input data represented in character-level one-hot encoding. The use of this network in a specific training-test assessment, allowed us to make considerations about the nucleosome organization in different eukaryotic organisms. In particular, it seems to emerge that the sequence features complexity of the DNA wrapped into nucleosomes increases with the complexity of the species.

References

1. Kornberg, R.D., Lorch, Y.: Twenty-five years of the nucleosome, fundamental particle of the eukaryote chromosome. Cell **98**(3), 285–294 (1999)
2. Mazina, M.Y., Vorobyeva, N.E.: The role of ATP-dependent chromatin remodeling complexes in regulation of genetic processes. Russ. J. Genet. **52**(5), 529–540 (2016)
3. Sala, A., et al.: Genome-wide characterization of chromatin binding and nucleosome spacing activity of the nucleosome remodelling ATPase ISWI. EMBO J. **30**(9), 1766–1777 (2011)
4. Mirabella, A.C., Foster, B.M., Bartke, T.: Chromatin deregulation in disease. Chromosoma **125**, 75–93 (2016)
5. Giancarlo, R., Lo Bosco, G., Pinello, L., Utro, F.: The three steps of clustering in the post-genomic era: a synopsis. In: Rizzo, R., Lisboa, P.J.G. (eds.) CIBB 2010. LNCS, vol. 6685, pp. 13–30. Springer, Heidelberg (2011). https://doi.org/10.1007/978-3-642-21946-7_2
6. Ciaramella, A., et al.: Interactive data analysis and clustering of genomic data. Neural Netw. **21**(2–3), 368–378 (2008)
7. Camastra, F., Di Taranto, M.D., Staiano, A.: Statistical and computational methods for genetic diseases: an overview. Comput. Math. Methods Med. **2015**, 954598 (2015)
8. Calcagno, G., et al.: A multilayer perceptron neural network-based approach for the identification of responsiveness to interferon therapy in multiple sclerosis patients. Inf. Sci. **180**(21), 4153–4163 (2010)
9. Di Taranto, D., et al.: Association of USF1 and APOA5 polymorphisms with familial combined hyperlipidemia in an Italian population. Mol. Cell. Probes **29**(1), 19–24 (2015)

10. Staiano, A., Di Taranto, M.D., Bloise, E., D'Agostino, M.N., et al.: Investigation of single nucleotide polymorphisms associated to familial combined hyperlipidemia with random forests. In: Apolloni, B., Bassis, S., Esposito, A., Morabito, F. (eds.) Neural Nets and Surroundings. Smart Innovation, Systems and Technologies, vol. 19, pp. 169–178. Springer, Heidelberg (2013). https://doi.org/10.1007/978-3-642-35467-0_18

11. Pinello, L., Lo Bosco, G., Yuan, G.-C.: Applications of alignment-free methods in epigenomics. Briefings Bioinform. **15**(3), 419–430 (2014)

12. Di Gesú, V., Lo Bosco, G., Pinello, L., Yuan, G.-C., Corona, D.F.V.: A multi-layer method to study genome-scale positions of nucleosomes. Genomics **93**(2), 140–145 (2009)

13. Struhl, K., Segal, E.: Determinants of nucleosome positioning. Nat. Struct. Mol. Biol. **20**(3), 267–273 (2013)

14. Yuan, G.-C.: Linking genome to epigenome. Wiley Interdisc. Rev.: Syst. Biol. Med. **4**(3), 297–309 (2012)

15. Hui, L., Ruichang, Z., Wei, X., Jihong, G., Ziheng, Z., Shuigeng, Z.: A comparative evaluation on prediction methods of nucleosome positioning. Briefings Bioinf. **15**(6), 1014–1027 (2014)

16. Lo Bosco, G.: Alignment free dissimilarities for nucleosome classification. In: Angelini, C., Rancoita, P.M.V., Rovetta, S. (eds.) CIBB 2015. LNCS, vol. 9874, pp. 114–128. Springer, Cham (2016). https://doi.org/10.1007/978-3-319-44332-4_9

17. Fici, G., Langiu, A., Lo Bosco, G., Rizzo, R.: Bacteria classification using minimal absent words. AIMS Med. Sci. **5**(1), 23–32 (2017)

18. Pinello, L., Lo Bosco, G., Hanlon, B., Yuan, G.-C.: A motif-independent metric for DNA sequence specificity. BMC Bioinf. **12**, 408 (2011)

19. Lo Bosco, G., Pinello, L.: A new feature selection methodology for K-mers representation of DNA sequences. In: di Serio, C., Liò, P., Nonis, A., Tagliaferri, R. (eds.) CIBB 2014. LNCS, vol. 8623, pp. 99–108. Springer, Cham (2015). https://doi.org/10.1007/978-3-319-24462-4_9

20. Rizzo, R., Fiannaca, A., La Rosa, M., Urso, A.: The general regression neural network to classify barcode and mini-barcode DNA. In: di Serio, C., Liò, P., Nonis, A., Tagliaferri, R. (eds.) CIBB 2014. LNCS, vol. 8623, pp. 142–155. Springer, Cham (2015). https://doi.org/10.1007/978-3-319-24462-4_13

21. Bengio, Y.: Learning deep architectures for AI. Found. Trends Mach. Learn. **2**(1), 1–127 (2009)

22. LeCun, Y., Bengio, Y., Hinton, G.: Deep learning. Nature **521**(7553), 436–444 (2015)

23. LeCun, Y., Bottou, L., Bengio, Y., Haffner, P.: Gradient-based learning applied to document recognition. Proc. IEEE **86**(11), 2278–2324 (1998)

24. Rizzo, R., Fiannaca, A., La Rosa, M., Urso, A.: A deep learning approach to DNA sequence classification. In: Angelini, C., Rancoita, P.M.V., Rovetta, S. (eds.) CIBB 2015. LNCS, vol. 9874, pp. 129–140. Springer, Cham (2016). https://doi.org/10.1007/978-3-319-44332-4_10

25. Lo Bosco, G., Di Gangi, M.A.: Deep learning architectures for DNA sequence classification. In: Petrosino, A., Loia, V., Pedrycz, W. (eds.) WILF 2016. LNCS (LNAI), vol. 10147, pp. 162–171. Springer, Cham (2017). https://doi.org/10.1007/978-3-319-52962-2_14

26. Fiannaca, A., et al.: Deep learning models for bacteria taxonomic classification of metagenomic data. BMC Bioinf. **19**, 198 (2018)

27. Lo Bosco, G., Rizzo, R., Fiannaca, A., La Rosa, M., Urso, A.: A deep learning model for epigenomic studies. In: 12th International Conference on Signal Image Technology & Internet Systems, SITIS 2016, pp. 688–692. IEEE, New York (2016)

28. Di Gangi, M.A., Gaglio, S., La Bua, C., Lo Bosco, G., Rizzo, R.: A deep learning network for exploiting positional information in nucleosome related sequences. In: Rojas, I., Ortuño, F. (eds.) IWBBIO 2017. LNCS, vol. 10209, pp. 524–533. Springer, Cham (2017). https://doi.org/10.1007/978-3-319-56154-7_47

29. Lo Bosco, G., Rizzo, R., Fiannaca, A., La Rosa, M., Urso, A.: Variable ranking feature selection for the identification of nucleosome related sequences. In: Benczúr, A. (ed.) ADBIS 2018. CCIS, vol. 909, pp. 314–324. Springer, Cham (2018). https://doi.org/10.1007/978-3-030-00063-9_30

30. Di Gangi, M., Lo Bosco, G., Rizzo, R.: Deep learning architectures for prediction of nucleosome positioning from sequences data. BMC Bioinf. **19**, 418 (2018)

31. Srivastava, N., Hinton, G.E., Krizhevsky, A., Sutskever, I., Salakhutdinov, R.: Dropout: a simple way to prevent neural networks from overfitting. J. Mach. Learn. Res. **15**(1), 1929–1958 (2014)

32. Hochreiter, S., Schmidhuber, J.: Long short-term memory. Neural Comput. **9**(8), 1735–1780 (1997)

33. Hochreiter, S., Bengio, Y., Frasconi, P., Schmidhuber, J.: Gradient flow in recurrent nets: the difficulty of learning long-term dependencies. In: Kremer, S.C., Kolen, J.F. (eds.) A Field Guide to Dynamical Recurrent Neural Networks. Wiley/IEEE, New York (2001)

34. Kingma, D.P., Ba, J.: Adam: a method for stochastic optimization. In: 3rd International Conference on Learning Representations. ICLR 2015, CoRR, abs/1412.6980 (2014)

35. Kaplan, N., et al.: The DNA-encoded nucleosome organization of a eukaryotic genome. Nature **458**, 362–366 (2009)

Modeling and Simulation Methods in System Biology

Searching for the Source of Difference: A Graphical Model Approach

Vera Djordjilović[1]([envelope]) [iD], Monica Chiogna[2] [iD], Chiara Romualdi[3] [iD], and Elisa Salviato[4] [iD]

[1] Department of Biostatistics, University of Oslo, Oslo, Norway
vera.djordjilovic@medisin.uio.no
[2] Department of Statistical Sciences, University of Bologna, Bologna, Italy
monica.chiogna2@unibo.it
[3] Department of Biology, University of Padova, Padua, Italy
chiara.romualdi@unipd.it
[4] IFOM, The FIRC Institute of Molecular Oncology, Milan, Italy
elisa.salviato@ifom.eu

Abstract. A growing body of evidence shows that when performing differential analysis it is highly beneficial to go beyond differences in the level of individual genes, and consider differences in their interactions as well. We propose an original statistical approach that identifies the set of variables driving the difference between two conditions under study. Our proposal, set within the framework of Gaussian graphical models, is implemented in the R package SourceSet, that also extends the analysis from a single to multiple pathways and provides several graphical outputs, including Cytoscape visualization to browse the results.

Keywords: Gaussian graphical models · Two-sample problem · Gene set analysis · Perturbation identification · Primary dysregulation identification

1 Introduction

A microarray experiment typically provides a list of differentially expressed genes, that represents the starting point of a highly difficult process of results interpretation. Biological interpretation becomes easier if differentially expressed genes show some similarity according to their functional annotation. Thus, in recent years, the interest has moved from the study of individual genes to that of groups of genes defined on the basis of functional categories, giving rise to the so-called Gene Set Analysis (GSA). For an extensive review see [8,9]. Being aimed at identifying groups of genes with possibly moderate, but coordinated, expression changes, GSA enables a better understanding of cellular processes involved in a biological process.

Latest approaches of GSA, namely, topology-based GSA methods, leverage information about the interconnections of genes (or other biomolecules) pictured

© Springer Nature Switzerland AG 2020
M. Raposo et al. (Eds.): CIBB 2018, LNBI 11925, pp. 131–138, 2020.
https://doi.org/10.1007/978-3-030-34585-3_12

in pathways, and offer improved performances – over methods based simply on clustering genes – in detecting dysregulations of functionally related genes across two or more biological conditions or groups of specimens (patients, cell lines, etc.). But once pathways which are perturbed in a given condition are identified, the question of how to get close to the possible source of dysregulation within a pathway is still an open one.

In this work, we present a new way to address the problem of identification of the possible source of dysregulation based on the new concept termed *source set*. Informally, the source set consists of genes that play crucial regulatory roles on the remaining genes in a pathway, and represents the minimal set of variables that explains the difference between two conditions. In other words, source set represents the source or the origin of the perturbation under study. Moreover, the set of genes affected by the source set through network propagation, i.e. the *secondary set* (see also [1]), is also identified. The method gives rise to a simple and fast estimating procedure implemented in SourceSet R package.

2   Scientific Background

A biological pathway can be described as a set of linked biological components interacting with each other to generate a single biological effect. It comprises a myriad of interactions, reactions, and regulations, often identified piecemeal over extended periods of time by a variety of researchers. Moreover, participants of one pathway can also be involved in other pathways, giving rise to intricate dependencies. As a result, pathway topology has to be considered a dynamic entity whose information is particularly challenging to compile and organize. Currently, there are a few comprehensive pathway repositories reporting pathway topologies as networks of functional interactions, see for example KEGG [6] or Reactome [5].

Among topological methods, [7] explicitly incorporates the dependence structure among genes provided by pathway topology into undirected Gaussian graphical models and compare two experimental conditions by testing equality of the associated Gaussian graphical distributions. The approach implemented in [7] converts a pathway into a graphical model $G = (V, E)$, where V are the nodes corresponding to the genes pictured in the graph and E are the edges of the graph (see [4] for further details) and assumes that observations are realizations of i.i.d. normal random vectors. In detail, for two conditions

$$\mathcal{M}_1(G) = \{Y_1 \sim N(\mu_1, \Sigma_1), \Sigma_1^{-1} \in S^+(G)\},$$

$$\mathcal{M}_2(G) = \{Y_2 \sim N(\mu_2, \Sigma_2), \Sigma_2^{-1} \in S^+(G)\},$$

where $S^+(G)$ is the set of symmetric positive definite matrices with null elements corresponding to the missing edges of G.

Dysregulation is detected by statistical tests on equality of the mean vectors and/or of the covariance matrices. The hypotheses of interest are therefore

$$H_0 : \Sigma_1 = \Sigma_2 \quad \text{vs} \quad \Sigma_1 \neq \Sigma_2. \tag{1}$$

and

$$H_0 : \mu_1 = \mu_2 \quad \text{vs} \quad H_1 : \mu_1 \neq \mu_2 \quad \text{subject to} \quad \Sigma_1 = \Sigma_2, \tag{2}$$

or

$$H_0 : \mu_1 = \mu_2 \quad \text{vs} \quad H_1 : \mu_1 \neq \mu_2 \quad \text{subject to} \quad \Sigma_1 \neq \Sigma_2. \tag{3}$$

When the global hypothesis of equality of the two multivariate distributions is rejected, the interest is usually in localizing the source of difference, which within the context of gene networks, corresponds to the task of identifying the source of dysregulation.

Within the approach of [7], identification of dysregulated subnetworks can be tackled by relying on the modular nature of the Gaussian graphical models: the graph is partitioned into smaller units, the so-called cliques, and the analysis is performed on each unit separately. Indeed, when the underlying graph is decomposable, clique-induced saturated marginal models can be tested separately. However, the cliques are not disjoint and overlapping subsets of variables among cliques, called separators, make identification of the source of dysregulation challenging.

3   Materials and Methods

The conceptual framework needed to define a source set is readily available and was introduced in the seminal work by Dawid and Lauritzen [2] under the name of Hyper Markov Laws. Indeed, one of the results that the Authors showed is that, in graphical models, sampling distributions of maximum likelihood estimators reflect separations in the graph, in a similar fashion as original variables, a property that can also be translated to statistical hypotheses and corresponding test statistics.

Our method exploits hyper Markov properties to decompose the testing problem into disjoint local testing problems. By disjoint local testing problems we mean independent hypotheses with no logical relations between them, i.e such that all combinations of true and false hypotheses are possible. As such decomposition is not unique, we suggest a way to deal with the non-uniqueness. Theoretical foundations of the proposed approach are covered in [3]; the focus of this work is on the implementation of the method available in the statistical software R SourceSet package.

The package, available at https://cran.r-project.org/web/packages/Source Set/, consists of four main functions. The main one, sourceSet, takes as an input a list of pathways and expression levels of genes in two conditions and estimates a source set for each input pathway (appropriately converted into an undirected decomposable graph). In particular, the function requires the following arguments:

graphs	A list of graphNEL objects representing the pathways to be analyzed
data	A matrix of expression levels with column names for genes and row names for samples

classes	A vector of length equal to the number of rows of data. Indicates the class (condition) of each statistical unit. Only two classes, labeled 1 and 2 are allowed;

⋮

The output is an object of the `sourceSetList` class. It contains as many lists as the input graphs, and provides the following information:

primarySet	A character vector containing the names of the variables belonging to the estimated source set (primary dysregulation);
secondarySet	A character vector containing the names of the variables affected by secondary dysregulation

⋮

In the above given output object, we adopted the terminology proposed in [1], that distinguishes between the *primary dysregulation* and the effect of the so-called network propagation, termed *secondary dysregulation*.

4   Experimental Results

Functions `infoSource`, `easyLookSource`, and `sourceSankeyDiagram` guide the user in interpreting results obtained from multiple pathways. Namely, although interpretation of the source set of a single pathway might be intuitive, the analysis of the whole collection of results obtained from multiple pathways might be complex. For this reason, we propose a guideline for the simultaneous analysis of multiple pathways, providing descriptive statistics and predefined plots. The final aim is that of providing researchers with tools permitting to zoom in on potential sources of differential behaviour that can subsequently be passed on for biological validation.

The `infoSource` provides a summary of the results by focusing on either genes or pathways. It supplies two different lists that are composed as follows:

$graph	
n.source	Number of genes belonging to the source set
n.secondary	Number of genes affected by dysregulation but not responsible for primary dysregulation (secondary dysregulation)
n.graph	Number of genes in the graph
n.cluster	Number of connected components in G

`primary.impact`	Relative size of the estimated source set. This index quantifies the portion of the graph affected by the primary dysregulation
`total.impact`	Relative size of the set of genes affected by some form of dysregulation. This index quantifies the portion of the graph affected by primary or secondary dysregulation
`p.value`	p-value for the hypothesis of equality of the two global distributions associated to the given pathway/graph

`$variable`

`n.graph`	Number of input pathways in which the gene is annotated
`specificity`	Percentage of input graphs containing the given gene with respect to the total number of input graphs
`primary.impact`	Percentage of input graphs, such that the given gene belongs to their estimated source set, with respect to the total number of input graphs in which the gene appears
`total.impact`	Percentage of input graphs, such that the given gene is affected by some form of dysregulation in the considered graph, with respect to the total number of input graphs in which the gene appears
`relevance`	Percentage of the input graphs such that the given variable belongs to their estimated source set, with respect to the total number of input graphs. It is a general measure of the importance of the gene based on the chosen pathways

The function `easyLookSource` allows to summarize the results of the analysis through a heatmap. The plot is composed of a matrix in which rows represent pathways and columns represent genes. An example is shown in Fig. 1.

Another way to summarize the results in a visual manner is through a Sankey diagram (Fig. 2). Sankey diagram highlights relationships between genes, pathways, and source sets. The layout is organized on three levels:

- first level (left) consists of genes that appear in at least one of the estimated source sets;
- second level (central) is made up of modules. A module is defined as a set of genes belonging to a connected subgraph of one pathway, that is also contained in the associated source set. A pathway can have multiple modules, and at the same time, one module can be contained within multiple pathways;
- third level (right) consists of pathways.

A link between two elements a and b is to be interpreted from left to right as "*element a is contained in element b*".

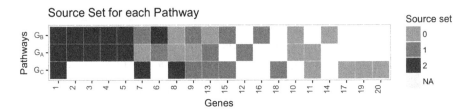

Fig. 1. easyLookSource function. The heat map highlights the role of individual genes in different pathways. For example, gene 7 is in the primary set of G_C (blue rectangle) and in the secondary set of G_A (light blue rectangle). Gene 10 belongs to G_A and G_B, but is not affected by dysregulation (gray rectangles). The same gene does not appear in G_C (white rectangle). (Color figure online)

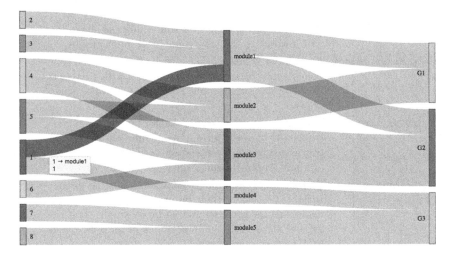

Fig. 2. sourceSankeyDiagram function output.

In addition, sourceCytoscape and sourceUnionCytoscape provide a connection with Cytoscape, a well known bioinformatics tool for visualizing, exploring, and manipulating biological networks. These functions, thanks to the connection with Cytoscape, allow the user to create a collection of graphs to be visualized in a unique session, while documenting interesting findings. A possible output is presented in Fig. 3.

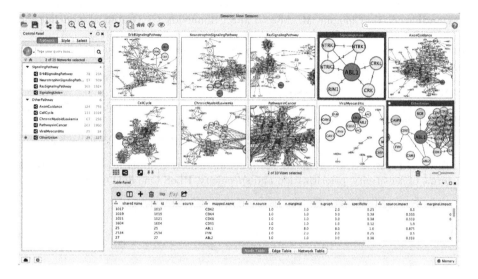

Fig. 3. Graphical output from `Cytoscape` interface.

5   Conclusions

We presented a novel computational tool, called `SourceSet`, for the identification of the origin of dysregulation in perturbed pathways. The theory behind the tool invokes Gaussian graphical models as a formal and meaningful language for representing biological pathways and hyper Markov laws for disentangling complexity and zooming in on potential sources of dysregulation. Such an approach has proved to be effective in recovering known sources of dysregulation in real data analysis; a guided example can be found in the `SourceSet` package vignette. In these examples, `SourceSet` takes two expression datasets, each from a different condition, for example, cancer and normal tissue, and identifies the genes that are likely to give rise and/or closely contribute to the disease associated dysregulation in chosen pathways. When results on multiple pathways are synthesized and unified through the specific functions offered to users, `SourceSet` provides statistical and biological investigators a powerful methodology for empirical research.

References

1. Ansari, S., Voichita, C., Donato, M., Tagett, R., Draghici, S.: A novel pathway analysis approach based on the unexplained disregulation of genes. Proc. IEEE **105**(3), 482–495 (2016)
2. Dawid, A.P., Lauritzen, S.: Hyper Markov laws in the statistical analysis of decomposable graphical models. Ann. Stat. **21**(3), 1272–1317 (1993)
3. Djordjilović, V., Chiogna, M.: Searching for a source of difference in Gaussian graphical models. arXiv preprint arXiv:1811.02503 (2018)

4. Djordjilović, V., Chiogna, M., Massa, M.S., Romualdi, C.: Graphical modeling for gene set analysis: a critical appraisal. Biom. J. **57**(5), 852–866 (2015)
5. Fabregat, A., et al.: The Reactome pathway knowledgebase. Nucleic Acids Res. **46**(D1), D649–D655 (2017)
6. Kanehisa, M., Goto, S.: KEGG: Kyoto encyclopedia of genes and genomes. Nucleic Acids Res. **28**(1), 27–30 (2000)
7. Massa, M.S., Chiogna, M., Romualdi, C.: Gene set analysis exploiting the topology of a pathway. BMC Syst. Biol. **4**(1), 121 (2010)
8. Mitrea, C., et al.: Methods and approaches in the topology-based analysis of biological pathways. Front. Physiol. **4**, 278 (2013)
9. Tarca, A.L., Bhatti, G., Romero, R.: A comparison of gene set analysis methods in terms of sensitivity, prioritization and specificity. PLoS One **8**(11), e79217 (2013)

A New Partially Segment-Wise Coupled Piece-Wise Linear Regression Model for Statistical Network Structure Inference

Mahdi Shafiee Kamalabad and Marco Grzegorczyk[✉]

Bernoulli Institute, Groningen University,
Nijenborgh 9, 9747 Groningen, The Netherlands
m.a.grzegorczyk@rug.nl
http://www.math.rug.nl/stat/People/Marco

Abstract. We propose a new non-homogeneous dynamic Bayesian network with partially segment-wise sequentially coupled network parameters. The idea is to infer the segmentation of a time series of network data using multiple changepoint processes, and to model the data in each segment by linear regression models. The conventional uncoupled models infer the network interaction parameters for each segment separately, without any systematic information-sharing among segments. More recently, it was proposed to couple the network interaction parameters sequentially among segments. The idea is to enforce the parameters of any segment to stay similar to those of the previous segment. This coupling mechanism can be disadvantageous, as it enforces coupling and does not feature any options to uncouple. We propose a new consensus model that infers for each individual segment whether it should be coupled to (or better should stay uncoupled from) the preceding one.

Keywords: Network structure learning · Dynamic Bayesian networks · Bayesian piece-wise linear regression · Partial segment-wise coupling

1 Introduction

In systems biology, dynamic Bayesian network models (DBNs) have become popular tools for learning the structures of regulatory networks from data. For temporal data, the usual assumption is that all regulatory interactions are subject to a time lag. There is then no acyclicity constraint on the network structure, and the parent nodes of each node can be learned separately. A common approach is then to make use of independent regression models to infer the parents (=covariates) of each individual network node, and to merge the individual parent sets in

This work was supported by the European Cooperation in Science and Technology (COST) [COST Action CA15109 European Cooperation for Statistics of Network Data Science (COSTNET)].

© Springer Nature Switzerland AG 2020
M. Raposo et al. (Eds.): CIBB 2018, LNBI 11925, pp. 139–152, 2020.
https://doi.org/10.1007/978-3-030-34585-3_13

form of a network; see Sect. 2.3 for a more detailed description. A shortcoming of dynamic Bayesian networks is that they are homogeneous models, so that the network parameters are not allowed to change over time. The same set of network parameters applies to all time points. For many applications, this assumption of homogeneity is unrealistic. For example, in cellular networks the strengths of the regulatory interactions often depend on unobserved external factors, such as cellular or experimental conditions, which do not necessarily stay constant over time. This renders the traditional dynamic Bayesian network models inappropriate for many biological applications. Therefore various non-homogeneous DBN models (NH-DBNs) have been proposed in the computational biology literature. The proposed NH-DBNs can be grouped into two classes: (i) NH-DBNs that allow only the network parameters to vary in time and (ii) NH-DBNs that allow the network structure and the network parameters to be time-dependent. We here focus on NH-DBNs that assume the network structure to be time-invariant (i), as this assumption is more realistic for our applications in Sect. 4.

When the network structure (i.e. the collection of covariate sets) does not change over time, while the network parameters (i.e. the regression coefficients) are time-dependent, piece-wise linear regression models can be used to model the data. A multiple changepoint process is applied to infer the segmentation of the data into disjoint segments, and the data within each segment are modelled by linear regression. The joint network structure, the number of changepoints and their locations, and the segment-specific network parameters (regression coefficients) are unknown and have to be inferred from the data. With the conventional uncoupled models (see, e.g., the work by Lèbre et al. [1]), the segment-specific network parameters have to be learned for each segment separately, without information-sharing among segments. For short time series, the segmentation leads to even shorter segments, sometimes containing a few data points only. This can lead to inflated inference uncertainties, so that the segment-specific network parameters cannot be properly learned from the data.

To address this issue, a model with fully sequentially coupled network parameters was proposed by Grzegorczyk and Husmeier [2]. The key idea is to allow for information-exchange among segments by a sequential coupling scheme. The posterior expectation of the network parameters of each segment h are used as prior expectation for the next segment $h + 1$. In the fully sequentially coupled model from Grzegorczyk and Husmeier [2], node-specific coupling strength parameters regulate the variance of the network parameter priors and so the effective coupling strength between all pairs $(h, h + 1)$ of neighboring segments. A disadvantage of this fully coupled model is that each segment $h \geq 2$ is enforced to be coupled to the preceding segment $h - 1$. That is, the regression coefficient prior distributions for each segment $h \geq 2$ are automatically centered around their posterior expectations from the preceding segment $h - 1$, and the coupling strength parameter only regulates the variance of these prior distributions. Low coupling strength parameters yield peaked prior distributions while high coupling strength parameters yield vague prior distributions around the posterior expectations from the preceding segment. In particular, in the fully sequentially coupled model from Grzegorczyk and Husmeier [2] there is no option to uncouple a segment $h \geq 2$ from the previous segment $h - 1$. Therefore, for networks with dissimilar segment-specific network parameters (regression coefficients), this information coupling scheme can become very counter-productive: Uncoupling can effectively only be achieved by making the network parameter prior distributions vague, so as to allow the regression coefficients to get dissimilar from their prior expectations (= the posterior expectations from the preceding segment).

In this paper, we address this disadvantage of the fully sequentially coupled model from Grzegorczyk and Husmeier [2]. We extend the fully coupled model by introducing a new option to uncouple segments. This yields a new model, which we refer to as the new partially segment-wise coupled model. The new models infers for each individual segment $h \geq 2$ whether it is coupled to (or uncoupled from) the preceding segment $h-1$. Hence, our new partially segment-wise coupled model can infer the best trade-off between the uncoupled model from Lèbre et al. [1] and the fully sequentially coupled model from Grzegorczyk and Husmeier [2]. The uncoupled model and the fully coupled model are the limiting cases, where either *all* segments are uncoupled or *all* segments are coupled.

In a complementary work [3], we have generalized the fully sequentially coupled model from Grzegorczyk and Husmeier [2] by introducing segment-specific coupling strength parameters. Although segment-specific coupling strengths increase the model flexibility, in this generalized fully coupled model [3] each segment stays coupled to the previous one and the uncoupled model from Lèbre et al. [1] cannot be reached as limiting case.

2   Methods

2.1   The New Partially Segment-Wise Coupled Model

Consider a piece-wise Bayesian linear regression model with response variable Y and covariate set $\pi = \{X_1, \ldots, X_k\}$. We assume that there are T temporal data points and that they can be divided into H disjoint segments with segment-specific regression coefficients. Let $\mathbf{y}_h$ be the response vector and $\mathbf{X}_h$ be the design matrix for segment h, where $\mathbf{X}_h$ includes a column of 1's for the intercept. For each segment $h = 1, \ldots, H$ we then have:

$$\mathbf{y}_h \sim \mathcal{N}(\mathbf{X}_h \mathbf{w}_h, \sigma^2 \mathbf{I}) \tag{1}$$

where $\mathbf{w}_h = (\mathbf{w}_{h,0}, \ldots, \mathbf{w}_{h,k})^\mathsf{T}$ is the regression coefficient vector for segment h, and σ^2 is the noise variance parameter, which we assume to have an inverse Gamma distribution:

$$\sigma^{-2} \sim \mathrm{GAM}(\alpha_\sigma, \beta_\sigma)$$

Onto the regression coefficient vectors $\mathbf{w}_h$ $(h = 1, \ldots, H)$ we impose the following novel prior distributions:

$$\mathbf{w}_h \sim \mathcal{N}(\boldsymbol{\mu}_h, \boldsymbol{\Sigma}_h) \text{ with } \boldsymbol{\mu}_h := \delta_h \tilde{\mathbf{w}}_{h-1} \text{ and } \boldsymbol{\Sigma}_h := \lambda_c{}^{\delta_h} \lambda_u{}^{1-\delta_h} \sigma^2 \mathbf{I} \tag{2}$$

where $\delta_1 := 0$, $\delta_h \in \{0,1\}$ for $h > 1$, $\tilde{\mathbf{w}}_0 := \mathbf{0}$. This yields:

$$
\mathbf{w}_h \sim \begin{cases} \mathcal{N}(\mathbf{0}, \quad \lambda_u \sigma^2 \mathbf{I}) & \text{if } h = 1 \text{ or } \delta_h = 0 \\ \mathcal{N}(\tilde{\mathbf{w}}_{h-1}, \lambda_c \sigma^2 \mathbf{I}) & \text{if } h > 1 \text{ and } \delta_h = 1 \end{cases} \tag{3}
$$

λ_c and λ_u are free hyperparameters, onto which we impose inverse Gamma prior distributions:

$$
\lambda_c^{-1} \sim \text{GAM}(\alpha_c, \beta_c)
$$
$$
\lambda_u^{-1} \sim \text{GAM}(\alpha_u, \beta_u)
$$

The newly introduced indicator variables $\delta_h \in \{0,1\}$ indicate whether segment $h \geq 2$ is coupled to segment $h - 1$ ($\delta_h = 1$) or not ($\delta_h = 0$), and $\tilde{\mathbf{w}}_{h-1}$ ($h \geq 2$) is the posterior expectation of $\mathbf{w}_{h-1}$; see Eq. (5) below.
The subscripts 'u' and 'c' indicate whether the hyperparameters apply to uncoupled ($\delta_h = 0$) or coupled ($\delta_h = 1$) segments. The posterior distribution (full conditional distribution) of $\mathbf{w}_h$ is:

$$
\mathbf{w}_h | (\sigma^2, \lambda_c, \lambda_u, \delta_h, \mathbf{y}_h) \sim \mathcal{N}(\tilde{\mathbf{w}}_h, \sigma^2 \tilde{\boldsymbol{\Sigma}}_h) \tag{4}
$$

where

$$
\tilde{\boldsymbol{\Sigma}}_h = \lambda_c^{-\delta_h} \lambda_u^{-(1-\delta_h)} \mathbf{I} + \mathbf{X}_h^{\mathsf{T}} \mathbf{X}_h
$$
$$
\tilde{\mathbf{w}}_h = \left(\tilde{\boldsymbol{\Sigma}}_h \right)^{-1} \left(\lambda_c^{-\delta_h} \lambda_u^{-(1-\delta_h)} \boldsymbol{\mu}_{h-1} + \mathbf{X}_h^{\mathsf{T}} \mathbf{y}_h \right) \tag{5}
$$

We assume that the new indicator variables $\delta_2, \ldots, \delta_H$ follow a Bernoulli distribution:

$$
\delta_h | p \sim \text{Ber}(p)
$$

where the probability hyperparameter $p \in [0,1]$ is Beta distributed:

$$
p \sim \text{Beta}(a, b)
$$

Our new model is then a consensus model between an uncoupled model and a fully coupled model:

- If $\delta_h = 0$ for all h, then $P(\mathbf{w}_h) = \mathcal{N}(\mathbf{0}, \lambda_u \sigma^2 \mathbf{I})$ for all h. This is an **uncoupled model** without information sharing among segments. The prior expectations of all regression coefficients $\mathbf{w}_{h,j}$ ($j = 0, \ldots, k$) are 0, and λ_u can be interpreted as a signal-to-noise ratio parameter.
- If $\delta_h = 1$ for $h \geq 2$, then $P(\mathbf{w}_h) = \mathcal{N}(\tilde{\mathbf{w}}_{h-1}, \lambda_c \sigma^2 \mathbf{I})$ for $h \geq 2$. This refers to the **fully sequentially coupled model** from [2]. The prior expectation of each individual regression coefficient $\mathbf{w}_{h,j}$ is its posterior expectation $\tilde{\mathbf{w}}_{h-1,j}$ from the previous segment ($j = 0, \ldots, k$), and λ_c can be interpreted as coupling (strength) parameter, where λ_c^{-1} refers to the coupling strength.
- Our new **partially segment-wise coupled model** infers the values of the binary variables δ_h ($h \geq 2$) from the data, so as to find the best trade-off between the uncoupled and the fully coupled model.

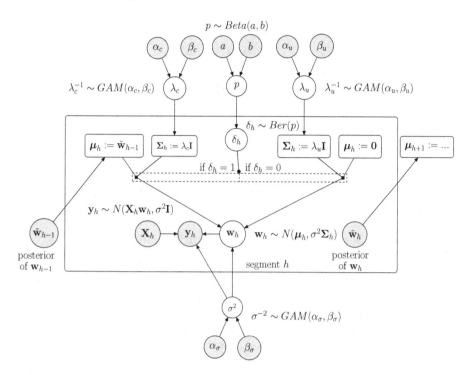

Fig. 1. Graphical representation of the new partially segment-wise coupled model. Parameters that have to be inferred are in white circles. The data and the fixed hyperparameters are in grey circles. The rectangles contain definitions that deterministically depend on the parent nodes. Everything within the plate is segment-specific.

Figure 1 shows a graphical model representation of our new partially coupled model and lists all prior distributions along with their hyperparameters.

For $\delta_h \sim \text{Ber}(p)$ with $p \sim \text{Beta}(a, b)$ the joint marginal density of $\{\delta_h\} := (\delta_2, \ldots, \delta_H)$ is:

$$p(\{\delta_h\}) = \int_0^1 p(p) \prod_{h=2}^H p(\delta_h|p) \, dp \tag{6}$$

$$= \frac{\Gamma(a+b)}{\Gamma(a)\Gamma(b)} \frac{\Gamma(a + \sum_{h=2}^H \delta_h)\Gamma(b + \sum_{h=2}^H (1 - \delta_h))}{\Gamma(a + b + (H-1))}$$

The marginal likelihood, $p(\mathbf{y}|\boldsymbol{\pi}, \boldsymbol{\tau}, \lambda_u, \lambda_c, \{\delta_h\})$, where $\boldsymbol{\pi} = \{X_1, \ldots, X_k\}$ is the covariate set and $\boldsymbol{\tau}$ denotes the data segmentation into the H segments $(h = 1, \ldots, H)$, can be computed analytically by applying the rule from Sect. 2.3

in Bishop [4]:

$$p(\mathbf{y}|\lambda_u, \lambda_c, \{\delta_h\}) = \frac{\Gamma(\frac{T}{2} + a_\sigma)}{\Gamma(a_\sigma)} \cdot \frac{\pi^{-T/2} \cdot (2b_\sigma)^{a_\sigma}}{\left(\prod\limits_{h=1}^{H} \det(\mathbf{C}_h)\right)^{1/2}} \cdot (2b_\sigma + \Delta^2)^{-(\frac{T}{2} + a_\sigma)} \quad (7)$$

where σ^2 and $\mathbf{w}_1, \ldots, \mathbf{w}_H$ have been integrated out, and

$$\mathbf{C}_h := \mathbf{I} + \mathbf{X}_h \boldsymbol{\Sigma}_h \mathbf{X}_h^\mathsf{T}$$

$$\Delta^2 := \sum_{h=1}^{H} (\mathbf{y}_h - \delta_h \mathbf{X}_h \tilde{\mathbf{w}}_{h-1})^\mathsf{T} \mathbf{C}_h^{-1} (\mathbf{y}_h - \delta_h \mathbf{X}_h \tilde{\mathbf{w}}_{h-1})$$

2.2   Covariate Set and Data Segmentation Learning

The data segmentation $\boldsymbol{\tau}$ and the covariate sets $\boldsymbol{\pi}$ are usually unknown and the major objective is to infer them from the data. Given N covariates, we assume that all subsets $\boldsymbol{\pi} \subset \{X_1, \ldots, X_N\}$ are equally likely a priori. For the data segmentation we use changepoint sets; i.e. we use a set of $H - 1$ changepoints $\boldsymbol{\tau} := \{\tau_1, \ldots, \tau_{H-1}\}$ to divide the temporal data points $\mathcal{D} = \{\mathcal{D}_1, \ldots, \mathcal{D}_T\}$ into H segments. Data point $\mathcal{D}_t$ is in segment h if $\tau_{h-1} < t \leq \tau_h$, where $\tau_0 := 0$ and $\tau_H := T$ are two pseudo-changepoints . We assume that the distances between neighboring changepoints are geometrically distributed with a fixed hyperparameter $\mathrm{p} \in (0, 1)$:

$$p(\boldsymbol{\tau}|\mathrm{p}) = \left(\prod_{h=1}^{H-1} (1 - \mathrm{p})^{\tau_h - \tau_{h-1} - 1} \mathrm{p}\right) (1 - \mathrm{p})^{\tau_H - \tau_{H-1} - 1} \quad (8)$$

$$= (1 - \mathrm{p})^{(T-1) - (H-1)} \mathrm{p}^{H-1}$$

The posterior distribution then takes the form:

$$p(\lambda_u, \lambda_c, \{\delta_h\}, \boldsymbol{\pi}, \boldsymbol{\tau}|\mathcal{D}) \propto p(\boldsymbol{\pi}) \cdot p(\boldsymbol{\tau}|\mathrm{p}) \cdot p(\lambda_u) \cdot p(\lambda_c) \cdot p(\{\delta_h\}) \quad (9)$$
$$\cdot p(\mathbf{y}|\boldsymbol{\pi}, \boldsymbol{\tau}, \lambda_u, \lambda_c, \{\delta_h\})$$

where the marginal likelihood $p(\mathbf{y}|\boldsymbol{\pi}, \boldsymbol{\tau}, \lambda_u, \lambda_c, \{\delta_h\})$ was defined in Eq. 7. Reversible Jump Markov Chain Monte Carlo (RJMCMC) simulations can be used to generate posterior samples from the posterior distribution in Eq. 9. In each RJMCMC iteration we re-sample the parameters in λ_u, λ_c and $\{\delta_h\}$ from their full conditional distributions (Gibbs sampling), and we perform Metropolis-Hastings moves to sample covariate sets $\boldsymbol{\pi}$ and changepoint sets $\boldsymbol{\tau}$.

Covariate Set Inference. For sampling covariate sets $\boldsymbol{\pi}$ we implement 3 move types: 'covariate additions (A)', 'covariate removals (R)', and 'covariate exchanges (E)'. Each move proposes to replace $\boldsymbol{\pi}$ by a new covariate set $\boldsymbol{\pi}^*$ having one covariate more (A) or less (R) or exchanged (E). When randomly selecting the move type and the involved covariate(s), we get the acceptance probability:

$$A(\pi \to \pi^*) = \min \left\{ 1, \frac{p(\mathbf{y}|\pi^*, \tau, \lambda_u, \lambda_c, \{\delta_h\})}{p(\mathbf{y}|\pi, \tau, \lambda_u, \lambda_c, \{\delta_h\})} \cdot \frac{p(\pi^*)}{p(\pi)} \cdot HR_\pi \right\}$$

where the move type (A, R, or E) specific Hastings ratios are:

$$HR_{\pi,A} = \frac{n - |\pi|}{|\pi^*|}, \quad HR_{\pi,R} = \frac{|\pi|}{n - |\pi^*|}, \quad HR_{\pi,E} = 1$$

Changepoint Set Inference. For sampling changepoint sets τ we also implement 3 move types: 'changepoint birth (B)', 'changepoint death (D)', and 'changepoint re-allocation (R)' moves. Each move proposes to replace τ by a new changepoint set τ^* having one changepoint added (B) or deleted (D) or re-allocated (R). We randomly select the move type, the involved changepoint and the new changepoint location. Changepoint moves also affect the numbers of parameters in the collection $\{\delta_h\}$. For each segment that stays unchanged we keep the old parameter. For altering segments we re-sample the corresponding parameters. To this end, we flip coins to get candidates for the involved δ_h's. This yields a new collection of binary variables $\{\delta_h\}^*$ and the Metropolis-Hastings acceptance probability is:

$$A([\tau, \{\delta_h\}] \to [\tau^*, \{\delta_h\}^*])$$
$$= \min \left\{ 1, \frac{p(\mathbf{y}|\pi, \tau^*, \lambda_u, \lambda_c, \{\delta_h\}^*)}{p(\mathbf{y}|\pi, \tau, \lambda_u, \lambda_c, \{\delta_h\})} \cdot \frac{p(\tau^*)}{p(\tau)} \cdot \frac{p(\{\delta_h\}^*)}{p(\{\delta_h\})} \cdot HR_\tau \right\}$$

where the move type specific (B, D, or R) Hastings ratios are

$$HR_{\tau,B} = \frac{T - 1 - |\tau^*|}{|\tau|} \cdot \frac{1}{2}, \quad HR_{\tau,D} = \frac{|\tau^*|}{T - 1 - |\tau|} \cdot 2, \quad HR_{\tau,R} = 1 \quad (10)$$

2.3   Network Structure Learning

If n variables $Z_1, \ldots, Z_n$ have been observed over time, the conventional dynamic Bayesian network assumption is that the regulatory interactions are subject to a time lag. A network edge $Z_i \to Z_j$ then indicates that Z_j at time point $t+1$ depends on the value of Z_i at time point t. The task of learning a network among the n variables can then be separated into n independent regression tasks. In the j-th regression model $Y := Z_j$ is the response and the other $N := n - 1$ variables $\{Z_1, \ldots, Z_{j-1}, Z_{j+1}, \ldots, Z_n\}$ are the potential covariates. Having inferred a covariate set π^j for each Z_j, a network can be built by merging the covariate sets in form of a network $\mathcal{N} := \{\pi^1, \ldots, \pi^n\}$. There is the edge $Z_i \to Z_j$ in the network $\mathcal{N}$ if and only if $Z_i \in \pi^j$.

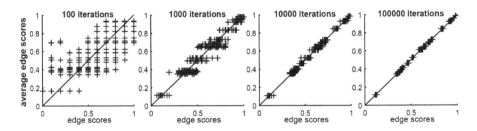

Fig. 2. Convergence diagnostic for the yeast data. For different MCMC simulation run length, $V \in \{100, 1000, 10000, 100000\}$, we performed 10 RJMCMC simulations with the new partially segment-wise coupled model. We used the hyperparameter p = 0.05 for the changepoint prior in Eq. 8. For each V there is a scatter plot where the simulation-specific edge scores (vertical axis) are plotted against the average scores for that V (horizontal axis).

For each network variable $Y := Z_j$ we generate a posterior sample from Eq. 9, $\{\lambda_u^{(j,w)}, \lambda_c^{(j,w)}, \{\delta_h\}^{(j,w)}, \boldsymbol{\pi}^{(j,w)}, \boldsymbol{\tau}^{(j,w)}\}_{w=1,...,W}$, and from the covariate sets we form a network sample: $\mathcal{N}^{(w)} = \{\boldsymbol{\pi}^{(1,w)}, \ldots, \boldsymbol{\pi}^{(n,w)}\}_{w=1,...,W}$.

The score $\hat{e}_{i,j} \in [0,1]$ of the edge $Z_i \to Z_j$ is the fraction of sampled networks that contain this edge; i.e. $\hat{e}_{i,j}$ is the estimated marginal posterior probability of $Z_i \to Z_j$.

If the true network is known and has M edges, we evaluate the network reconstruction accuracy as follows: for each threshold $\xi \in [0,1]$ we extract the n_ξ edges whose scores $\hat{e}_{i,j}$ exceed ξ, and we count the number of true positives T_ξ among them. Plotting the precisions $P_\xi := T_\xi/n_\xi$ against the recalls $R_\xi := T_\xi/M$, gives the precision-recall curve. We refer to the area under the precision-recall curve as AUC ('area under curve') value. The higher the AUC value, the higher the network reconstruction accuracy.

3   Hyperparameter Settings and RJMCMC Simulation Run Lengths

For the empirical cross-model comparison, we re-use the hyperparameters from the earlier works by Lèbre et al. [1] and Grzegorczyk and Husmeier [2]:

$$\sigma^{-2} \sim GAM(\alpha_\sigma = \nu, \beta_\sigma = \nu)$$

with $\nu = 0.005$, and

$$\lambda_u^{-1} \sim GAM(\alpha_u = 2, \beta_u = 0.2)$$
$$\lambda_c^{-1} \sim GAM(\alpha_c = 3, \beta_c = 3)$$

For our new partially segment-wise coupled model we use the same hyperparameters with the extension: $\delta_h \sim BER(p)$ with $p \sim BETA(a = 1, b = 1)$.

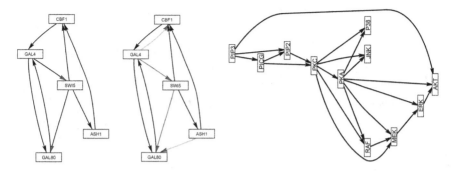

Fig. 3. Network topologies. *Left*: The true yeast network with $n = 5$ nodes and $M = 8$ edges. *Centre*: Network prediction inferred with the new partially segment-wise coupled model. We extracted the $M = 8$ edges with the highest scores. The two grey solid edges correspond to false positives. The two grey dotted edges refer to false negatives. *Right*: RAF pathway with $n = 11$ nodes and $M = 20$ edges.

For each of the three models we run RJMCMC simulation for $V = 100,000$ iterations. Setting the burn-in phase to $0.5V$ (50%) and thinning out by the factor 10 during the sampling phase, yields $W = 0.5V/10 = 5000$ samples from each posterior distribution. To check for convergence, we compared the samples of independent simulations, using standard trace plot diagnostics as well as scatter plots of the estimated edge scores. For the data sets, analyzed here, the diagnostics indicated almost perfect convergence already after $V = 10,000$ iterations; see Fig. 2 for some example scatter plot diagnostics.

4   Results

4.1   Synthetic RAF Protein Pathway Data

We generate synthetic RAF pathway data and assume the data segmentation to be known, i.e. we keep the changepoints in τ fixed. The RAF pathway [5] has $n = 11$ nodes and $M = 20$ edges, as shown in the right panel of Fig. 3. We generate data with $H = 4$ segments having 10 data points each. For each node Z_j and its parent nodes in π^j we sample the regression coefficients for $h = 1$ from standard Gaussian distributions and collect them in a vector $\mathbf{w}_1^j$, which we normalize to Euclidean norm 1, $\mathbf{w}_1^j \leftarrow \mathbf{w}_1^j/|\mathbf{w}_1^j|$. For the segments $h = 2, 3, 4$ we use: $\mathbf{w}_h^j = \mathbf{w}_{h-1}^j$ ($\delta_h = 1$, coupled) or $\mathbf{w}_h^j = -\mathbf{w}_{h-1}^j$ ($\delta_h = 0$, uncoupled). The design matrices $\mathbf{X}_h^j$ contain a first column of 1's for the intercept and the segment-specific values of the parent nodes, shifted by one time point. To the values of Z_j: $\mathbf{z}_h^j = \mathbf{X}_h^j \mathbf{w}_h^j$ we add Gaussian noise with standard deviation $\sigma = 0.05$. For all eight coupling scenarios $(\delta_2, \delta_3, \delta_4) \in \{0, 1\}^3$, we generate 25 data sets with different regression coefficients.

Figure 4 compares the network reconstruction accuracies in terms of average AUC value differences. For 6 out of 8 scenarios the AUC differences are in

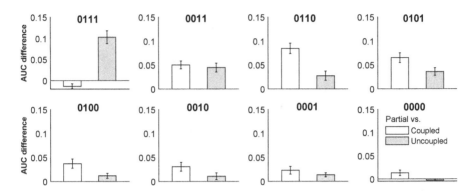

Fig. 4. Network reconstruction improvements for RAF data. For each scenario $(\delta_2, \delta_3, \delta_4)$ there are two bars of AUC differences between the new and the uncoupled model (white), and the new and the coupled model (grey). Positive values are in favor of the new model. The error bars give 95% t-test confidence intervals.

favor of the new partially coupled model. Only when all segments $h \geq 2$ are coupled ('111') or all segments are uncoupled ('000'), the new model performs slightly worse than the fully coupled or the uncoupled model, respectively. For the new partially coupled model, Fig. 5 shows the posterior probabilities that the segments $h = 2, 3, 4$ are coupled. The trends are in good agreement with the true coupling scenarios, i.e. the new model correctly infers if the regression coefficients are similar (identical) or different (opposite signs).

4.2   Saccharomyces Cerevisiae Gene Expression Data

Cantone et al. [6] synthetically designed a network in *S. cerevisiae* (yeast) with $n = 5$ genes and $M = 8$ edges, and then measured gene expression data under galactose- and glucose-metabolism: 16 measurements were taken in galactose and 21 measurements were taken in glucose. This is an ideal benchmark data set, as the network structure is known, so the network reconstruction accuracies can be cross-compared on real wet-lab data. The true network topology is shown in the left panel of Fig. 3. We pre-process the data as described in [2]. Here we assume the changepoint(s) to be unknown, so that we infer them from the data. As the number of changepoints grows with the hyperparameter p of the changepoint prior in Eq. 8, we implement the models with different values p. The average AUC scores of the models are shown in Fig. 6. The uncoupled model is consistently inferior to the new partially coupled model. The new partially coupled model also performs better than the coupled model. One exemption occurs for $p = 0.1$, where the coupled model is slightly superior.

The center panel of Fig. 3 shows a network prediction that was obtained with the new partially segment-wise coupled NH-DBN model.

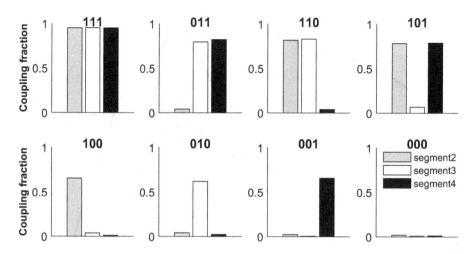

Fig. 5. Diagnostic plot for the new partially segment-wise coupled model on RAF pathway data. For each scenario $(\delta_2, \delta_3, \delta_4) \in \{0, 1\}^3$ there is a bar chart of the posterior probabilities that segment h is coupled to segment $h - 1$ $(h = 2, 3, 4)$. It can be clearly seen that the trends are in good agreement with the true underlying coupling scenarios. There are high coupling fractions for the coupled segments (with $\delta_h = 1$) and low coupling fractions for the uncoupled segments (with $\delta_h = 0$).

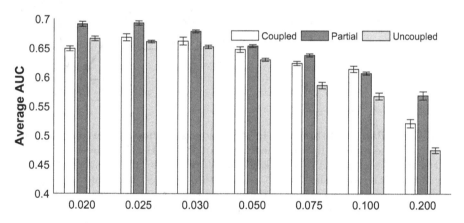

Fig. 6. Network reconstruction accuracy for *S. cerevisiae* network. We implemented the models with seven different hyperparameters p of the changepoint prior in Eq. 8. There is a bar chart for each $p \in \{0.02, , 0.025.0.03, 0.05, 0.075, 0.1, 0.2\}$, and the bars show the model-specific average AUC scores; the error bars indicate standard deviations.

4.3   Arabidopsis Thaliana Gene Expression Data

The circadian clock network in *A. thaliana* optimizes the gene regulatory processes w.r.t. the daily dark:light cycles (photo periods). In four experiments Ara-

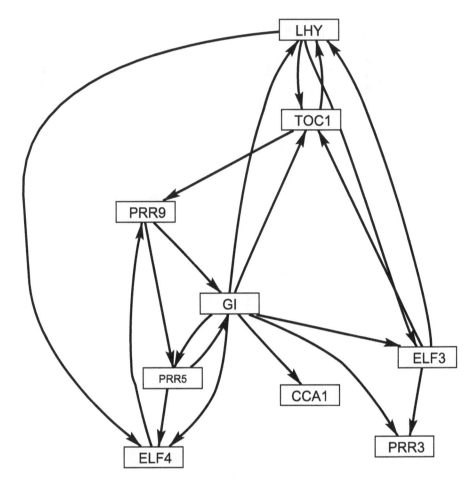

Fig. 7. Circadian clock network in *Arabidopsis thaliana*, inferred with the newly proposed partially segment-wise coupled NH-DBN model. We have implemented the model with the hyperparameter p = 0.1 for the changepoint process prior in Eq. 8. The figure shows the 20 edges with the highest marginal edge posterior probabilities (edge scores). Many of the inferred edges are consistent with the plant biology literature; see main text for further details.

bidopsis plants were entrained in different dark:light cycles, before gene expressions were measured under constant light condition over 24- and 48-hours. We follow earlier studies [2], and merge the four time series to one single data set with $T = 47$ data points and focus our attention on the $n = 9$ core genes: LHY, TOC1, CCA1, ELF4, ELF3, GI, PRR9, PRR5, and PRR3.

Here we cannot objectively cross-compare the network reconstruction accuracies, as the true underlying network topology is not known. Figure 7 shows the structure of the network that was inferred with the new partially segment-wise coupled model, using p = 0.1 for the changepoint process prior in Eq. 8. For obtaining the prediction in Fig. 7, we extracted the 20 edges with the highest marginal edge posterior probabilities (edge scores). A proper biological evaluation of the predicted network topology is beyond the scope of this paper, but we note that many of the inferred edges are consistent with the plant biology literature. For example, the feedback loop between the two genes LHY and $TOC1$ is one of the most important key features of the circadian clock network; see, e.g., the early work by Locke et al. [8]. Many of the other predicted network edges have been reported in more recent works. For example, the five edges $LHY \to ELF3$, $LHY \to ELF4$, $GI \to TOC1$, $ELF3 \to PRR3$ and $ELF4 \to PRR9$ can all be found in the Arabidopsis circadian clock network of Herrero et al. [7].

5 Conclusion

We have proposed a new partially segment-wise coupled non-homogeneous dynamic Bayesian network model (NH-DBN). Our new model is a consensus model between the standard uncoupled NH-DBN model and the fully coupled non-homogeneous dynamic Bayesian network model from [2]; see Fig. 1 for a graphical model representation.

Our empirical results on synthetic RAF pathway data (see Fig. 4) and on *S. cerevisiae* gene expression data (see Fig. 6) show that the new partially segment-wise coupled model reaches higher network reconstruction accuracies than its competitors, namely the uncoupled NH-DBN and the fully sequentially coupled NH-DBN. We have also seen that the new partially coupled model can correctly infer from the data whether the parameters of neighboring segments are similar or dissimilar; see Fig. 5 for a diagnostic plot.

In the third and last empirical case study, we have applied the new partially segment-wise coupled model to gene expression data from the circadian clock network in *Arabidopsis thaliana*. The inferred network topology (shown in Fig. 7) is consistent with the plant biology literature.

References

1. Lèbre, S., Becq, J., Devaux, F., Lelandais, G., Stumpf, M.P.H.: Statistical inference of the time-varying structure of gene-regulation networks. BMC Syst. Biol. 4, Article 130 (2010)
2. Grzegorczyk, M., Husmeier, D.: A non-homogeneous dynamic Bayesian network with sequentially coupled interaction parameters for applications in systems and synthetic biology. Stat. Appl. Genet. Mol. Biol. (SAGMB) **11**(4), Article 7 (2012)
3. Shafiee Kamalabad, M., Grzegorczyk, M.: Improving nonhomogeneous dynamic Bayesian networks with sequentially coupled parameters. Stat. Neerl. **72**(3), 281–305 (2018)
4. Bishop, C.M.: Pattern Recognition and Machine Learning, 1st edn. Springer, Singapore (2006)
5. Sachs, K., Perez, O., Pe'er, D., Lauffenburger, D.A., Nolan, G.P.: Causal protein-signaling networks derived from multiparameter single-cell data. Science **308**, 523–529 (2005)

6. Cantone, I., et al.: A yeast synthetic network for in Vivo assessment of reverse-engineering and modeling approaches. Cell **137**, 172–181 (2009)
7. Herrero, E., et al.: EARLY FLOWERING4 recruitment of EARLY FLOWERING3 in the nucleus sustains the Arabidopsis circadian clock. Plant Cell Online **24**(2), 428–443 (2012)
8. Locke, J.C.W., et al.: Experimental validation of a predicted feedback loop in the multi-oscillator clock of Arabidopsis thaliana. Mol. Syst. Biol. **2**(1), online article (2006)

Inhibition of Primed Ebola Virus Glycoprotein by Peptide Compound Conjugated to HIV-1 Tat Peptide Through a Virtual Screening Approach

Ahmad Husein Alkaff⬥, Mutiara Saragih⬥,
Mochammad Arfin Fardiansyah Nasution⬥,
and Usman Sumo Friend Tambunan$^{(\boxtimes)}$⬥

Bioinformatics Research Group, Department of Chemistry,
Faculty of Mathematics and Natural Sciences, Universitas Indonesia,
Kampus UI, Depok 16424, Indonesia
usman@ui.ac.id

Abstract. A higher prevalence of Ebola hemorrhagic fever is caused by Ebola virus (EBOV). It enters into the host cell through macropinocytosis mechanism. During the entry process, the primed viral glycoprotein (GPcl) interacts with a lysosomal cholesterol transporter, Niemann Pick C1 (NPC1), leading to the fusion of the viral envelope and the host endosomal membrane. Hence, disrupting the interaction between EBOV GPcl and host NPC1 is a promising way to prevent the viral nucleocapsid content entering the cytoplasm. In this study, a virtual screening approach has been used to investigate peptide compounds conjugated to HIV-1 Tat peptide as drug lead candidate inhibiting EBOV GPcl. About 50,261 peptides from NCBI PubChem database, which acts as ligands, were subjected to initial toxicological screening to omit ligands with undesired properties. The remaining ligands underwent a pharmacophore search, rigid docking, and flexible docking simulation to discover ligands with favorable inhibition activities. Calfluxin, SNF 8906, grgesy, phosphoramidon, and endothelin (16-21) were five ligands which have lower $\Delta G_{binding}$ value compared to the standard ligand. The chosen ligands were subjected to absorption, distribution, metabolism, excretion, and toxicity (ADME-Tox) analysis, which was accomplished by pkCSM software. Subsequently, they were conjugated to HIV-1 Tat peptide to accumulate them inside the endosome. The inhibition activity was reevaluated by the second flexible molecular docking simulation. As a result, only C-Calfluxin showed improved affinity while managing minimal conformational changes in protein-peptide interaction compared to its respective unconjugated ligand.

Keywords: Ebola · GPcl · NPC1 · Peptide · HIV-1 Tat · Virtual pharmacophore screening · Flexible molecular docking simulation

1 Introduction

Ebola virus (EBOV) is classified under the Ebolavirus genus in Filoviridae family. It was identified in 1976 as an agent that caused a deadly hemorrhagic fever [1]. In the 2013–2016 Ebola outbreak, EBOV caused more than 28,646 cases and took the lives of

© Springer Nature Switzerland AG 2020
M. Raposo et al. (Eds.): CIBB 2018, LNBI 11925, pp. 153–165, 2020.
https://doi.org/10.1007/978-3-030-34585-3_14

11,323 people [2]. Recently, the outbreak occurred in the Republic Democratic of the Congo on April-May 2018, leading to 50 reported cases; 25 of which ended in death [3]. Currently, no approved drug to cure EBOV infection, even though the rVSV-ZEBOV vaccine has shown promising protection from EBOV infection [4, 5]. Therefore, the drug development for the treatment of EBOV infection is urgently needed.

2   Scientific Background

EBOV enters cells via the binding of viral glycoprotein (GP) spike on the surface of the viral envelope to receptors on the surface of the host cell. After EBOV GP attachment to the cell surface protein such as T cell immunoglobulin and mucin domain-containing proteins, EBOV is internalized into the endosome via macropinocytosis mechanism. The furin-like proteases cleave the GP spike into GP1 and GP2 subunits, which still intact via an inter-subunit disulfide bond and non-covalent interactions. Subsequently, Cathepsin L and Cathepsin B remove about 60% of amino acids from GP1, resulting in a 19 kDa primed GP (GPcl) [6]. The GPcl, unlike full-length GP, could bind to Niemann Pick C1 (NPC1), a lysosomal cholesterol transporter, triggering the conformation changes on GP2 which results in the fusion of EBOV envelope and host endosomal membrane [7]. Hence, preventing the interaction between GPcl and NPC1 is a promising approach to block cytoplasmic delivery of the viral nucleocapsid content.

Recently, more than 7,000 naturally occurring peptides have been identified. It is known to have unique pharmacological properties due to the high specificity and selectivity, as well as low toxicity. As a drug, the peptide is considered as relatively safe and well tolerated. Therefore, there is a growing interest in pharmaceutical research on the peptide-based drug [6]. In this research, we investigated the potency of peptide compounds as an inhibitor of EBOV GPcl through virtual screening approach of pharmacophore-based virtual screening and molecular docking simulations.

3   Materials and Methods

The three-dimensional structure of EBOV GPcl, which selected as the target protein, was acquired from RCSB Protein Data Bank (PDB) with PDB ID: 5JNX. The peptide compounds acquired from NCBI PubChem database were chosen as the ligand, while HIV-1 Tat peptide sequence was obtained from Sigma-Aldrich. The initial toxicological screening was performed using DataWarrior v4.5.2 software. Molecular Operating Environment (MOE) 2014.09 was used to carry out the pharmacophore search, rigid docking, and flexible docking simulation. The absorption, distribution, metabolism, excretion, and toxicity (ADME-Tox) properties were determined by utilizing pkCSM software. Conjugation of the selected ligand with HIV-1 Tat peptide was done using ACDLabs Chemsketch v12.01 software followed by the second flexible docking simulation to determine the effect of HIV-1 Tat peptide to ligand-protein binding and interaction.

4   Experimental Results

4.1   Initial Toxicological Screening

About 50,261 peptide compounds were obtained from the NCBI PubChem database. The peptides from this database not only consist of linear structure but also the branched and cyclical structure. Peptides with the additional functional group and unique amino acids such as glycopeptides, lipopeptides, peptoids, and peptaibols are also included. In term of function, it ranges from antimicrobial, cell-penetrating, and signaling peptides. The hormones and neuropeptides are comprised as well. Peptides with undesired mutagenic, tumorigenic, reproductive effect, and irritant properties were omitted through the initial toxicological screening by using the DataWarrior v4.5.2 software. From this step, 44,582 peptides proceeded to the next analysis as the selected ligands.

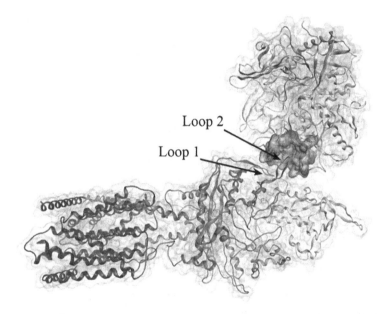

Fig. 1. EBOV GPcl (magenta) in complex with NPC1. The NPC1 is constructed of 13 transmembrane segments (blue) and three lumenal domain A (yellow), C (orange), and I (green). The green and purple area on the binding site indicate the hydrophobic and hydrophilic region, respectively. (Color figure online)

4.2   Pre-docking Preparation

The human NPC1 is composed of 1,278 amino acids and consisted of four domains: 13 transmembrane segments, luminal domain A, C, and I. EBOV GPcl binds to the NPC1 by anchoring the hydrophobic cavity of EBOV GPcl head to loop 1 and 2 of NPC1 luminal domain C (see Fig. 1) [7]. The 3D protein structure of EBOV GPcl in complex with human NPC1 was prepared using MOE 2014.09 software. The unwanted water and glucose molecules were eliminated. Also, 1,271 of 1,278 amino acid sequence of NPC1 was deleted with only leaving the protruding loop two (NPC1-L2) which consist of Gly500, Asp501, Asp502, Phe503, Phe504, Val505, and Tyr506 amino acid residue. AMBER 10: EHT with R-field solvation was selected as forcefield parameter. AMBER 10: EHT is a parameterized forcefield for proteins and nucleic acids while R-field is a solvent model that disable implicit solvation energy (E_{sol}) and uses the reaction field form of the electrostatic energy (E_{ele}) equation. The default protocol of 'LigX' feature was employed to add missing hydrogen and perform energy minimization process.

The selected ligands and standards were also prepared using MOE 2014.09 software. MMFF94x, which suitable for preparing small organic molecules, with R-field solvation was selected as the forcefield parameter. The default protocol of wash and energy minimization procedure was selected to optimize the molecular structure of ligand and standard peptides. The molecular chirality was preserved because of its essential role in affecting the molecular activity.

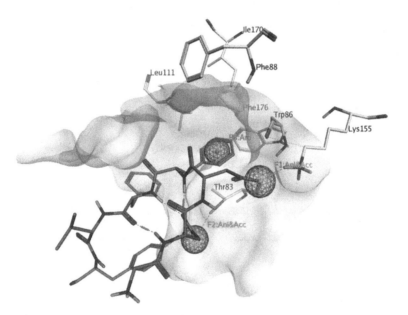

Fig. 2. Pharmacophore feature of NPC1-L2 (orange) in the EBOV GPcl biding site. The blue and orange sphere represents the 'anion and hydrogen bond acceptor' and 'aromatic' pharmacophore features, in addition to the green and purple area of the binding site indicate the hydrophobic and hydrophilic region, respectively. (Color figure online)

4.3   Pharmacophore-Based Virtual Screening

A pharmacophore is a three-dimensional arrangement that carries the essential features that enable a molecule to exert a particular biological activity [8]. The pharmacophore features are built based on the essential protein-ligand interaction from the known structure of the protein-ligand complex. In this research, the pharmacophore features were generated based on the interaction of EBOV GPcl with NPC1-L2 peptide (see

Fig. 2). The 'Ani&Acc' pharmacophore feature, which stands for 'anion and hydrogen bond acceptor,' on Asp501 and Asp502 were generated because it generates essential hydrogen bond with Thr83 and Lys155, respectively. Also, 'Aro' pharmacophore feature, which is an 'aromatic' centroid annotation located at the geometric center of the benzene ring,' was generated on Phe503 which lies inside the hydrophobic cavity of EBOV GPcl binding site consisting of nonpolar residues Trp 86, Phe88, Leu111, Ile170, and Phe176. "Pharmacophore Search" feature was utilized to select ligands which have the constructed pharmacophore features and expected for having biological activity similar to NPC1-L2. From about 44,582 ligands, only 660 ligands pass the pharmacophore-based virtual screening.

Table 1. The $\Delta G_{binding}$ and RMSD value of best ligands and the standard against EBOV GPcl.

Ligand name	PubChem CID	Flexible docking	
		$\Delta G_{binding}$	RMSD
Calfluxin	348609099	−10.4847	0.8261
SNF 8906	319153670	−8.4813	0.7998
Grgesy	135261596	−8.1038	1.8910
Phosphoramidon	319155145	−7.7372	0.7403
Endothelin (16-21)	348629755	−7.3627	0.8255
NPC1-L2*	N/A	−7.0571	1.5523

Note: *Standard.

4.4 Molecular Docking Simulation of the Ligand

The 660 selected ligands underwent rigid molecular docking followed by flexible molecular docking simulation, a method to assess the position, orientation, and conformation of a ligand that interacts with the protein target binding site [9]. 'Rigid Receptor' protocol with the retain value of 30 repetitions was used as rigid docking method, a docking mode in which the binding site is rigid while the ligand moves flexibly following lock and key theorem. Then, 'Induced Fit' protocol, which enables both the ligand and binding site more flexible in finding the most suitable conformation, was performed with the retain value of 100 repetitions as a flexible docking method. Default 'Pharmacophore' placement with 'London dG' rescoring and 'Forcefield' refinement with GBVI/WSA dG rescoring was selected as the calculation method for both molecular docking method. The ligands which have a root-mean-square deviation (RMSD) value refine lower than 2.0 Å and lower Gibbs binding energy ($\Delta G_{binding}$) than the standards are considered as a potential ligand. RMSD account for the acceptable magnitudes of conformational changes generated from the simulation while $\Delta G_{binding}$ estimates the affinity and conformational stability of the protein-ligand complex. After the first and second simulation, only five ligands were regarded as a potential inhibitor of EBOV GPcl (Table 1) based on the $\Delta G_{binding}$ and RMSD value.

Table 2. Chemical properties of best peptide ligands and the standard against EBOV GPcl.

Ligand name	Molecular weight	TPSA	H-Don	H-Acc	Rotatable bond	Log P
Calfluxin	1541.46	811	27	32	52	−13.4
SNF 8906	1100.20	407	12	18	31	−2.0
Grgesy	667.68	351	12	13	21	−6.6
Phosphoramidon	543.51	211	8	11	11	−0.2
Endothelin (16–21)	759.94	291	10	11	23	−0.1
NPC1-L2*	844.89	320	8	19	24	−5.5

Note: *Standard, molecular weight in g/mol, and TPSA in Å^2.

4.5   ADME-Tox Prediction

The pharmacokinetics and toxicity properties dictate whether a drug candidate can be progressed into a drug for delivering its therapeutic effects. The molecular weight (MW), topological polar surface area (TPSA), H-bond donor (H-Don), H-bond acceptor (H-Acc), rotatable bond count, and octanol/water partition coefficient (Log P) are the physicochemical parameters associated with the oral bioavailability (Table 2) [10]. The high properties value from each parameter indicates the high solubility but low intestinal permeability of the ligands, which indicates the low oral bioavailability. The peptide as the drug faces a challenge of rapid degradation due to its proneness to hydrolysis and oxidation, especially in the gastrointestinal tract [6]. Hence, the peptide-based drug is usually delivered through intermittent intravenous, intramuscular, pulmonary, or subcutaneous administration to overcome its weaknesses [11].

The best ligands also subjected to further ADME analysis using the pkCSM software (Table 3). The software calculates the general compound properties and distance-based graph signatures to train and test machine learning-based predictors [12]. As a result, the ADME properties represented by P-gp substrate/inhibitor, fraction unbound, human cytochrome P450 (CYP) substrate/inhibitor, and total clearance was predicted.

Table 3. The best ligands and the standard ADME properties predicted using pkCSM software.

Ligand name	Absorption			Distribution	Metabolism		Excretion
	P-gp substrate	P-gp inhibitor I	P-gp inhibitor II	Fraction unbound	CYP substrate	CYP inhibitor	Total clearance
Calfluxin	No	No	No	0.375	No	No	−0.984
SNF 8906	Yes	No	No	0.295	No	No	−0.063
Grgesy	Yes	No	No	0.511	No	No	0.644
Phosphoramidon	Yes	No	No	0.344	No	No	0.248
Endothelin (16–21)	Yes	No	No	0.379	No	No	0.108
NPC1-L2*	Yes	No	No	0.271	CYP3A4	No	0.426

Note: *Standard and total clearance in log (mL/min/Kg).

P-glycoprotein (P-gp) is an efflux membrane transporter. It is commonly scattered throughout the body and is in charge of constricting cellular uptake and the distribution of xenobiotics and toxic substances [13]. P-gp is located in the luminal membrane of entire intestine from the duodenum to the rectum, with high expression in the enterocytes of the small intestine. Grgesy, SNF 8906, and endothelin indicated as a substrate of P-gp. This property may reduce the oral bioavailability of the candidate drug. All of the ligands did not inhibit the P-gp, meaning that they were poorly permeable. Ideally, the inhibition of P-gp is mainly accomplished in order to increase the delivery of the therapeutic agent [13].

The CYP is a family of enzymes that catalyze the metabolism of a huge variety of compounds such as fatty acids, steroids, drugs, xenobiotics, and environmental toxins [14]. According to the presented result, no ligands was acting as substrates. It indicates that all of the ligands are not substrates for CYP. Hence, the ligands are not converted to reactive species that covalently bind to CYP isoenzyme leading to their inactivation [15]. Moreover, there was no ligand determined as the inhibitor of CYP. The compounds which inhibit any isoform of CYP leads to the malfunctioning of the drug metabolism and elevation of toxicity [16].

After the drugs were absorbed and entry into the systemic circulation either by intravascular injection or by absorption from any of various extravascular site, then the drug is distributed to cells or tissues. One of the critical factors in the distribution of drug efficacy in pharmacokinetic and pharmacodynamic studies is fraction unbound. Commonly the unbound drug is capable of interacting with pharmacological target proteins such as enzymes, channels, and receptors. Also, it is able to diffuse amidst plasma and tissues [17]. Nearly all drugs bind to plasma and tissue proteins, ensuing in decline in pharmacologically free active drug concentrations. The fraction unbound analyzed in this simulation depict the ratio of free against total drug concentration in plasma [18]. Hence, the higher fraction unbound value indicates the higher concentration of free active drug concentrations which ready to be distributed to the site of action. All ligands and standards demonstrated the fraction unbound value lower than 0.500 implying their low free drug concentration, except for Grgesy which have fraction unbound of 0.511.

Drug clearance is defined as the volume of the body compartment from which drug is removed in unit time [19]. Total drug clearance is the sum of the drug elimination processes. In this simulation, this process includes hepatic clearance (metabolism in the liver and biliary clearance) and renal clearance (excretion via the kidneys). Total clearance is related to the elimination half-life and distribution volume. The increase in total clearance provided distribution volume is constant, results in a decrease in elimination rate half-life [20]. Therefore, the negative total clearance value of calfluxin and SNF 8906 implies the rapid clearance of the ligands.

Table 4. The best ligands and the standard toxicity properties predicted using pkCSM software.

Ligand name	Ames toxicity	hERG I	Hepatoxicity
Calfluxin	No	No	No
SNF 8906	No	No	Yes
Grgesy	No	No	Yes
Phosphoramidon	No	No	No
Endothelin (16–21)	No	No	Yes
NPC1-L2*	No	No	Yes

Note: *Standard.

The pkCSM software was also used to determine the toxicity properties of the best ligands, which represented in the Ames toxicity, the human ether-a-go-go-related gene (hERG), and hepatoxicity (Table 4). The Ames test is a short-term bacterial reverse mutation assay specifically arranged to identify a broad range of chemical substances which can promote genetic damage that leads to genetic mutation [21]. In the Ames test, the mutagen substances or compounds may be detected by the liability of histidine-dependent strains of Salmonella typhimurium which carrying different mutations in various genes in the histidine operon [21, 22]. If the compounds are a mutagen, it may be caused by the mutations or base-pair substitutions of the Salmonella typhimurium [22]. Grgesy, SNF 8906, calfluxin, and endothelin were negative in accordance with the Ames test because they do not have any tendency to become mutagenic properties.

The hERG encodes a trimeric potassium channel. hERG plays an important role in cardiac potential, which is related to long QT syndrome and may cause avoidable sudden cardiac death [23]. In human, hERG1 is expressed in various tissues and cell types, including the retina, brain, cardiac, and smooth muscle. Some compounds such as terfenadine and grepafloxacin which are known as an inhibitor of the hERG1 channel may cause a reduction in the polarizing current by blocking the central cavity of hERG and promoting ventricular arrhythmia [24]. Neither ligands nor standard ligand inhibited the hERG1 channel.

The liver plays a vital role in the maintenance, performance, and regulating homeostasis of the body. Hepatotoxin is defined as a drug that caused liver damage [25]. Hepatotoxicity is the injury or liver damage caused by exposure of hepatotoxin. Ggresy, SNF 8906, calfluxin, and endothelin have positive value on the hepatoxicity properties. It indicates that all of the ligands may disrupt liver performance. Based on the hepatoxicity prediction, phosphoramidon has no potential to damage the liver.

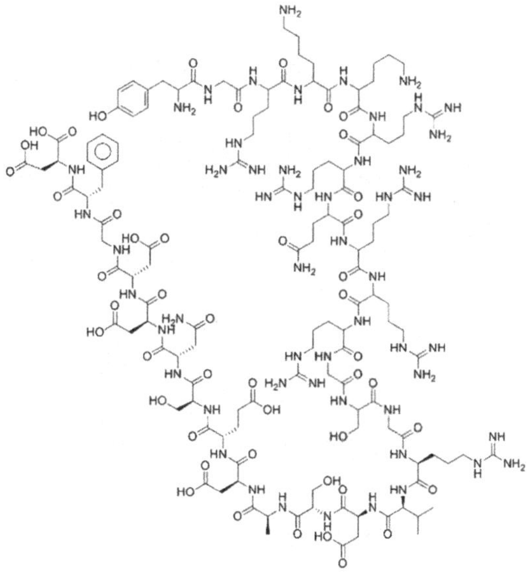

Fig. 3. The structure of C-calfluxin which constructed from HIV-1 Tat peptide (red), linker (green), and calfluxin (blue). (Color figure online)

4.6 Molecular Docking Simulation of the Conjugated Ligand

The binding site of EBOV GPcl is exposed and binds to NPC1 inside the endosome. Hence, HIV-1 Tat peptide, a carrier peptide which accumulates in the endosome, was conjugated to the selected ligands to enhance the probability of the ligands binds with EBOV GPcl binding site. The HIV-1 Tat peptide sequence is constructed of Tyr-Gly-Arg-Lys-Lys-Arg-Arg-Gln-Arg-Arg-Arg amino acid residues. Also, the linker, which consists of Gly-Ser-Gly amino acid residues, was attached between the HIV-1 Tat peptide C-terminal and the N-terminal of the ligands. The linker was used to prevent the ligand activity from being disturbed by the HIV-1 Tat peptide and vice versa [26]. Calfluxin, Grgesy, and Endothelin (16-21) have their N-terminal protrude outside the binding site which makes them suitable for HIV-1 Tat peptide conjugation, producing three conjugated ligands (C-ligand) C-Calfluxin (see Fig. 3), C-Grgesy, and C-Endothelin (16-21). SNF 8906 is a branched peptide which has three N-terminal, two of which protrude outside the binding site, producing C1-SNF 8906 and C2-SNF 8906. On the other hand, phosphoramidon cannot be conjugated with HIV-1 Tat peptide because of the phosphate molecule attached to its N-terminal.

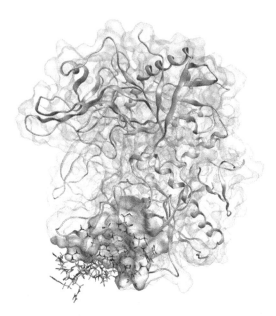

Fig. 4. Full-length C-calfluxin (blue) binds EBOV GPcl. The green and purple area of the binding site indicate the hydrophobic and hydrophilic region, respectively. (Color figure online)

The flexible molecular docking simulation with the 'Induced Fit' protocol, the retain value of 300 repetitions and the same calculation method as the previous simulation was employed to assess the effect of HIV-1 Tat peptide conjugation to the protein-ligand interaction. From five C-ligand, only C-Calfluxin (see Fig. 4) showed improvement in binding affinity with lower $\Delta G_{binding}$ value (−19.8885 kcal/mol) compared to the unconjugated ligand, calfluxin (−10.4847 kcal/mol), while maintaining minimal conformational changes in calfluxin-EBOV GPcl interaction. The enhanced binding affinity might due to the attachment of polar HIV-1 Tat peptide to the hydrophilic surface of the EBOV GPcl.

From the flexible molecular docking simulation, it is revealed that C-calfluxin is bound to the EBOV GPcl binding site in the same position as NPC1-L2. The comparison of C-calfluxin and calfluxin interaction in the EBOV GPcl binding site is presented in Fig. 5. The benzene ring from the C-calfluxin Phe27 and calfluxin Phe13 resides in the same position as a hydrophobic cavity surrounded by Trp86, Phe88, Leu111, Ile170, and Phe176. The vital hydrogen bond interaction with Thr183 is established by an Asp28 backbone of C-calfluxin, while calfluxin is failed to do the same. C-calfluxin is managed to form two essential hydrogen bonds with Lys155 via its Asp25 side chain and Asp20 backbone, while calfluxin can only generate one hydrogen bond via its Asp3 side chain. Also, C-calfluxin forms a hydrogen bond with Thr534 by its Asp17 backbone. In general, whether or not calfluxin is conjugated to HIV-1 Tat peptide, it can bind to the designated binding site to form the desired interactions.

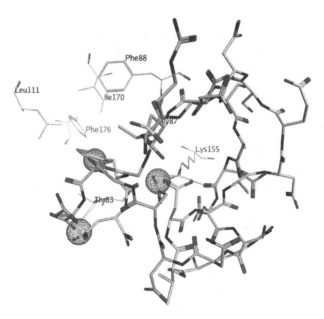

Fig. 5. Calfluxin (yellow) and C-Clfluxin (blue) conformation in the EBOV GP binding site. The magenta-colored amino acid represents the essential residues of EBOV GP. (Color figure online)

5   Conclusions

The result showed that all five selected ligands have lower $\Delta G_{binding}$ energy and better interaction with EBOV GPcl compared to NPC1-L2. The ADME-Tox analysis of the ligands exhibited relatively comparable pharmacochemical characteristics compared to the standards. The exceptionally negative Log P and total clearance value of calfluxin signify its high solubility and rapid clearance in human. After the conjugation of the ligand to the HIV-1 Tat, only C-calfluxin showed minimal conformational changes in the binding site compared to the other C-ligands. Considering the improvement in $\Delta G_{binding}$ energy and molecular interaction, C-calfluxin was expected to interact with EBOV GPcl more spontaneously and formed a more stable protein-ligand complex compared to the unconjugated calfluxin. Also, C-calfluxin was more likely to have better antiviral activity than calfluxin and the standards with its tendency to accumulate inside the endosome. Hence, it indicated that calfluxin, a gonadotropic neuropeptide, which conjugated with HIV-1 Tat peptide, has potential as a drug lead for treating Ebola by inhibiting the EBOV GPcl. Finally, the protein-ligand crystallography analysis must be carried out to investigate C-calfluxin interaction with EBOV GPcl before proceeded to in vitro and in vivo analysis where the compound is assessed under a more complex biological condition.

Acknowledgments. This research is financially supported by the Ministry of Research, Technology, and Higher Education (Kemenristekdikti) through Hibah Penelitian Kerjasama Luar Negeri Tahun 2018 No: 538/UN2.R3.1/HKP.05.00/2018. All authors were responsible for conducting the research and writing the approved final version of the manuscript. The authors declare that there is no conflict of interest regarding this manuscript.

References

1. Falasca, L., et al.: Molecular mechanisms of Ebola virus pathogenesis: focus on cell death. Cell Death Differ. **22**, 1250–1259 (2015)
2. WHO—Ebola situation reports: archive. WHO (2016)
3. Barry, A., et al.: Outbreak of Ebola virus disease in the Democratic Republic of the Congo, April–May, 2018: an epidemiological study. Lancet **392**, 213–221 (2018)
4. Smith, D.R., Holbrook, M.R., Gowen, B.B.: Animal models of viral hemorrhagic fever. Antiviral Res. **112**, 59–79 (2014)
5. Henao-Restrepo, A.M., et al.: Efficacy and effectiveness of an rVSV-vectored vaccine in preventing Ebola virus disease: final results from the Guinea ring vaccination, open-label, cluster-randomised trial (Ebola Ça Suffit!). Lancet **389**, 505–518 (2017)
6. Fosgerau, K., Hoffmann, T.: Peptide therapeutics: current status and future directions. Drug Discov. Today **20**, 122–128 (2015)
7. Wang, H., et al.: Ebola viral glycoprotein bound to its endosomal receptor Niemann-Pick C1. Cell **164**, 258–268 (2016)
8. Dror, O., Shulman-Peleg, A., Nussinov, R., Wolfson, H.: Predicting molecular interactions in silico: I. a guide to pharmacophore identification and its applications to drug design. Curr. Med. Chem. **11**, 71–90 (2004)
9. Zoete, V., Grosdidier, A., Michielin, O.: Docking, virtual high throughput screening and in silico fragment-based drug design. J. Cell Mol. Med. **13**, 238–248 (2009)
10. Lipinski, C.A.: Lead profiling lead- and drug-like compounds : the rule-of-five revolution. Drug Discov. Today Technol. **I**, 337–341 (2004)
11. Sehgal, A.: Peptides 2006 - New Applications in Discovery, Manufacturing, and Therapeutics (2006)
12. Pires, D.E.V., Blundell, T.L., Ascher, D.B.: pkCSM: predicting small-molecule pharmacokinetic and toxicity properties using graph-based signatures. J. Med. Chem. **58**, 4066–4072 (2015)
13. Amin, L.: P-glycoprotein inhibition for optimal drug delivery. Lib. Acad. **7**, 27–34 (2013)
14. Zhou, S.-F.: Drugs behave as substrates, inhibitors and inducers of human cytochrome P450 3A4. Curr. Drug Metab. **9**, 310–322 (2008)
15. Badyal, D.K., Dadhich, A.P.: Cytochrome P450 and drug interactions. Indian J. Pharmacol. **33**, 248–259 (2001)
16. Malik, A., Manan, A., Mirza, U.: Molecular docking and in silico ADMET studies of silibinin and glycyrrhetic acid anti-inflammatory activity. Trop. J. Pharm. Res. **16**, 67–74 (2017)
17. Watanabe, R., et al.: Predicting fraction unbound in human plasma from chemical structure: improved accuracy in the low value ranges. Mol. Pharm. **15**, 5302–5311 (2018)
18. Gonzalez, D., Schmidt, S., Derendorf, H.: Importance of relating efficacy measures to unbound drug concentrations for anti-infective agents. Clin. Microbiol. Rev. **26**, 274–288 (2013)

19. Barber, H.E., Petrie, J.C.: Today's treatment elimination of drugs. Br. Med. J. **282**, 809–810 (1981)
20. Rang, H.P., Ritter, J., Flower, R.J., Rod J., Henderson, G.: Rang & Dale's Pharmacology. Churchill Livingstone (2015)
21. Mortelmans, K., Zeiger, E.: The Ames Salmonella/microsome mutagenicity assay. Fundam. Mol. Mech. Mutagen. **455**, 29–60 (2000)
22. Hansen, K.: Benchmark data set for in silico prediction of Ames mutagenicity. J. Chem. Inf. Model. **49**, 2077–2081 (2009)
23. Zhang, C., et al.: In silico prediction of hERG potassium channel blockage by chemical category approaches. Toxicol. Res. (Camb) **5**, 570–582 (2016)
24. Durdagi, S., Subbotina, J., Guo, J., Duff, H.J., Noskov, S.Y.: Insights into the molecular mechanism of hERG1 channel activation and blockade by drugs. Curr. Med. Chem. **17**, 3514–3532 (2010)
25. Pandit, A., Sachdeva, T., Bafna, P.: Drug-induced hepatotoxicity: a review. J. Appl. Pharm. Sci. **02**, 233–243 (2012)
26. Higgins, C.D., Koellhoffer, J.F., Chandran, K., Lai, J.R.: C-peptide inhibitors of Ebola virus glycoprotein-mediated cell entry: effects of conjugation to cholesterol and side chain-side chain crosslinking. Bioorg. Med. Chem. Lett. **23**, 5356–5360 (2013)

Pharmacophore Modelling, Virtual Screening, and Molecular Docking Simulations of Natural Product Compounds as Potential Inhibitors of Ebola Virus Nucleoprotein

Mochammad Arfin Fardiansyah Nasution ⓘ,
Ahmad Husein Alkaff ⓘ, Ilmi Fadhilah Rizki ⓘ, Ridla Bakri ⓘ,
and Usman Sumo Friend Tambunan$^{(\boxtimes)}$ ⓘ

Bioinformatics Research Group, Department of Chemistry,
Faculty of Mathematics and Natural Sciences, Universitas Indonesia,
Kampus UI, Depok 16424, Indonesia
usman@ui.ac.id

Abstract. Ebola virus (EBOV) prevails as a serious public health issue which infected at least 27,000 people and claimed the lives of about 11,000 people in the latest Ebola outbreak in 2014. Although the virus has been known for almost 40 years, currently there is no approved drug for this virus. Hence, the development of a new drug candidate for Ebola is required to anticipate the future outbreak that may happen. In this research, about 229,538 natural product (NP) compounds were retrieved and screened using a computational approach against EBOV nucleoprotein (NP). In the beginning, all NP compounds were screened throughout computational toxicity and druglikeness prediction tests, followed by pharmacophore-based virtual screening and molecular docking simulation to identify their binding affinity and molecular interaction in the RNA-binding groove of EBOV NP. All of the results were compared to 18β-glycyrrhetinic acid, the standard molecule of EBOV NP. In the end, about five NP compounds (UNPD213871, UNPD199951, UNPD124962, UNPD139843, and UNPD147202) were identified to have exciting activities against EBOV NP. Therefore, based on the results of this study, these compounds appeared to have potential inhibition activities against EBOV NP and can be proposed for further in silico and in vitro studies.

Keywords: Ebola virus · Natural product compounds · Ebola virus nucleoprotein · Inhibitors · Pharmacophore-based virtual screening · Molecular docking

1 Introduction

Zaire ebolavirus is a species of lipid-enveloped, non-segmented, and negatively stranded RNA virus which was classified in the *Filoviridae* family [1]. Ebola virus (EBOV) has caused several large Ebola epidemics and become the most virulent species of *Filoviridae*. The EBOV nucleoprotein (NP) is the largest nucleoprotein of

© Springer Nature Switzerland AG 2020
M. Raposo et al. (Eds.): CIBB 2018, LNBI 11925, pp. 166–178, 2020.
https://doi.org/10.1007/978-3-030-34585-3_15

the non-segmented negative-stranded RNA viruses which constructed of 739 amino acid residues [2]. EBOV possesses a nucleocapsid to facilitate genomic RNA encapsidation to form viral ribonucleoprotein complex (RNP) together with NP and polymerase (L), which plays an essential role in the virus life cycle [3]. The RNP acts as the template for replication that produces viral genomic RNA to be packaged in the virion. The blockade of RNP formation and its function would provide a great opportunity for antiviral development.

2 Scientific Background

Identification of the antiviral mechanisms from these natural products has shed light on where they interact with the viral life cycle, such as viral entry, replication, assembly, and release, as well as on the targeting of virus-host-specific interactions [4]. Indonesia, for example, is a home of more than 17,000 islands and 30,000 floral species, which make this archipelago country has the second-highest biodiversity worldwide after Brazil [5]. Many of these tropical plants have medicinal value and commonly used in the drug discovery and development field [6]. However, the cost of random screening of huge database can be overpriced, and it makes sense to use in silico screening where possible to filter down the number of compounds used. This research used pharmacophore-based virtual screening method to generate the pharmacophore and use it as a template for screening is to find molecules (hits) that have chemical features similar to those of the template [7, 8]. The pharmacophore models are suitable for finding compounds that fit into the number of constraints, which makes the analysis is significantly faster than using other simulation methods, such as docking simulation. Thus, pharmacophore searching is ideal for screening gigantic databases of chemical structures [9]. Furthermore, the virtual screening was combined with molecular docking to find the best potential ligand to inhibit EBOV NP from the chosen NP database.

3 Materials and Methods

The methodology of this research was taken and modified based on the previously established research [10, 11]. This research used several offline software: DataWarrior v.4.6.1 [12], Molecular Operating Environment (MOE) 2014.09 [13], and PerkinElmer ChemBioDraw Ultra. For online access, Research Collaboratory for Structural Bioinformatics Protein Data Bank (RCSB-PDB) [14], and Universal Natural Product Database (UNPD) [15] were used. Additionally, 18β-glycyrrhetinic acid, a pentacyclic triterpenoid compound which has been studied earlier by Fu et al. in 2016 to have a binding affinity with EBOV NP in its RNA-binding groove, was selected as the standard ligand for this study [16]. The complete flowchart of this research is depicted in Fig. 1.

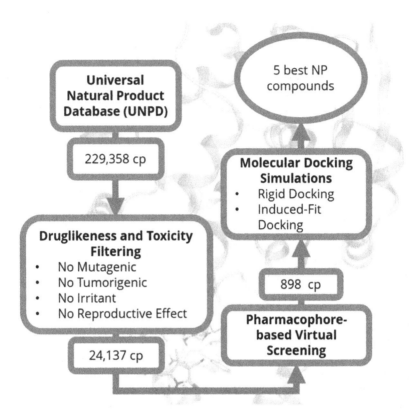

Fig. 1. Research flowchart in this study; which represents the standard procedure of computer-aided drug design and development protocols, including computational ADME-Tox filtering, pharmacophore-based virtual screening, and molecular docking simulations.

3.1 Pre-docking Preparation

In the beginning, both 18β-glycyrrhetinic acid and NP ligands, which collected from the UNPD database, were screened through a computational ADME-Tox scheme using DataWarrior v.4.6.1 to identify their molecular properties, toxicity potency, and druglikeness score. Followed by the optimization and minimization of the remaining NP compounds using the standard protocol of MOE 2014.09 software. Furthermore, the 3D structure of EBOV NP was also optimized using MOE 2014.09 software, using the default protocol of preparing the 3D structure of the protein.

3.2 Pharmacophore Modelling and Pharmacophore-Based Virtual Screening

In this study, the pharmacophore model of EBOV NP in its RNA-binding groove was generated by using MOE 2014.09 software, which based on the molecular interaction of 18β-glycyrrhetinic acid in the RNA-binding groove of this protein. Since there was no 3D structure of this ligand bound with the protein that currently available, the interaction of 18β-glycyrrhetinic acid was identified and predicted after the molecular docking simulation had been performed, using the 'Induced-Fit' protocol, with the addition of standard procedure of PLIF (Protein-Ligand Interaction Fingerprints) method from MOE 2014.09 software. The generated pharmacophore model from this step underwent pharmacophore-based virtual screening, using the remaining NP compounds that retrieved from the previous step.

3.3   Molecular Docking Simulations

In this study, the four steps of rigid docking and induced-fit docking were applied to the remaining compounds that matched with the pharmacophore model from the previous screening. These simulations were done using MOE 2014.09 software. In this simu-lations, about five protein-ligand complexes were chosen as the best ligands, based on the Gibbs free binding energy ($\Delta G_{binding}$) value, Root-Mean-Square Deviance (RMSD) score and molecular interactions compared to 18β-glycyrrhetinic acid as the standard ligand.

4   Experimental Results

4.1   Preparation of the Standard Ligand and the Indonesia Natural Product

In this study, about 229,358 NP compounds were collected from the UNPD database as ligand database; this database was constructed of NP compounds, including flavonoids, terpenoids, alkaloids, steroids, and another group of natural product compounds that contained in microorganisms, plants, and animals from several databases [15]. Many of these NP have medicinal value and commonly used in the drug discovery and devel-opment field, many of which are anti-viral agents [6]. All of these NP compounds were subjected into computational ADME-Tox screening test to verify their molecular properties (e.g., molecular weight, hydrogen bond donor/acceptor, logP, TPSA, and a number of rotatable bonds), toxicity properties (e.g., mutagenic, tumorigenic, irritant, and reproductive effective), and druglikeness. In this research, both MOE 2014.09 and DataWarrior v4.7.2 software were utilized to perform the computational ADME-Tox screening test, any compounds that have druglikeness score lower than 0, have any toxicity properties or not sufficient enough to be absorbed through oral administration according to Lipinski's Rule of Five (RO5) were eliminated from this study. Conse-quently, only 24,137 NP compounds passed these screening tests were remained and prepared for the next phase.

The remaining NP ligands and 18β-glycyrrhetinic acid were prepared to further on, by minimizing and optimizing their molecular structure by using MOE 2014.09 soft-ware. The default protocols of 'Wash,' 'Partial Charge,' and 'Energy Minimize' were selected and conducted, while the 'MMFF94x' forcefield in the 'Gas Phase' solvation was chosen as the main parameters, along with 'RMS gradient' of 0.001 kcal/mol.Å. Also, the molecular chirality of corresponding compounds was kept and conserved by clicking 'Preserve Existing Chirality' box during the minimized process occurred within the 'Energy Minimize' feature.

4.2   Preparation and Optimization of EBOV NP 3D Structure

In this study, the 3D structure of EBOV NP was represented through the PDB ID: 4Z9P, which has been previously studied [3]. It consists of 309 amino acid residues that lie from Ala35 to Ala310. At first, the EBOV NP structure that retrieved previously from RCSB-PDB database was prepared first by removing any water molecules from the protein structure, followed by performing the default protocol of 'LigX' on MOE 2014.09 software, with 'Amber10:EHT' forcefield and 'Gas Phase' solvation were selected with the Tether value of 100,000. This process was mandatory in this study to add any missing hydrogen atoms that are not contained in the initial protein structure. Finally, the optimized EBOV NP structure was saved in .moe file format.

4.3   Pharmacophore Generation, Validation, and Database Screening

In this study, the 3D structure of EBOV NP was represented through the PDB ID: 4Z9P, which has been previously studied. The binding site of EBOV NP, known as an RNA-binding groove, holds an important role in EBOV proliferation process in the host cells and viral RNA encapsidation [16]. This site contains six positive-charged, highly-conserved residues, namely Lys160, Lys171, Gln238, Lys248, Arg298, and His310. Thus, the inhibition of this site may play an important part to prevent the proliferation process of EBOV. In this study, the pharmacophore-based screening was performed to identify the NP compounds that specifically bind with the important residues of EBOV NP RNA-binding groove. First, 18β-glycyrrhetinic acid was subjected to induced-fit based molecular docking simulations to recognize the binding affinity, along with the molecular interactions during the formation of EBOV NP-18β-glycyrrhetinic acid complex. The docking simulation itself was conducted by selecting 'Triangle Matcher – London dG' as the 'Placement' method, along with 'Forcefield – GBVI/WSA dG' as the 'Refinement' method, with the retain value of 300 and 10, respectively. Furthermore, the EBOV NP-18β-glycyrrhetinic acid complex that has the lowest $\Delta G_{binding}$ was selected. EBOV NP-18β-glycyrrhetinic acid complex that has a $\Delta G_{binding}$ value of -6.2489 kcal/mol was further analyzed. Moreover, the 18β-glycyrrhetinic acid was also observed to form hydrogen bond interactions with Lys160, Lys171, and Arg174 within the carboxylate group of the respective ligand, as it is shown in Fig. 2A. Hence, the pharmacophore query of EBOV NP was built according to the final pose of EBOV NP-18β-glycyrrhetinic acid, which consisted of five pharmacophore features; four hydrophobic regions and one anion, lone pairs, and H-bond acceptors region, which is resembled by the carboxylate group of 18β-glycyrrhetinic acid. The pharmacophore features of EBOV NP in its RNA-binding groove is displayed in Fig. 2b.

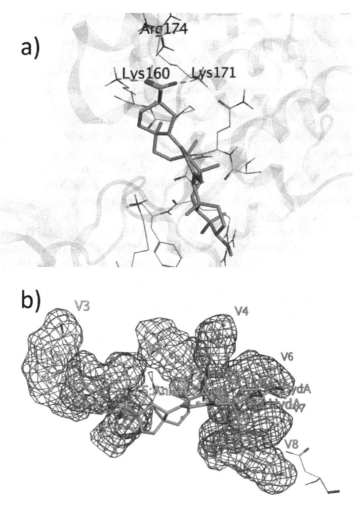

Fig. 2. (a) Molecular interaction of 18β-glycyrrhetinic acid in the RNA-binding groove of EBOV NP (b) Pharmacophore feature of EBOV NP

After the pharmacophore feature had been constructed, the pharmacophore-based virtual screening was performed on the remaining 24,137 NP compounds that have been screening previously against the RNA-binding groove of EBOV NP using 'Pharmacophore Search' feature on MOE 2014.09 software. In this step, the pharmacophore feature that obtained from the prior step was used as the pharmacophore template, with the default protocol was selected during the screening process. Only 898 NP compounds that matched with the preferred pharmacophore features of EBOV NP in the RNA-binding groove. Thus, these compounds were kept and saved in .mdb file format to be prepared further on for the next pharmacophore-based molecular docking simulations.

4.4   Pharmacophore-Based Molecular Docking Simulation

About 898 NP compounds from previous pharmacophore-based virtual screening underwent the series of molecular docking simulations, which include the rigid molecular docking and flexible molecular docking simulation. The parameters from the previous simulation were selected, with the exception of 'Pharmacophore – London dG' parameter as the 'Placement' method instead of 'Triangle Matcher – London dG.' In general, the docking simulations were conducted four times; the first two by rigid molecular docking simulations and followed by two-times flexible molecular docking simulations. Within each simulation, the ligands which have RMSD value lower than 2.0 Å and higher $\Delta G_{binding}$ value than 18β-glycyrrhetinic acid would be omitted from this study. Additionally, the molecular interactions between the NP compound in the RNA-binding groove of EBOV NP were also observed as well in the result analysis of final flexible docking simulation. In the end, only five ligands that considered as the potential EBOV NP inhibitors, according to their $\Delta G_{binding}$ and RMSD value. They were UNPD213871, UNPD199951, UNPD124962, UNPD139843, and UNPD147202, as it displayed in Table 1 and Fig. 3. From the final docking results, the UNPD213871 ligand was deemed as the most potent inhibitor among all, $\Delta G_{binding}$ value wise, with -9.1024 kcal/mol. Followed by UNPD199951, UNPD124962, UNPD139843, and UNPD147202, with $\Delta G_{binding}$ value of -7.6399, -7.5248, -7.1958, and -6.8425 kcal/mol, respectively. These results showed that all five NP compounds have lower $\Delta G_{binding}$ value than 18β-glycyrrhetinic acid, which predicted to have $\Delta G_{binding}$ value of -6.2489 kcal/mol.

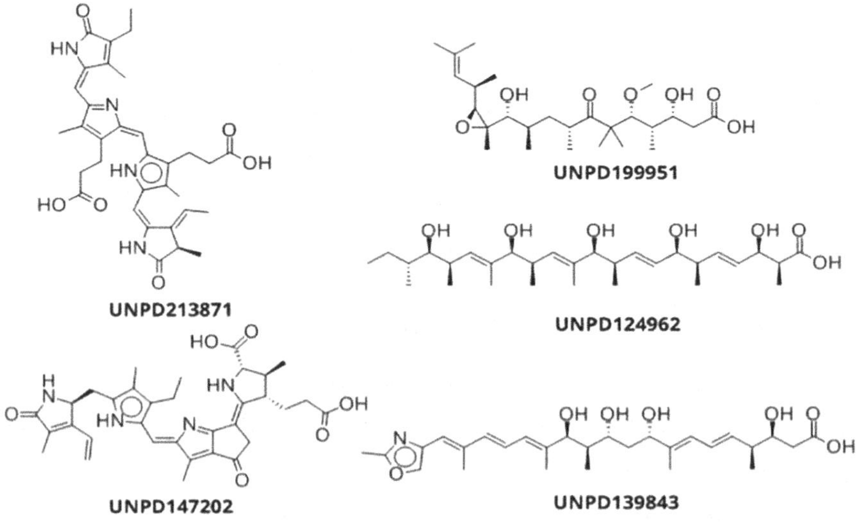

Fig. 3. The chemical structure of five best natural product compounds to inhibit EBOV NP

Table 1. List of the best five natural product and the standard compounds to inhibit EBOV NP in the RNA-binding groove, alongside with their $\Delta G_{binding}$ and RMSD values

No	Ligand Name	$\Delta G_{binding}$ (kcal/mol)	RMSD (Å)	Binding Interactions
1	UNPD213871	-9.1024	0.9947	Lys160, Lys171, Lys171, Arg174, Arg174, Gly243
2	UNPD199951	-7.6399	0.7270	Lys160, Lys171, Arg174, Arg174, Lys248, Lys160
3	UNPD124962	-7.5248	1.0207	Lys160, Lys171, Gln238, Arg174
4	UNPD139843	-7.1958	0.5515	Arg174, Gln238, Lys248, Lys160, Lys171, Lys171
5	UNPD147202	-6.8425	0.6746	Lys160, Lys171, Arg174, Lys248, Arg174, Lys248
S1	18β-glycyrrhetinic acid	-6.2489	-	Lys160, Lys171, Arg174, Arg174

Note: Blue color in the 'Hydrogen Bond Interactions' marks the important amino acid residue in RNA-binding groove of EBOV NP, the underlined residues mark the ionic interaction, and the italic residues mark the cationic pi-pi Interaction.

According to the docking simulation result from Table 1, all of the five best ligands interacted with Lys160, Lys171, and Arg174, forming either a hydrogen bond or ionic bond, depends on each ligand pose that generated during the formation of ligand-EBOV NP complex. Additionally, except UNPD213871, all best NP ligands were also formed an additional significant interaction with one of the essential RNA-binding residues, either Gln238 or Lys248. Thus, their inhibition activity should be taken into consideration even though the $\Delta G_{binding}$ value of those ligands were not as low as UNPD213871. The in-depth analysis of the molecular interactions that formed during the docking simulations will be explained further in the next subsection.

4.5   Binding Mode Interaction Analysis from Docking Simulation

EBOV NP-UNPD213871

UNPD213871, as it previously mentioned, was the best NP ligand that interacted with EBOV NP in the RNA-binding groove, in terms of $\Delta G_{binding}$ value. However, the $\Delta G_{binding}$ value that formed during the docking simulation was not the sole factor to determine the best ligand of all; the intermolecular interaction which created during the simulation was identified as well. According to the docking results, which depicted in Fig. 4, the UNPD213871 ligand formed five hydrogen bonds with the EBOV NP. In details, both Lys171 and Arg174 residues interacted with UNPD213871 as hydrogen bond donor, which formed between the amine functional groups (-NH$_2$) of the corresponded residues with the carboxylate functional groups (-COOH) of the ligand. Moreover, the carbonyl functional group (C = O), which occurred in the ligand in the form of γ-lactam functional group, was interacted with the hydrogen atom in the backbone of Gly243 (-NH) through hydrogen bond interaction as well. Additionally, other amino acid residues such as Pro159, Lys160, Val162, Val164, Gln238, Arg240, Phe241, Ser242, Leu245, Lys248, and Asp252 also interacted with the UNPD213871 ligand through van der Waals interactions.

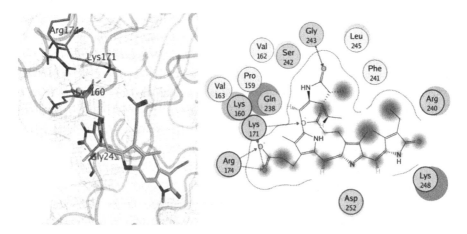

Fig. 4. The binding mode interaction of UNPD213871 in the RNA-binding groove of EBOV NP in two-dimensional (left) and three-dimensional (right) figures

EBOV NP-UNPD199951

UNPD199951 have five hydrogen bond interactions, one ionic interaction, and seven residues that interacted through van der Waals interactions, according to the docking result. Similar to the UNPD213871, the amine functional groups ($-NH_2$) in the side chain of Lys160, Lys171, and Arg174 residues interacted with the carboxylate functional group (-COOH) in the ligand. Moreover, the hydroxyl group (-OH) in the ligand acted as hydrogen bond acceptor of Lys248, which interacted via its hydrogen atom in its amine functional group. Interestingly, Lys160 formed ionic interaction with the carboxylate functional group of the ligand as well, which may happen since Lys160, and the ligand existed in the human body in the form of their cationic ($-NH_3^+$) and anionic ($-COO^-$) forms. The molecular interaction during the formation of the EBOV NP-UNPD199951 complex is depicted in Fig. 5.

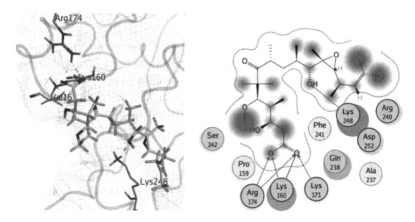

Fig. 5. The binding mode interaction of UNPD199951 in the RNA-binding groove of EBOV NP in two-dimensional (left) and three-dimensional (right) figures

EBOV NP-UNPD124962

The third-best scored NP compound, UNPD124962, according to the docking simulation result that depicted in Fig. 6, formed four hydrogen bond interactions (Lys160, Lys171, Gln238, and Lys248), one ionic interaction (Arg174), and seven van der Waals interaction (Pro159, Arg240, Phe241, Leu245, Thr249, Asp252, and His253) in the RNA-binding groove of EBOV NP. Similar with other two NP ligands, the carboxylate functional group in the ligand was responsible for the formation of hydrogen bond interactions between the ligand and both Lys160 and Lys171, while this functional group, since it exists in ionic form, may also form ionic interaction as well, which was shown in the formation of ionic interaction between the respective ligand and Arg174. Moreover, the hydroxyl functional groups in the ligand were responsible for the formation of hydrogen bond interactions of Gln238 and Lys248.

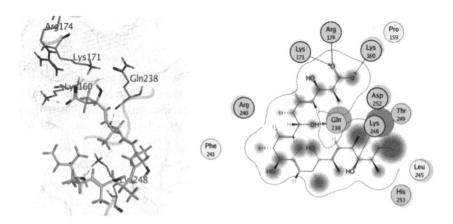

Fig. 6. The binding mode interaction of UNPD124962 in the RNA-binding groove of EBOV NP in two-dimensional (left) and three-dimensional (right) figures

EBOV NP-UNPD139843

The intermolecular interaction of UNPD139843 ligand in the binding site of EBOV NP was relatively different from the other four NP ligands. Despite this ligand interacted with Lys160, Lys171, and Arg174, followed by Gln238 and Lys248, which almost rather the same as other ligands, the mechanism of interaction of these ligands were quite different. In general, Lys160 and Lys171 interacted with this ligand via ionic interactions, which may happen since these ligands had amine functional group in the form of their cationic form ($-NH_3^+$) and formed the ionic interactions with the carboxylate functional group in the anionic form ($-COO^-$). Besides, Arg174, Gln238, and Lys248 interacted with UNPD139843 ligand through hydrogen bond interactions, and seven other amino acid residues interacted through van der Waals interaction. The molecular interactions of EBOV NP-UNPD139843 complex are displayed in Fig. 7.

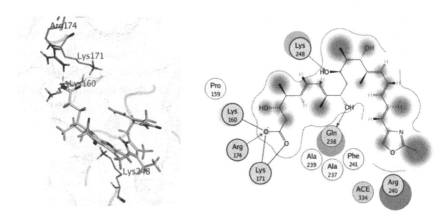

Fig. 7. The binding mode interaction of UNPD139843 in the RNA-binding groove of EBOV NP in two-dimensional (left) and three-dimensional (right) figures

EBOV NP-UNPD147202

UNPD147202, surprisingly, has one of the most interesting ligand-EBOV NP complex interaction, as it displays in Fig. 8, despite being the ligand with the highest $\Delta G_{binding}$ value. This ligand directly interacted with nine amino acid residues; forming H-bond interaction with Lys160, Lys171, Arg174, and Lys248, ionic interaction with Arg174, pi-pi interaction with Lys248, and van der Waals interaction with Gln238, Arg240, Phe241, and Asp252. Thus, this compound may be considered to be one of the most potent inhibitors among all NP ligands, although its complex stability should be determined further through molecular dynamics simulations.

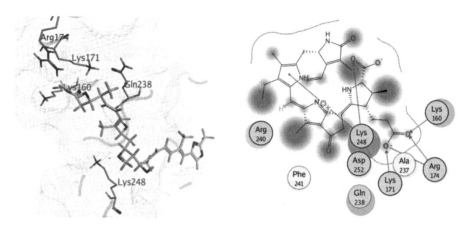

Fig. 8. The binding mode interaction of UNPD147202 in the RNA-binding groove of EBOV NP in two-dimensional (left) and three-dimensional (right) figures

5   Conclusions

The computational methods have become a significant tool to support the identification of the potential compounds in the early stage of drug design and development. In this study, the effectiveness of the EBOV NP as the potential antiviral target was evaluated using a series of the computational ADME-Tox screening test, pharmacophore-based virtual screening, and molecular docking simulations. After the computational methods were performed throughout the research, it was revealed that five natural product compounds could be considered as a potential drug candidate for EBOV targeting NP in its RNA-binding groove. Despite having decent inhibition activities, according to their molecular interactions and $\Delta G_{binding}$ energy, the initial computational ADME-Tox screening test also discovered that all five ligands had good druglikeness and possessed no toxicity properties. Thus, these compounds should be investigated further through additional in silico experiments, such as pharmacological predictions and molecular dynamics simulations, to predict their bioactivities, health effect, and toxicity properties, as well as their complex stability under real environment.

Acknowledgements. The research and the publication costs are entirely funded by Kementerian Riset, Teknologi, and Pendidikan Tinggi (Kemenristekdikti), Republic of Indonesia, through Hibah Penelitian Kerjasama Luar Negeri Tahun 2018 No: 538/UN2.R3.1/HKP.05.00/2018. All authors were responsible equally for writing the manuscript, conducting the experiments, as well as approving the final version of the manuscript. Finally, none conflict of interest is declared.

References

1. Dowall, S.D., et al.: Antiviral screening of multiple compounds against Ebola virus. Viruses **8**, 1–17 (2016)
2. Sanchez, A., Kiley, M.P., Holloway, B.P., McCormick, J.B., Auperin, D.D.: The nucleoprotein gene of Ebola virus: Cloning, sequencing, and in vitro expression. Virology **170**, 81–91 (1989)
3. Dong, S., et al.: Insight into the Ebola virus nucleocapsid assembly mechanism: crystal structure of Ebola virus nucleoprotein core domain at 1.8 Å resolution. Protein Cell **6**, 351–362 (2015)
4. Lin, L.T., Hsu, W.C., Lin, C.C.: Antiviral natural products and herbal medicines. J. Tradit. Complement. Med. **4**, 24–35 (2014)
5. Von Rintelen, K., Arida, E., Häuser, C.: A review of biodiversity-related issues and challenges in megadiverse Indonesia and other Southeast Asian countries. Res. Ideas Outcomes **3**, 1–16 (2017)
6. Orbell, J., Coulepis, T.: Medicinal plants of Indonesia. Asia Pac. Biotech News **11**, 726–743 (2007)
7. Harvey, A.L.: Natural products in drug discovery. Drug Discov. Today **13**, 894–901 (2008)
8. Yang, S.-Y.: Pharmacophore modeling and applications in drug discovery: challenges and recent advances. Drug Discov. Today **15**, 444–450 (2010)
9. Young, D.C.: Computational Drug Design: A Guide for Computational and Medicinal Chemists (2009)
10. Rizki, I.F., Nasution, M.A.F., Siregar, S., Ekawati, M.M., Tambunan, U.S.F.: Screening of Sonic Hedgehog (Shh) inhibitors in the Hedgehog signaling pathway from Traditional Chinese Medicine (TCM) database through structure-based pharmacophore design. In: Zhang, F., Cai, Z., Skums, P., Zhang, S. (eds.) ISBRA 2018. LNCS, vol. 10847, pp. 179–184. Springer, Cham (2018). https://doi.org/10.1007/978-3-319-94968-0_16
11. Nasution, M.A.F., Toepak, E.P., Tambunan, U.S.F.: Flexible docking-based molecular dynamics simulation of natural product compounds and Ebola virus Nucleocapsid (EBOV NP): a computational approach to discover new drug for combating Ebola 19, 2011 (2018)
12. Sander, T., Freyss, J., Von Korff, M., Rufener, C.: DataWarrior: an open-source program for chemistry aware data visualization and analysis. J. Chem. Inf. Model. **55**, 460–473 (2015)
13. Vilar, S., Cozza, G., Moro, S.: Medicinal chemistry and the molecular operating environment (MOE): application of QSAR and molecular docking to drug discovery. Curr. Top. Med. Chem. **8**, 1555–1572 (2008)
14. Rose, P.W., et al.: The RCSB protein data bank: integrative view of protein, gene and 3D structural information. Nucl. Acids Res. gkw1000 (2016)
15. Gu, J., Gui, Y., Chen, L., Yuan, G., Lu, H.Z., Xu, X.: Use of natural products as chemical library for drug discovery and network pharmacology. PLoS ONE **8**, 1–10 (2013)
16. Fu, X., et al.: Novel chemical ligands to Ebola virus and Marburg virus nucleoproteins identified by combining affinity mass spectrometry and metabolomics approaches. Sci. Rep. **6**, 29680 (2016)

Global Sensitivity Analysis
of Constraint-Based Metabolic Models

Chiara Damiani[1] ![ORCID], Dario Pescini[3,4(✉)] ![ORCID], and Marco S. Nobile[2,3] ![ORCID]

[1] Department of Biotechnology and Biosciences,
University of Milano-Bicocca, 20126 Milan, Italy
[2] Department of Informatics, Systems and Communication,
University of Milano-Bicocca, 20126 Milan, Italy
nobile@disco.unimib.it
[3] SYSBIO.IT Centre of Systems Biology, 20126 Milan, Italy
[4] Department of Statistics and Quantitative Methods, University of Milano-Bicocca,
20126 Milan, Italy
dario.pescini@unimib.it

Abstract. In the latter years, detailed genome-wide metabolic models have been proposed, paving the way to thorough investigations of the connection between genotype and phenotype in human cells. Nevertheless, classic modeling and dynamic simulation approaches—based either on differential equations integration, Markov chains or hybrid methods—are still unfeasible on genome-wide models due to the lack of detailed information about kinetic parameters and initial molecular amounts. By relying on a steady-state assumption and constraints on extracellular fluxes, constraint-based modeling provides an alternative means—computationally less expensive than dynamic simulation—for the investigation of genome-wide biochemical models. Still, the predictions provided by constraint-based analysis methods (e.g., flux balance analysis) are strongly dependent on the choice of flux boundaries. To contain possible errors induced by erroneous boundary choices, a rational approach suggests to focus on the pivotal ones. In this work we propose a novel methodology for the automatic identification of the key fluxes in large-scale constraint-based models, exploiting variance-based sensitivity analysis and distributing the computation on massively multi-core architectures. We show a proof-of-concept of our approach on core models of relatively small size (up to 314 reactions and 256 chemical species), highlighting the computational challenges.

Keywords: Flux Balance Analysis · Constraint-Based Modeling ·
Global sensitivity analysis · MPI · Linear Programming

1 Scientific Background

Computational models of metabolism are increasingly being reconstructed and simulated with the aim of connecting genotype with phenotype. In view of their

© Springer Nature Switzerland AG 2020
M. Raposo et al. (Eds.): CIBB 2018, LNBI 11925, pp. 179–186, 2020.
https://doi.org/10.1007/978-3-030-34585-3_16

potential in unraveling the fragility points of complex pathological diseases in which a rearrangement of metabolism plays an essential role, such as cancer, diabetes, or neurodegenerative disorders, as well as of their extensive applications within metabolic engineering to increase the cells' production of a certain substance, genome-wide metabolic models have been reconstructed for many organisms, spanning from prokaryotes to *Homo sapiens*. These reconstructions include virtually all the reactions that can be catalyzed by the enzymes encoded by a given genome. Despite the recent advancements in dynamic simulation [1], large-scale biochemical models can still be challenging as kinetic parameters of rate laws are largely undetermined. For instance, the human metabolic network includes more than 7.000 reactions [2]. For this reason, these networks are typically simulated by means of constraint-based methods, and in particular of Flux Balance Analysis (FBA), as reviewed in [3]. Although constraint-based models are not appropriate for the simulation of molecular networks in general, they are well suited for metabolic networks, as the concentration of intracellular metabolites in time can be reasonably approximated as constant. Despite neglecting information on transient dynamics, these approaches allow the identification of key features of metabolism such as growth yield, network robustness, and gene essentiality. In the case of unicellular organisms, steady-state extracellular fluxes (e.g., consumption rate of carbon and nitrogen) can be promptly derived from chemostat experiments.

Given such constraints, FBA has proven able to correctly predict the expected growth yield [4]. In the case of human metabolism, it is more difficult to properly constrain the FBA solution, especially when the aim is to investigate the metabolic program of human cells *in vivo*. The latest curated version of human metabolic network Recon 2.2 [2] considers 747 nutrients that can be exchanged with the environment, of which 701 can be taken up by the cell. If we let the influx of all these metabolites unbound, the resulting phenotype could be very different from biological reality. On the other hand it is unrealistic to provide a correct bound for each of them. It is therefore of paramount importance to be able to assess the relative influence of these boundaries on the model outputs, not only to identify the extracellular fluxes that should be tightly constrained, but also to understand which are the inputs that most correlate with output. In the case of cancer metabolism, this would allow to investigate which nutrients mostly affect cancer growth, providing indications for treatment.

Ranking model parameters according to their contribution is the typical aim of sensitivity analysis (SA), which computes a sensitivity coefficient for each parameter. The most widely used approaches to SA are defined as local, because they assess the variability of one parameter at the time within a neighborhood. These methods cannot be applied to genome-wide metabolic models, unless a baseline parametrization is provided, according to experimental measures. Because estimation of extracellular fluxes for each possible metabolite is hardly available, global methods that explore the entire parameter space are required. At this aim, variance-based methods have been proposed to measure the output variance explained by each input, which are also able to explore the effect of interactions among parameters [5]. A few applications of variance-based methods on small dynamic models have been proposed (e.g., [4]). Conversely, to our knowledge, no attempt has been made to assess the sensitivity of the inputs of constraint-based models. At this purpose, we here propose a workflow to be extended to any constraint-based model. As a proof of principle, we show a preliminary application of the approach to reduced model of human metabolism, which focus on central carbon metabolism. Specifically, we leverage a Sobol variance-based sensitivity analysis [5,6], coupled to the Saltelli's scheme for low discrepancy random sampling [7].

2 Materials and Methods

In this section we summarize the methods exploited in this work: Constraint-Based Modeling (CBM) of biochemical systems and variance-based sensitivity analysis.

2.1 Constraint-Based Modeling

The starting point of a CBM is the stoichiometric matrix S, i.e., a matrix defined by the stoichiometric values of the reactants and products of each reaction in the metabolic network: each row corresponds to a metabolite, while columns correspond to reactions. In Fig. 1 an example of formal specification of a toy metabolic model and of the steady-state assumption is provided. Because metabolic networks typically include more reactions (hence fluxes) than metabolites, the stoichiometric constraints and the steady-state assumption alone lead to an underdetermined system in which a bounded solution space of all feasible flux distributions can be identified. Additional constraints (e.g., reactions' irreversibility) should be incorporated to further restrict the solution space, by specifying the maximum and minimum values of the flux through any given reaction. Capacity constraints are generally set according to experimental data. On top of CBM, FBA [8] allows to determine an optimal flux distribution within the obtained feasible solution space with respect to a given objective function, modeling all assumptions about cell's behavior with respect to a specified natural objective. Since all constraints, boundaries, and the objective function itself are linear, Linear Programming methods can be leveraged to determine the optimal fluxes solution. Given the vector v_{min} and v_{max} specifying the lower and upper bound

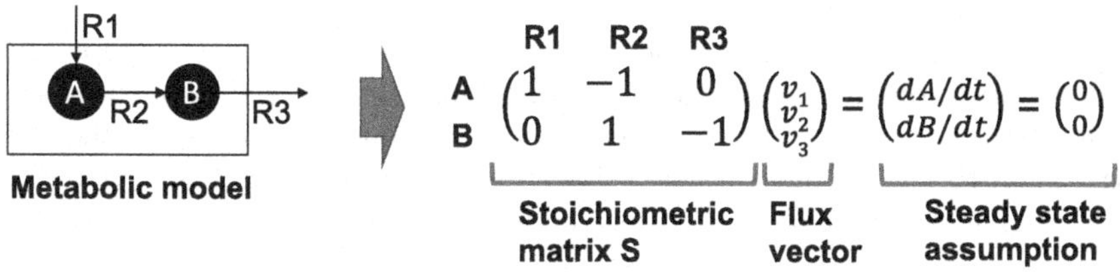

Fig. 1. Example of the steady-state assumption for a toy metabolic network, represented as a stoichiometric matrix S.

respectively for each flux v_i of the vector of fluxes $\boldsymbol{v}$, the linear programming problem is postulated as follows:

$$\text{maximise} \sum w_i v_i$$

$$\text{subject to } Sv = \mathbf{0}, v_{min} \leq v \leq v_{max}, \tag{1}$$

where w_i is the objective coefficient of flux v_i. Differently from mechanistic models, where the detailed temporal behavior of each metabolite is explicitly calculated by means of ODEs or Markov chains [3], FBA allows to efficiently determine the optimal fluxes configuration at steady-state. Thus, FBA is computationally less expensive than classic simulation—since the latter usually requires advanced implicit integration methods, relying on the systematic evaluation of massive Jacobian matrices associated to the systems of coupled differential equations, or the explicit calculation of the effect of every single reaction in the case of stochastic simulation—paving the way to the analysis of metabolism at the genome-wide scale.

2.2 Sensitivity Analysis

SA investigates how the uncertainty in some output of a given mathematical model can be apportioned to different sources of uncertainty in its inputs [9]. The outcome of a SA run is generally a ranking of the sensitivity coefficients associated to the varied inputs [10,11]. Many SA methods can be employed for the computational investigation of biological models (e.g., Morris' elementary-effects [12,13], variance-based sensitivity [5,7], derivative-based sensitivity [14]).

In this work we are considering how the uncertainty of flux boundaries on a FBA model affects the objective function, in this case the production of biomass. Although SA represents a powerful means to understand the model's behavior, a proper investigation of the multi-dimensional boundaries space implies a combinatorial explosion of configurations to be tested and optimized. However, all linear programming optimizations are mutually independent, so that huge computational effort can be mitigated by means of a parallel or distributed architecture.

The SA is performed with Sobol's variance-based method [5], as implemented by the SALib library [15]. Given V variables in the system, the method calculates the sensitivity indexes using on $N \times (V \times 2 + 2)$ independent random samples of the parameters space determined using Saltelli's approach [7], where N is a user-defined number of samples.

We here compute the first order index S_i (as per Eq. 2), which represents the main effect contribution of the input factor X_i to the variance of the output Y, and the total effect index S_{T_i} (as per Eq. 3), which accounts for the total contribution to the output variation due to each factor, i.e its first-order effect plus all higher-order effects due to interactions [16].

$$S_{T_i} = \frac{Var[E(Y|x_i)]}{Var(Y)} \tag{2}$$

$$S_{T_i} = \frac{Var[E(Y|x_{-i})]}{Var(Y)} \tag{3}$$

3 Results

As a preliminary test, we performed the SA on the simple model ENGRO1 (characterized by 84 reactions and 67 metabolites) [17], which was designed to evaluate the contribution of glucose and glutamine to biomass formation in cancer cells, and for which extensive scan of the few network parameters and investigation of model dynamics was previously reported [17]. Metabolite uptake is therefore allowed for glutamine, glucose and oxygen only. Although, in order to achieve mass balance, uptake reactions are included also for metabolites whose synthesis is not accounted in the network (tetrahydrofolate and methionine) and for arginine, which is required for ornithine production, the utilization of the carbon of this substrates is not allowed by ENGRO1 stoichiometry, thus their consumption will directly correlate to the growth capabilities of the network. The number of variables is thus $V = 6$ and we assumed $N = 2500$, corresponding to a total of 65 000 independent FBA runs.

Figure 2 shows the results of a SA performed on the ENGRO1 core model: on the top we report the sorted first-order sensitivity indices, while on the bottom we report the sorted total sensitivity indices.

The sensitivity indices portray a picture that is consistent with the results reported in [17], yet of way more immediate interpretation. According to previous results, it is indeed apparent in Fig. 2 that oxygen is by far the most sensitive nutrient, whereas glutamine and glucose show similar values for the sensitivity index, with the former slightly less sensitive than the latter. Worth of note, the sensitivity of glutamine equals that of glucose when the total effect is considered.

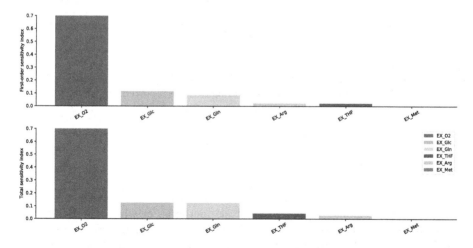

Fig. 2. Results of the SA on the ENGRO1 model. Both the first-order sensitivity index (top panel) and the total effect index (bottom figure) are reported.

We then applied the SA methodology to a less characterized model: the HMRcore model introduced in [18] and updated in [19–21]. Although the number of intake reactions is still limited, the model is more complex than ENGRO1, as it takes into account the compartmentalization of nutrients. Two compartments are modeled: mitochondria and cytoplasm. The model includes 314 reactions and 256 metabolites. The number of variables for which we assessed the sensitivity in this case is $V = 12$ and we assumed $N = 2500$, corresponding to a total of 35 000 independent FBA runs. Figure 3 shows the results of the SA performed on the HMRcore model. According to our results, also for this case, the highest sensitivity is associated, not surprisingly, to oxygen. Interestingly, differently from the ENGRO1 model, the importance of glucose greatly outperforms that of glutamine. Another non trivial result is that the amino acids arginine ranks higher than glutamine. This result suggests that the role of the compartments and of the set of reactions considered in the model interfere with the favored nutrient and that arginine may represent a valide alternative to glutamine as a source of carbon and nitrogen.

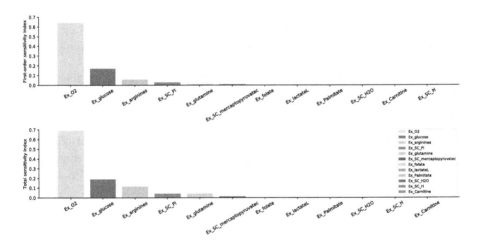

Fig. 3. Results of the SA on the HMRcore model. Both the first-order sensitivity index (top panel) and the total effect index (bottom figure) are reported.

4   Conclusion

In this work we presented a novel framework for the analysis of constraint-based models, based on global sensitivity analysis methods, in which the perturbed input variables are the uptake fluxes. We tested the methodology on two core models: the simple ENGRO model [17] and the more complex HMRcore model [18, 19].

Our results showed how the sensitivity indexes provide a straightforward ranking of the importance of the parameters under study, without the need to perform many laborious analyses. Moreover, we showed that, especially as the model complexity increases, results are not obvious, thus highlighting the importance of performing a sensitivity analysis of large models in order to identify those nutrients for which a careful setting of the linear programming problem constraints is required.

The number of uptake fluxes to be perturbed for the calculation of sensitivity indexes for the genome-wide model Recon2.2 [2] would be 701, so that approximately 3.5 million FBA runs would be necessary. According to our preliminary tests, a full sequential SA on the HMR model requires approximately 370 s to be completed on a workstation equipped with a CPU Intel i7-7700 with clock 2.81 GHz. However, since all FBA runs are mutually independent, the process can be accelerated by distributing the optimizations on multiple cores. We are currently implementing a parallel version of the algorithm based on MPI [22], in order to leverage massively multicore supercomputers and tackle genome-wide sensitivity analysis. Tier-0 supercomputers like Marconi, maintained by the Italian supercomputing consortium CINECA, provide access to several computing nodes. Specifically, the computing nodes belonging to Marconi's A2 partition are equipped with 68-cores Intel Xeon Phi 7250 accelerators, so that more than 240 000 cores could, in principle, be exploited to distribute the calculations and strongly reduce the overall running time. Even though CINECA imposes some limitations on the number of simultaneous computing nodes that can be requested—affecting the level of parallelism that can be actually achieved—we estimate that approximately 500 batches of 7000 parallel FBAs would be sufficient to perform the whole analysis. Since a single batch requires the same amount of time of a single FBA run (MPI communication overhead is negligible), the results of a genome-wide SA would be collected in less than 8 h.

References

1. Tangherloni, A., Nobile, M., Besozzi, D., Mauri, G., Cazzaniga, P.: LASSIE: simulating large-scale models of biochemical systems on GPUs. BMC Bioinform. **18**(1), 246 (2017)
2. Swainston, N., et al.: Recon 2.2: from reconstruction to model of human metabolism. Metabolomics **12**(7), 1–7 (2016)
3. Cazzaniga, P., et al.: Computational strategies for a system-level understanding of metabolism. Metabolites **4**, 1034–1087 (2014)
4. Bordbar, A., Monk, J.M., King, Z.A., Palsson, B.O.: Constraint-based models predict metabolic and associated cellular functions. Nat. Rev. Genet. **15**(2), 107 (2014)
5. Sobol, I.M.: Global sensitivity indices for nonlinear mathematical models and their Monte Carlo estimates. Math. Comput. Simul. **55**(1–3), 271–280 (2001)
6. Saltelli, A.: Making best use of model evaluations to compute sensitivity indices. Comput. Phys. Commun. **145**(2), 280–297 (2002)
7. Saltelli, A., Annoni, P., Azzini, I., Campolongo, F., Ratto, M., Tarantola, S.: Variance based sensitivity analysis of model output. Design and estimator for the total sensitivity index. Comput. Phys. Commun. **181**(2), 259–270 (2010)

8. Orth, J.D., Thiele, I., Palsson, B.Ø.: What is flux balance analysis? Nat. Biotechnol. **28**(3), 245–248 (2010)

9. Saltelli, A., Ratto, M., Tarantola, S., Campolongo, F.: Sensitivity analysis for chemical models. Chem. Rev. **105**, 2811–2827 (2005)

10. Damiani, C., Filisetti, A., Graudenzi, A., Lecca, P.: Parameter sensitivity analysis of stochastic models: application to catalytic reaction networks. Comput. Biol. Chem. **42**, 5–17 (2013)

11. Nobile, M.S., Mauri, G.: Accelerated analysis of biological parameters space using GPUs. In: Malyshkin, V. (ed.) PaCT 2017. LNCS, vol. 10421, pp. 70–81. Springer, Cham (2017). https://doi.org/10.1007/978-3-319-62932-2_6

12. Morris, M.D.: Factorial sampling plans for preliminary computational experiments. Technometrics **33**(2), 161–174 (1991)

13. Campolongo, F., Cariboni, J., Saltelli, A.: An effective screening design for sensitivity analysis of large models. Environ. Model. Softw. **22**(10), 1509–1518 (2007). Modelling, computer-assisted simulations, and mapping of dangerous phenomena for hazard assessment

14. Sobol, I.M., Kucherenko, S.: Derivative based global sensitivity measures and their link with global sensitivity indices. Math. Comput. Simul. **79**(10), 3009–3017 (2009)

15. Usher, W., Herman, J., Whealton, C., Hadka, D.: SALib/SALib: Launch! (2016)

16. Saltelli, A., et al.: Global Sensitivity Analysis: The Primer. Wiley, Hoboken (2008)

17. Damiani, C., et al.: A metabolic core model elucidates how enhanced utilization of glucose and glutamine, with enhanced glutamine-dependent lactate production, promotes cancer cell growth: the WarburQ effect. PLoS Comput. Biol. **13**(9), e1005758 (2017a)

18. Di Filippo, M., et al.: Zooming-in on cancer metabolic rewiring with tissue specific constraint-based models. Comput. Biol. Chem. **62**, 60–69 (2016)

19. Damiani, C., Di Filippo, M., Pescini, D., Maspero, D., Colombo, R., Mauri, G.: popFBA: tackling intratumour heterogeneity with flux balance analysis. Bioinformatics **3**(14), i311–i318 (2017)

20. Graudenzi, A., et al.: Integration of transcriptomic data and metabolic networks in cancer samples reveals highly significant prognostic power. J. Biomed. Inform. **87**, 37–49 (2018)

21. Damiani, C.: Integration of single-cell RNA-seq data into population models to characterize cancer metabolism. PLoS Comput. Biol. **15**(2), e1006733 (2019)

22. Gabriel, E., et al.: Open MPI: goals, concept, and design of a next generation MPI implementation. In: Kranzlmüller, D., Kacsuk, P., Dongarra, J. (eds.) EuroPVM/MPI 2004. LNCS, vol. 3241, pp. 97–104. Springer, Heidelberg (2004). https://doi.org/10.1007/978-3-540-30218-6_19

Efficient and Settings-Free Calibration of Detailed Kinetic Metabolic Models with Enzyme Isoforms Characterization

Niccolò Totis[1,2] , Andrea Tangherloni[3,4,5,6] , Marco Beccuti[1] ,
Paolo Cazzaniga[7,8] , Marco S. Nobile[3,8] , Daniela Besozzi[3] ,
Marzio Pennisi[9] , and Francesco Pappalardo[10(✉)]

[1] Department of Computer Science, University of Torino, Torino, Italy
beccuti@di.unito.it
[2] Bio- & Chemical Systems Technology, Reactor Engineering and Safety (CREaS),
KU Leuven, Leuven, Belgium
niccolo.totis@kuleuven.be
[3] Department of Informatics, Systems and Communication,
University of Milano-Bicocca, Milano, Italy
{marco.nobile,daniela.besozzi}@unimib.it
[4] Department of Haematology, University of Cambridge, Cambridge, UK
at860@cam.ac.uk
[5] Wellcome Trust Sanger Institute, Hinxton, UK
[6] Medical Research Council Cambridge Stem Cell Institute, Cambridge, UK
[7] Department of Human and Social Sciences, University of Bergamo, Bergamo, Italy
paolo.cazzaniga@unibg.it
[8] SYSBIO.IT Centre for Systems Biology, Milano, Italy
[9] Department of Mathematics and Computer Science, University of Catania,
Catania, Italy
mpennisi@dmi.unict.it
[10] Department of Drug Sciences, University of Catania, Catania, Italy
francesco.pappalardo@unict.it

Abstract. Mathematical modeling and computational analyses are essential tools to understand and gain novel insights on the functioning of complex biochemical systems. In the specific case of metabolic reaction networks, which are regulated by many other intracellular processes, various challenging problems hinder the definition of compact and fully calibrated mathematical models, as well as the execution of computationally efficient analyses of their emergent dynamics. These problems especially occur when the model explicitly takes into account the presence and

N. Totis and A. Tangherloni—Equal contribution.

Electronic supplementary material The online version of this chapter (https://doi.org/10.1007/978-3-030-34585-3_17) contains supplementary material, which is available to authorized users.

The original version of this chapter was revised: The name of the author "Andrea Tangherloni" was incorrect. The correction to this chapter is available at https://doi.org/10.1007/978-3-030-34585-3_31

the effect of different isoforms of metabolic enzymes. Since the kinetic characterization of the different isoforms is most of the times unavailable, Parameter Estimation (PE) procedures are typically required to properly calibrate the model. To address these issues, in this work we combine the descriptive power of Stochastic Symmetric Nets, a parametric and compact extension of the Petri Net formalism, with FST-PSO, an efficient and settings-free meta-heuristics for global optimization that is suitable for the PE problem. To prove the effectiveness of our modeling and calibration approach, we investigate here a large-scale kinetic model of human intracellular metabolism. To efficiently execute the large number of simulations required by PE, we exploit LASSIE, a deterministic simulator that offloads the calculations onto the cores of Graphics Processing Units, thus allowing a drastic reduction of the running time. Our results attest that estimating isoform-specific kinetic parameters allows to predict how the knock-down of specific enzyme isoforms affects the dynamic behavior of the metabolic network. Moreover, we show that, thanks to LASSIE, we achieved a speed-up of $\sim$30$\times$ with respect to the same analysis carried out on Central Processing Units.

Keywords: Metabolic reaction networks $\cdot$ GPU-powered simulations $\cdot$ Parameter Estimation

1   Introduction

Metabolic networks are inherently complex systems characterized by thousands of reactions and metabolites, regulated by a number of other intra- and extracellular processes [4,26]. Kinetic models of metabolism exploit mechanistic biological information to provide a detailed description of the reaction network, as well as to simulate or predict its dynamic behavior in different conditions. To this aim, both the structure of the network (i.e., how metabolites interact with enzymes, operating either as reactants, products, inhibitors or activators in each reaction) and the kinetic parameters need to be properly defined. Unfortunately, due to the high costs and complexity of experimental procedures, these data are scarce or incomplete, and the resulting model is usually undetermined [8]. In addition, the modeling task is further complicated by the fact that the flux of metabolites through the network strongly depends on many interrelated processes like transcription, translation, post-translational modifications, and allosteric control [26]. As a matter of fact, many sources of indetermination can affect models of metabolic systems, such as uncharacterized enzyme isoform mixtures [37].

Enzyme isoforms, also called isozymes, are structurally similar but nonidentical protein complexes that are able to catalyze the same biochemical reaction. Every cell type presents specific mixtures of isoforms of its metabolic enzymes and, in order to quantify their abundance, experimental techniques (e.g., proteomics analyses) can be used [42]. The structural differences of these proteins result in different kinetic behaviors of the catalytic process [33,45]. However, the main databases that collect information about enzyme kinetics and the associated parameters, such as SABIO-RK [44] and BRENDA [36], generally lack details about the differences among isozymes.

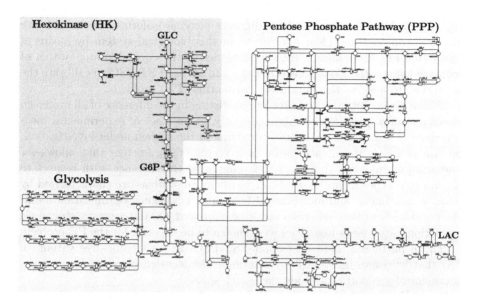

Fig. 1. SSN model of human intracellular core metabolic pathways (hexokinase (green), glycolysis (pink), and pentose phosphate pathway (yellow)) in a red blood cell. GLC: glucose; G6P: glucose-6-phosphate; LAC: lactate. (Color figure online)

The knowledge so far available in the literature thus prevents modelers from formulating a precise mathematical representation of the kinetic differences of isozymes. As a result, the vast majority of published metabolic models do not explicitly take these kinetic differences into account [37]. Instead, the average behavior of the whole isoform mixture is reproduced by using a single set of kinetic parameters, either taken from the literature or inferred through global optimization methods [29]. The main drawback of this assumption emerges, in particular, when the two following conditions superimpose: (i) the isozymes of the modeled metabolic system display significant kinetic variations; (ii) in the modeled experimental condition the abundance of isozymes differs from the condition in which the kinetic parameters of the whole mixtures were originally estimated. These issues should be carefully taken into consideration since the expression levels of metabolic enzymes change due to genetic knock-downs, therapeutic interventions, as well as various environmental stimuli [26]; therefore, the investigations of metabolic systems in perturbed conditions would highly benefit from the explicit representation of enzyme isoforms.

In this work, we present a computational approach that integrates the descriptive power of a parametric and compact extension of Petri Nets, called Stochastic Symmetric Nets (SSNs) [11], with Fuzzy Self-Tuning Particle Swarm Optimization (FST-PSO) [28], an efficient and settings-free meta-heuristics suitable for solving global optimization problems. As a case study, we apply our approach to investigate a large-scale kinetic model of the central energy-producing pathways in human cells, in which different enzyme isoforms are defined. We provide a compact and clear description of this biological system by means of a SSN (see Fig. 1)—which is then used to derive the corresponding system of Ordinary Differential Equations (ODEs)—and we use FST-PSO to calibrate the model, i.e., to address the Parameter Estimation (PE) problem.

PE is a global optimization problem consisting in the inference of all unknown parameters of a given model. More precisely, given a set of experimental measurements of some chemical species occurring in the system under investigation, the aim of PE is the identification of a vector of parameters that allows for generating a simulated dynamics that minimizes the distance with respect to the available experimental data. Since biochemical systems are characterized by complex, non-linear, and multi-modal behaviors, traditional optimization methods (e.g., Gradient Descent [15]) cannot be employed. On the contrary, stochastic population-based meta-heuristics were shown to be effective for the PE problem [14,29]; in particular, here we exploit FST-PSO [28], a settings-free variant of PSO that leverages fuzzy logic to determine the most effective settings for the movement of the particles inside the search space.

Population-based optimization algorithms, like FST-PSO, are iterative processes requiring the execution of a large number of fitness evaluations. In the case of PE, every fitness evaluation is generally based on a simulation of the system dynamics with a given model parameterization. This task can be computationally challenging for large-scale models; however, the simulation of mathematical models can be efficiently parallelized by using High-Performance Computing infrastructures [31]. Here, we take advantage of LASSIE [39], a deterministic simulator capable of offloading onto the Graphics Processing Unit (GPU) all the calculations necessary to numerically integrate a system of ODEs. As such, LASSIE can drastically reduce the running time to execute high numbers of simulations.

This work is structured as follows. In Sect. 2, we introduce the SSN formalism, the FST-PSO algorithm to tackle the PE problem, the main characteristics of LASSIE, and the intracellular metabolic model employed as a case study to show the applicability and effectiveness of our approach. In Sect. 3, we show the outcome of the PE analysis, and a comparison of the computational performance of LASSIE with respect to LSODA [34], a state-of-the-art ODE solver running on CPU. Finally, in Sect. 4, we provide some final remarks and directions for future works.

2   Materials and Methods

In this section, we introduce the SSN formalism, suitable for the description of large-scale kinetic models; we present the parameter estimation problem and the GPU-powered strategy leveraged to reduce the computational effort of the whole methodology. Finally, we describe the intracellular human metabolic network, used as a case study to assess the effectiveness of our approach.

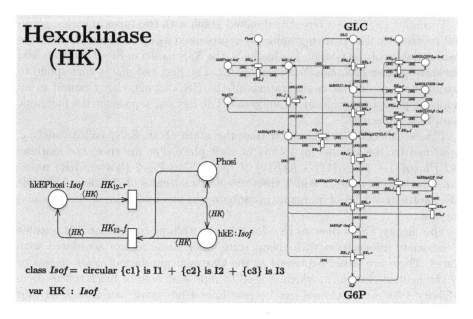

Fig. 2. Hexokinase (HK) submodel. On the left, an expanded portion of the submodel, used to present an example of the nomenclature and visualization of SSN components. At the bottom of the figure, the formal definitions of a color class and a color variable are shown. For the model presented in this work, one color class, called *Isof*, further partitioned into three static subclasses *I1*, *I2* and *I3* was defined. For the color class *Isof* one color variable, *HK*, was specified. The arcs of the transitions that involve places containing colored tokens of class *Isof* are labeled with functions ⟨*HK*⟩ that describe the specific color variable, *HK*, associated to those places. GLC: glucose; G6P: glucose-6-phosphate.

2.1 The SSN Formalism

Petri Nets (PNs) and their extensions are a family of formalism that are well suited for modeling in a natural way interactions among system components (i.e., synchronization, sequentiality, concurrency, and conflict) [27]. In the literature, PNs were successfully exploited to study a wide range of applications, ranging from chemical processes to man-made systems (e.g., communication networks, computational distributed systems, manufacturing systems). In particular, the first application of PNs to model biological pathways was published by Reddy *et al.* in [35]; afterward, many other research works highlighted the advantages of using PNs to model biological systems [12,19,20,41].

In this work, we focus on SSNs [3,11], a PN formalism that provides a more compact and parametric description of the system, improving the performance of ordinary PNs. A brief introduction of SSNs is provided hereafter using the net in Fig. 2 (bottom left), which describes the processes of phosphorylation and dephosphorylation of the metabolic enzyme hexokinase (HK).

Formally, an SSN is a bipartite directed graph with two types of nodes: *places* and *transitions*. The places, graphically represented as circles, coincide with the state variables of the system. For instance, the SSN model in Fig. 2 (bottom left) has three places: *hkE*, *hkEPhosi*, and *Phosi*. The first two places correspond to two different molecular configurations of the HK enzyme, either bound to an inorganic phosphate molecule or unbound. The last one represents the inorganic phosphate molecules.

Places can contain tokens, so that the state of an SSN, called *marking*, is defined by the number of tokens in each place. For instance, the marking $hkEPhosi(4) + Phosi(12) + hkE(2)$ of the SSN in Fig. 2 (bottom left) represents the system state in which there are four molecules of the bound enzyme HK, twelve molecules of inorganic phosphate and two molecules of the unbound enzyme.

Specifically, SSNs allow us to color tokens within places so that it is possible to associate information with them; thus, *color domains* are associated with places. These are either expressed as the Cartesian product of *color classes* or by the neutral element ε, which is used to represent *neutral* black tokens as in ordinary PNs. Color classes can be partitioned into *static subclasses*, so that colors in a class encode entities of the same nature (i.e., the same enzyme), but only colors within the same static subclass behave similarly (i.e., the same isoform).

In our example, a single color class (i.e., *Isof*) is introduced to distinguish between the isoforms of HK enzyme. Indeed, *Isof* is partitioned into three *static subclasses* $I1$, $I2$, and $I3$, whose related colors ($c1$, $c2$, and $c3$) represent the specific isoforms of HK. It is important to stress that while the color domain of places *hkE* and *hkEPhosi* is *Isof*, the domain of place *Phosi* is ε. According to this, a possible marking of the system, in which the integer coefficients express the number of molecules of the related compounds, is the following: $hkEPhosi$ $(4 \cdot \langle I1 \rangle + 2 \cdot \langle I3 \rangle) + Phosi(12) + hkE(2 \cdot \langle I2 \rangle)$. This expression represents the system state in which there are four molecules of the isoform $I1$ and two molecules of isoform $I3$ of the bound enzyme *hkEPhosi*, twelve molecules of inorganic phosphate *Phosi*, and two molecules of isoform $I2$ of the unbound enzyme *hkE*.

Differently from places, transitions are nodes that are graphically represented as boxes, and correspond to the events of the system. For instance, the transition HK_{12_f} models the phosphorylation of a HK molecule, while the transition HK_{12_r} models its reverse reaction.

Places and transitions are connected by arcs, labelled by *arc functions*, which specify both the conditions that enable a transition to occur, expressed as requirements that the marking of the system has to satisfy, and how the marking then changes as a consequence of the transition firing.

The stochastic behavior of an SSN model is characterized by the assumption that the firing of any enabled transition occurs after a random delay sampled from a negative exponential distribution. A function ω is associated with each transition and defines its firing rate according to the set of colored tokens involved in the transition firing. For instance, the firing rate value of transition HK_{12_f}

in the human intracellular metabolic model can be defined as follows: (i) 2.0, if the isoform $I1$ of $hkEPhosi$ is involved in the reaction; (ii) 3.0, if the reaction involves the isoform $I2$ of $hkEPhosi$; (iii) 0.5, for any other involved isoform of $hkEPhosi$.

Thus, these stochastic firing delays, sampled from a negative exponential distribution, allow us to automatically derive the associated stochastic process, i.e., a Continuous Time Markov Chain (CTMC), which describes the dynamics of the SSN model. In detail, the CTMC states are identified with SSN markings, and the state changes of the CTMC correspond to the marking changes of the SSN model.

In the literature, different techniques are proposed to solve the underlying CTMC, e.g., Gillespie's Stochastic Simulation Algorithm [40]. Unfortunately, the calculation of a trajectory of the system using exact or approximate stochastic approaches can be computationally challenging, especially in the case of very complex models. Under these circumstances, a deterministic approach [24] can be efficiently leveraged. In particular, given an SSN model, we can derive a deterministic process [3], described through a system of ODEs, which well approximates its behavior.

2.2   GPU-Powered Parameter Estimation

The dynamics of mathematical models of biological systems can be accurately simulated only when a precise parameterization is available, e.g., when the rates of the reactions driving the emergent behavior of the system are known. Unfortunately, these kinetic parameters are difficult or even impossible to measure by means of *in vivo* experiments, and the lack of these values limits the execution of computational investigations. This leads to the PE issue, which is a non-linear, non-convex, and multi-modal optimization problem that can be effectively tackled by means of Computational Intelligence methods.

According to [14], one of the most effective techniques for PE is PSO [23], a population-based meta-heuristic belonging to the class of Swarm Intelligence methods, designed to deal with real-valued optimization problems. In PSO, a *swarm* of randomly generated candidate solutions (called *particles*) moves inside a D-dimensional bounded search space, cooperating to identify the optimal solution to the given problem. During each iteration, the position of each particle changes following two attractors, that is, the best position found by the swarm so far, and the best position found by the particle so far. The social $c_{soc} \in \mathbb{R}^+$ and the cognitive $c_{cog} \in \mathbb{R}^+$ parameters are used to balance the aforementioned attractors, respectively. Moreover, an inertia factor $w \in \mathbb{R}^+$ is exploited to weigh the velocity of the particles, thus avoiding chaotic behaviors in the swarm. The velocity of a particle on the d-th component of the search space, with $d = 1, \ldots, D$, can also be limited below a maximum value $v_{max_d} \in \mathbb{R}^+$, or above a minimum velocity $v_{min_d} \in \mathbb{R}^+$, to prevent the loss of diversity in the swarm [25].

The functioning settings of PSO strongly affect its optimization performance. Thus, they should be, in principle, carefully crafted according to the fitness landscape of the problem under investigation. To work around the settings determination task, in this work we employ FST-PSO [28], a settings-free version of PSO where each particle automatically adjusts its own settings by means of fuzzy rules during the optimization. In particular, we employ the specific variant of FST-PSO presented in [29], as it was shown to be one of the most effective meta-heuristics for the PE problem.

The most time consuming task of PE is the fitness calculation that, in this work, consists in simulating the dynamics of the metabolic model by using the kinetic parameters encoded by every particle in the swarm, and comparing the outcome with a target dynamics [30,32] (specifically, the amounts of GLC and LAC). To accelerate the calculation of the fitness functions, we rely on an improved version of the GPU-powered deterministic simulator LASSIE [39], designed to simulate large-scale biochemical models. Such version of LASSIE introduces the possibility of running in parallel multiple simulations of large-scale models, achieving both thread- and block-level parallelism on the GPU, allowing to (i) perform many simulations of the same system of ODEs (characterized by different parameterizations) in a parallel fashion, and (ii) accelerate the numerical integration of each instance of the system of ODEs by distributing the required calculations on the available GPU cores. LASSIE takes as input a model formalized through SSNs and translated into the equivalent formalism of reaction-based models [5], and generates the corresponding system of ODEs under the assumption of mass action kinetics [10]. A particular advantage of LASSIE is that it is able to automatically recognize and simulate biochemical systems characterized by stiffness: a system of biochemical reactions is considered stiff when there are two well-separated dynamical modes, determined by fast and slow reactions [17].

2.3   Human Intracellular Metabolic Network

In this work, we consider the model of the red blood cell metabolism presented in [22], which consists in a fully parameterized set of reactions following the mass action rate law [10]. The model contains 92 metabolites and 94 reactions describing the central pathways involved in carbohydrate metabolism—namely, glycolysis and the pentose phosphate pathway—in a human red blood cell. For our analyses, we neglected the tissue specificity and we used it as a model for a generic human cell. Moreover, differently from [22], we did not consider the uptake of extracellular substrates. Instead, we considered the different isoforms of the enzyme hexokinase (HK), the first enzyme of the glycolythic pathway that converts glucose (GLC) into glucose-6-phosphate (G6P). We decided to explicitly represent the three isoforms of HK that are the most abundant across different cell types and are known to have different kinetic properties [43]. Thus, the network is modelled by relying on the SSN formalism, in which, as shown in Fig. 2, colors are used to encode the different isoforms of HK (i.e., the color class *Isof* divided in the three subclasses *I1*, *I2* and *I3*).

The corresponding SSN model comprises 92 places and 174 transitions, and is summarized by the schemes in Fig. 1. Out of the 92 places, 11 are colored and involved in 26 transitions that represent the elementary steps of HK reaction.

In details, the model can be divided into three submodels describing the three main involved pathways, which are highlighted with green, pink and yellow boxes. In Fig. 1, the green box contains the colored places and transitions associated with hexokinase reaction, the pink box contains all the reactions modeling the glycolythic pathway, while the yellow box contains the reactions involved in the pentose phosphate pathway. The compactness of the SSN representation can be appreciated considering that a non-colored SPN with an identical behavior would have included two redundant replicates for each of the colored places and related transitions, resulting in a SPN with 114 places and 226 transitions.

3   Results and Discussion

In this section we present the results of the PE carried out by integrating Great-SPN [1]—a framework for the analysis of Discrete Event Dynamic Systems described through the PN formalisms—with FST-PSO [28], and LASSIE [39]. In order to show the effectiveness of FST-PSO for the PE problem with respect to other state-of-the-art methods, we compared its performance against Covariance Matrix Adaptation Evolution Strategy (CMA-ES) [18], Differential Evolution (DE) [13], and Genetic Algorithms (GAs) [6,21]. The mutation strategy of DE used in this work is the DE/rand/1/bin; the crossover probability was set to $CR = 0.25$; the differential weight was set to $F = 1.0$. For GAs, we used a Gaussian mutation with probability $p_{mu} = 0.2$; two-point crossover with probability $p_{cr} = 0.99$; tournament selection, with tournament size equal to 3 (additional details can be found in [13]). All methods were implemented in Python using the Distributed Evolutionary Algorithms (DEAP) framework (v. 1.0.2) [16]. In all tests, the initial population of each meta-heuristic was randomly generated by sampling the parameters of each individual using a log-uniform distribution [9,29], restricting the values in the boundaries of the search space. All simulations were performed using an absolute error tolerance equal to 10^{-8}, and a relative error tolerance equal to 10^{-4}. All tests were repeated 30 times to collect statistical information and assess the Average Best Fitness (ABF) value, in order to discuss the average behavior of each algorithm.

The SSN model presented in this work (Fig. 1) represents an extension of the model proposed by Jamshidi and Palsson [22], in which we increased the level of detail by specifying the three main isoforms of the HK enzyme (i.e., HK-I, HK-II and HK-III). By so doing, we included new reactions corresponding to new transitions in the SSN model, whose kinetic constants are unknown. In our tests, we also assumed the relative abundances of HK-I, HK-II, and HK-III respectively equal to 0.6, 0.3, and 0.1, which represent a *possible* condition for the metabolic network.

To investigate this extended model, we considered the following workflow:

- the approach described in [22] is used to build a baseline large-scale kinetic metabolic model with mass action kinetics;
- the model is defined by means of a SSN, where colored places represent enzyme isoforms of biological relevance, thus obtaining a compact and clear description of the system;
- experimental data are gathered and exploited to infer the isoform-specific kinetic parameters by means of FST-PSO coupled with LASSIE;
- the model is finally used to predict the behavior of the system when isoform-specific modifications are introduced.

The experimental data needed to solve the PE problem are (i) time-course measurements of the main upstream and downstream metabolites, and (ii) a quantification of the proportion of isozymes in the mixture. The former data can be produced by biochemical assays or Mass Spectrometry techniques [7,38], while the latter can be derived from proteomics experiments [42] or estimated from gene-expression data [2]. However, to test the effectiveness of the approach presented here, we relied on synthetic experimental data, generated by simulating the baseline model in which no isoforms are specified.

The model extended with different isoforms includes 78 new transitions, 3 for each reaction, whose kinetic constants are unknown. The values of the corresponding 78 mass-action kinetic parameters were then inferred searching the best fit with the synthetic experimental measurements of glucose (GLC) and lactate (LAC) concentrations in a 50 h time window. The ranges of feasible parameter values were defined by considering the 26 original parameter values used to generate the synthetic data, and allowing for a variation of at most 3 orders of magnitude above and below these values.

Figure 3 shows the comparison between the experimental data (DTTS, red dots) and the simulation of GLC and LAC obtained with the best parameterizations found by CMA-ES (blue line), DE (green line), GAs (orange line), and FST-PSO (pink line). We observe that all meta-heuristics except CMA-ES achieved a perfect fitting at the end of the optimization process. Figure 4 reports the ABF calculated according to the results of 3 independent PE repetitions, showing that FST-PSO is capable of outperforming the other meta-heuristics for this specific task, confirming the results presented in [29].

From the computational perspective, LASSIE strongly reduced the running time required by the PE task, achieving a 30× speed-up running on a NVIDIA GeForce Titan X (3072 cores, clock 1.075 GHz and RAM 12 GB) with respect to the same analysis carried out with the ODE solver LSODA [34] running on a CPU Intel Core i7-2600 (3.4 GHz, RAM 8 GB).

We then explored how the model could be used to reproduce the effect of an isoform-specific modification of the system. This scenario can represent a gene knock-down experiment, the effect of a drug with an isoform-specific target, or a change in isozyme expression after the cell is exposed to an environmental

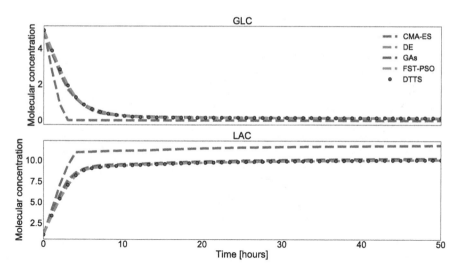

Fig. 3. Dynamics of the GLC (*top*) and LAC (*bottom*) species of the human intracellular metabolic network, obtained with the best parameterization found by CMA-ES, DE, GAs and FST-PSO, compared with the experimental data (DTTS). (Color figure online)

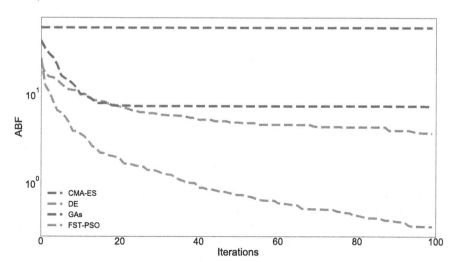

Fig. 4. ABF calculated running CMA-ES, DE, GA and FST-PSO for 20 independent repetitions. FST-PSO clearly outperforms the competitor methods.

stimulus [26]. In Fig. 5 we show that 50%, 75%, and 100% reductions of the concentration of HK-I isoform, the isoform with the highest abundance (0.6), affect the dynamics of key metabolites in the network in a 10 h time window. Noteworthy, from these results we can see that a complete knock-out of one isoform is not necessarily detrimental for intracellular energy-producing path-

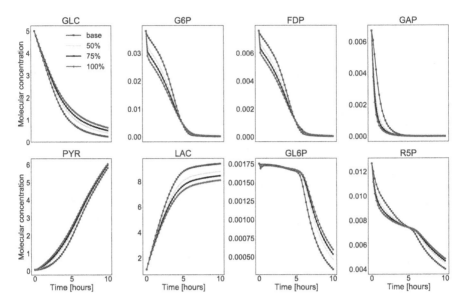

Fig. 5. Dynamic profiles of metabolite concentration simulated with the base-line model, and with increasing $(50, 75, 100\%)$ knock-down interventions on the HK I isoform, in a 10 h time window. Glucose (GLC) and glucose-6-phosphate (G6P) are upstream metabolites with respect to HK. Fructose-diphosphate (FDP), glyceraldehyde-3-phosphate (GAP), pyruvate (PYR) and lactate (LAC) are down-stream metabolites with respect to HK. 6-Phosphogluconolactone (GL6P) and ribose 5-phosphate (R5P) are involved in the pentose phosphate pathway.

ways, as other isoforms with efficient kinetics can provide alternative catalytic routes. The simulations of the model in a 50 h time window indeed show that the 50%, 75%, and 100% knock-down interventions only result in negligible effects on steady state metabolite concentrations (see Fig. 6).

The dynamics of GLC and LAC apparently display the most evident differences in a 50 h window; however, it should be considered that this result was indeed expected as these two metabolites represent, respectively, the "source" and the "sink" entities in the network (see Fig. 1), and no influx/efflux reactions were modeled. If we compare the GLC and LAC dynamics, we can observe that altering the activity of enzymes like HK, which is known to exert high control over the whole pathway [37], produces proximal effects (upstream metabolite GLC is directly degraded by HK) and distal effects (9 metabolic reactions separate HK from the downstream metabolite LAC) on a comparable scale.

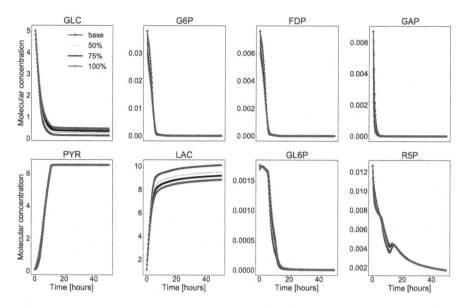

Fig. 6. Dynamic profiles of metabolite concentration simulated with the baseline model, and with increasing $(50, 75, 100\%)$ knock-down interventions on the HK I isoform, in a 50 h time window. Glucose (GLC) and glucose-6-phosphate (G6P) are upstream metabolites with respect to HK. Fructose-diphosphate (FDP), glyceraldehyde-3-phosphate (GAP), pyruvate (PYR) and lactate (LAC) are downstream metabolites with respect to HK. 6-Phosphogluconolactone (GL6P) and ribose 5-phosphate (R5P) are involved in the pentose phosphate pathway.

4   Conclusions

When mathematical models are used to investigate the effects produced by the different kinetic properties of metabolic isozymes, issues of model representation and efficient simulation of the dynamics often arise. In this work, we showed how the compact and parametric representation achieved with SSNs can be combined with FST-PSO coupled with LASSIE to effectively and efficiently solve the PE problem, thanks to the parallelization of the computations on the GPU. The approach presented in this work is particularly well suited to deal with complex metabolic models that, for instance, include many reactions alternatively catalyzed by various isozymes with unknown kinetic parameters. As a matter of fact, we successfully estimated the 78 missing parameters related to HK isozymes. These allowed us to correctly reproduce the network behavior and to perform new *in silico* experiments. Thanks to the GPU-powered simulator exploited in this work, we achieved a 30× speed-up with respect to the same methodology running on CPU and exploiting LSODA as ODE solver.

As a future extension, we will apply our methodology to expand other portions of the intracellular metabolic network, so that the effects of combined modifications on multiple enzymes could be analyzed. In addition, we plan to employ multi-GPU systems to further improve the overall computational performances of our approach.

Acknowledgments. This work was conducted in part using the resources of the Advanced Computing Center for Research and Education at Vanderbilt University, Nashville, TN, USA.

The work of MB was partially supported by Fond. CRT - "Experimentation and study of models for the evaluation of the performance and the energy efficiency of C3S."

References

1. Babar, J., Beccuti, M., Donatelli, S., Miner, A.S.: GreatSPN enhanced with decision diagram data structures. In: Lilius, J., Penczek, W. (eds.) Application and Theory of Petri Nets. PETRI NETS 2010, Lecture Notes in Computer Science, vol. 6128, pp. 308–317. Springer, Berlin, Heidelberg (2010). https://doi.org/10.1007/978-3-642-13675-7_19
2. Barker, B.E., et al.: A robust and efficient method for estimating enzyme complex abundance and metabolic flux from expression data. Comput. Biol. Chem. **59**, 98–112 (2015)
3. Beccuti, M., et al.: From symmetric nets to differential equations exploiting model symmetries. Comput. J. **58**(1), 23–39 (2015)
4. Bennett, M.R., et al.: Metabolic gene regulation in a dynamically changing environment. Nature **454**(7208), 1119 (2008)
5. Besozzi, D.: Reaction-based models of biochemical networks. In: Beckmann, A., Bienvenu, L., Jonoska, N. (eds.) CiE 2016. LNCS, vol. 9709, pp. 24–34. Springer, Cham (2016). https://doi.org/10.1007/978-3-319-40189-8_3
6. Besozzi, D., Cazzaniga, P., Mauri, G., Pescini, D., Vanneschi, L.: A comparison of genetic algorithms and particle swarm optimization for parameter estimation in stochastic biochemical systems. In: Pizzuti, C., Ritchie, M.D., Giacobini, M. (eds.) Evolutionary Computation, Machine Learning and Data Mining in Bioinformatics. Lecture Notes in Computer Science, vol. 5483, pp. 116–127. Springer, Berlin Heidelberg (2009). https://doi.org/10.1007/978-3-642-01184-9_11
7. Bordbar, A., Yurkovich, J.T., Paglia, G., Rolfsson, O., Sigurjónsson, Ó.E., Palsson, B.O.: Elucidating dynamic metabolic physiology through network integration of quantitative time-course metabolomics. Sci. Rep. **7**, 46249 (2017)
8. Cazzaniga, P., et al.: Computational strategies for a system-level understanding of metabolism. Metabolites **4**, 1034–1087 (2014)
9. Cazzaniga, P., Nobile, M., Besozzi, D.: The impact of particles initialization in PSO: parameter estimation as a case in point. In: Proceedings of Conference on Computational Intelligence in Bioinformatics and Computational Biology, pp. 1–8. IEEE (2015)
10. Chellaboina, V., Bhat, S.P., Haddad, W.M., Bernstein, D.S.: Modeling and analysis of mass-action kinetics. IEEE Control Syst. Mag. **29**(4), 60–78 (2009)
11. Chiola, G., Dutheillet, C., Franceschinis, G., Haddad, S.: Stochastic well-formed coloured nets for symmetric modelling applications. IEEE Trans. Comput. **42**(11), 1343–1360 (1993)
12. Cordero, F., et al.: Multi-level model for the investigation of oncoantigen-driven vaccination effect. BMC Bioinform. **14**(Suppl. 6) (2013). Article number S11

13. Das, S., Suganthan, P.: Differential evolution: a survey of the state-of-the-art. IEEE Trans. Evol. Comput. **15**(1), 4–31 (2011)
14. Dräger, A., Kronfeld, M., Ziller, M., Supper, J., Planatscher, H., Magnus, J.: Modeling metabolic networks in *C. glutamicum*: a comparison of rate laws in combination with various parameter optimization strategies. BMC Syst. Biol. **3**, 5 (2009)
15. Fletcher, R.: Practical Methods of Optimization. Wiley, Hoboken (2013)
16. Fortin, F., De Rainville, F., Gardner, M., Parizeau, M., Gagné, C.: DEAP: evolutionary algorithms made easy. J. Mach. Learn. Res. **13**, 2171–2175 (2012)
17. Gillespie, D.T.: Stochastic simulation of chemical kinetics. Annu. Rev. Phys. Chem. **58**, 35–55 (2007)
18. Hansen, N., Ostermeier, A.: Adapting arbitrary normal mutation distributions in evolution strategies: the covariance matrix adaptation. In: 1999 IEEE Congress on Evolutionary Computation (CEC), pp. 312–317. IEEE (1996)
19. Herajy, M., Fei, L., Rohr, C., Heiner, M.: Coloured hybrid Petri Nets: an adaptable modelling approach for multi-scale biological networks. Comput. Biol. Chem. **76**, 87–100 (2018)
20. Hofestädt, R.: A Petri Net application of metabolic processes. J. Syst. Anal. Model. Simul. **16**, 113–122 (1994)
21. Holland, J.: Adaptation in Natural and Artificial Systems: An Introductory Analysis with Applications to Biology, Control and Artificial Intelligence. MIT Press, Cambridge (1992)
22. Jamshidi, N., Palsson, B.Ø.: Mass action stoichiometric simulation models: incorporating kinetics and regulation into stoichiometric models. Biophys. J. **98**(2), 175–185 (2010)
23. Kennedy, J., Eberhart, R.C.: Particle swarm optimization. In: Proceedings of the International Conference on Neural Networks, vol. 4, pp. 1942–1948. IEEE (1995)
24. Kurtz, T.G.: Solutions of ordinary differential equations as limits of pure jump Markov processes. J. Appl. Probab. **1**(7), 49–58 (1970)
25. Liu, H., Abraham, A., Zhang, W.: A fuzzy adaptive turbulent particle swarm optimisation. Int. J. Innov. Comput. Appl. **1**(1), 39–47 (2007)
26. Metallo, C.M., Vander Heiden, M.G.: Understanding metabolic regulation and its influence on cell physiology. Mol. Cell **49**(3), 388–398 (2013)
27. Murata, T.: Petri Nets: properties, analysis and applications. Proc. IEEE **77**(4), 541–580 (1989)
28. Nobile, M.S., Cazzaniga, P., Besozzi, D., Colombo, R., Mauri, G., Pasi, G.: Fuzzy self-tuning PSO: a settings-free algorithm for global optimization. Swarm Evol. Comput. **39**, 70–85 (2018)
29. Nobile, M.S., et al.: Computational intelligence for parameter estimation of biochemical systems. In: 2018 IEEE Congress on Evolutionary Computation (CEC). IEEE (2018)
30. Nobile, M.S., Besozzi, D., Cazzaniga, P., Mauri, G., Pescini, D.: Estimating reaction constants in stochastic biological systems with a multi-swarm PSO running on GPUs. In: Proceedings of the 14th Annual Conference Companion on Genetic and Evolutionary Computation, pp. 1421–1422. ACM (2012)
31. Nobile, M.S., Cazzaniga, P., Tangherloni, A., Besozzi, D.: Graphics processing units in bioinformatics, computational biology and systems biology. Brief. Bioinform. **18**(5), 870–885 (2016)
32. Nobile, M.S., Tangherloni, A., Besozzi, D., Cazzaniga, P.: GPU-powered and settings-free parameter estimation of biochemical systems. In: 2016 IEEE Congress on Evolutionary Computation (CEC), pp. 32–39. IEEE (2016)

33. O'Brien, J., Kla, K.M., Hopkins, I.B., Malecki, E.A., McKenna, M.C.: Kinetic parameters and lactate dehydrogenase isozyme activities support possible lactate utilization by neurons. Neurochem. Res. **32**(4–5), 597–607 (2007)
34. Petzold, L.: Automatic selection of methods for solving stiff and nonstiff systems of ordinary differential equations. SIAM J. Sci. Stat. Comput. **4**(1), 136–148 (1983)
35. Reddy, V., Mavrovouniotis, M., Liebman, M.: Petri Net representation in metabolic pathways. In: Proceedings of International Conference on Intelligent Systems for Molecular Biology, pp. 328–336 (1993)
36. Schomburg, I., et al.: BRENDA, the enzyme database: updates and major new developments. Nucl. Acids Res. **32**(suppl_1), D431–D433 (2004)
37. Smallbone, K., et al.: A model of yeast glycolysis based on a consistent kinetic characterisation of all its enzymes. FEBS Lett. **587**(17), 2832–2841 (2013)
38. Sriyudthsak, K., Shiraishi, F., Hirai, M.Y.: Identification of a metabolic reaction network from time-series data of metabolite concentrations. PLoS ONE **8**(1), e51212 (2013)
39. Tangherloni, A., Nobile, M.S., Besozzi, D., Mauri, D., Cazzaniga, P.: LASSIE: simulating large-scale models of biochemical systems on GPUs. BMC Bioinform. **18**(1), 246 (2017)
40. Tangherloni, A., Nobile, M.S., Cazzaniga, P., Besozzi, D., Mauri, G.: Gillespie's stochastic simulation algorithm on MIC coprocessors. J. Supercomput. **73**(2), 676–686 (2017)
41. Totis, N., Follia, L., Riganti, C., Novelli, F., Cordero, F., Beccuti, M.: Overcoming the lack of kinetic information in biochemical reactions networks. SIGMETRICS Perform. Eval. Rev. **44**(4), 91–102 (2017)
42. Uhlén, M., Fagerberg, L., Hallström, B.M., Lindskog, C., Oksvold, P., Mardinoglu, A., Sivertsson, Å., Kampf, C., Sjöstedt, E., Asplund, A., et al.: Tissue-based map of the human proteome. Science **347**(6220), 1260419 (2015)
43. Wilson, J.E.: Isozymes of mammalian hexokinase: structure, subcellular localization and metabolic function. J. Exp. Biol. **206**(12), 2049–2057 (2003)
44. Wittig, U., et al.: SABIO-RK-database for biochemical reaction kinetics. Nucl. Acids Res. **40**(D1), D790–D796 (2011)
45. Wuntch, T., Chen, R.F., Vesell, E.S.: Lactate dehydrogenase isozymes: kinetic properties at high enzyme concentrations. Science **167**(3914), 63–65 (1970)

Computational Models in Health Informatics and Medicine

Automatic Discrimination of Auditory Stimuli Perceived by the Human Brain

Angela Serra[1,2] , Antonio della Pietra[1], Marcus Herdener[3] ,
Roberto Tagliaferri[1(✉)] , and Fabrizio Esposito[4]

[1] NeuRoNeLab, DISA-MIS University of Salerno, 84084 Fisciano, SA, Italy
robtag@unisa.it
[2] Faculty of Medicine and Health Technology, Tampere University, Tampere, Finland
[3] Department of Psychiatry, Psychotherapy and Psychosomatics,
University Hospital for Psychiatry Zurich, Lenggstr. 31, 8032 Zurich, Switzerland
[4] Department of Medicine, Surgery and Dentistry "Scuola Medica Salernitana",
University of Salerno, Via S. Allende, 84081 Baronissi, Salerno, Italy

Abstract. Humans are able to perceive small difference of sound frequency but it is still unknown how the difference in frequency information is represented at the level of the primary sensory cortex. Indeed, analysis of fMRI imaging identified tonotopic maps through the auditory pathways to the primary sensory cortex. These maps are unfortunately too coarse to show ultra-fine discrimination. Then, the hypothesis is that this small frequency differences are recognised thanks to the information coming from a large set of auditory neurons. To investigate this possibility, a multi-voxel pattern discriminating analysis of the response of BOLD-fMRI in the bilateral auditory cortex to tonal stimuli with different shift in frequency was performed. Our results suggest that small shifts in the frequency are easily classified compared with big shifts and that multiple areas of the auditory cortex are involved in the tone recognition.

Keywords: Tonotonic maps · Auditory cortex · BOLD-fMRI

1 Introduction

Humans can resolve just-noticeable differences of sound frequency (in the range of 2% frequency difference) that are substantially smaller than the spectral resolution of peripheral sensors in the cochlea [1], where sound frequencies are represented topographically (cochleotopic representations) but at lower resolution. More specifically, the just-noticeable differences that can be perceived are around 30 times smaller than the bandwidth of peripheral filters [1]. It is unclear, however, how this fine-grained frequency information can be derived (or perceived) from the sensory representations of sound frequency in primary sensory cortices. Using functional magnetic resonance imaging (fMRI) and voxel-based analyses of blood oxygen level dependent (BOLD) signals, it has been shown that sound frequency is encoded in (relatively coarse) tonotopic maps that are

© Springer Nature Switzerland AG 2020
M. Raposo et al. (Eds.): CIBB 2018, LNBI 11925, pp. 205–211, 2020.
https://doi.org/10.1007/978-3-030-34585-3_18

preserved throughout the auditory pathway up to primary sensory cortex [8]. However, the isofrequency contours within fMRI maps usually appear too coarse to underlie ultra-fine frequency discrimination. To investigate the possibility that just-noticeable frequency differences could be derived by integrating information over a large and distributed population of auditory neurons, a multi-voxel pattern discriminative analysis of BOLD-fMRI responses to pairs of tonal stimuli with either bigger or smaller frequency differences was performed in the bilateral auditory cortex of three healthy volunteers. Multivariate processing and machine learning were used to learn and automatically classify the directions of big and small frequency changes in auditory stimulus transitions between different tones. To the best of our knowledge this is the first attempt to perform this kind of analysis using fMRI data. Thus, the goal of this work is to perform a pilot study to understand the type and amount of information available and if there is room for further investigations and speculations.

2   Materials and Methods

2.1   Dataset

Experiments were performed on the fMRI data acquired on three volunteers (SUBJ1, SUBJ2 and SUBJ3). When lying into the scanner the subjects heard two different tones of alternating frequency. The objective of the study was to investigate if the human brain is able to recognise the shift in frequency. Each subject heard two kinds of stimuli, one with a big and one with a small change in the frequency. Two experimental setups were used, one with 500–525 Hz for the big (called 500big) and 500–510 Hz for the small (called 500small). The other with 1000–1025 Hz for the big (called 1000big) and 1000–1010 Hz for the small (called 1000small). Due to the nature of the experiments, only the auditory cortex was taken into consideration, by using a brain mask that considered around 2×10^3 voxels (See Fig. 1). For each subject 49 trials were acquired, 24 trials have a shift from a higher to a lower threshold (labelled as $+1$), and 25 have a shift from a lower to a higher threshold (labelled as -1). The focus of the study here is to identify the shift in the frequencies between all the subjects, this is why we created four data sets, by concatenating the stimulus of the three subjects, for the different frequency shifts. Before concatenation, each dataset is first standardised to have zero mean and standard deviation equal to one. When combined, quantile normalisation [2] is performed to make the empirical distribution of the three datasets approximately equal. Furthermore, the voxels containing null values were removed from the analyses.

2.2   fMRI Data Acquisition

Three adult subjects with no history of neurological or hearing disorders participated to an fMRI study. During data acquisition, subjects were asked to keep eyes open and to listen to the sounds. FMRI data were acquired on a 3 T

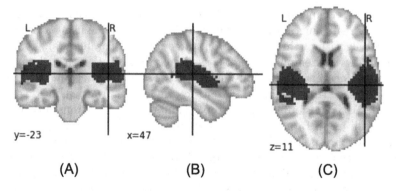

Fig. 1. Mask of the auditory cortex in the coronal (A), sagittal (B) and horizontal (C) planes.

clinical MRI scanner (Siemens Magnetom Trio TIM, Germany) equipped with 12-channel head coil. A structural T1-weighted high-resolution ($1 \times 1 \times 1\,\mathrm{mm}^3$) data set covering the whole brain was collected for each subject with a three-dimensional magnetization prepared rapid acquisition gradient echo (MPRAGE) sequence. Functional images were acquired using a low-impact-noise acquisition fMRI sequence [9] which increases the dynamic range of the blood oxygenation level-dependent (BOLD) signal in response to acoustic stimuli and allows for a time-efficient mapping of acoustic properties in auditory cortex. Functional volumes were positioned parallel to the lateral sulcus and acquired using the following parameters: gradient recalled (GRE) echo-planar imaging (EPI) sequence with 12 image slices, 3 mm slice thickness, 1398 msec volume repetition time (TR), $20 \times 20\,\mathrm{cm}^2$ field of view (FOV), 96×96 matrix size, 48 msec echo time (TE), 80° flip angle (FA). Functional scans were acquired during the continuous presentation of acoustic stimuli consisting of two alternating frequencies in a blocked design (ABABAB...). Separate runs were acquired with different alternating frequencies (500–510 Hz, 500–525 Hz, 1000–1010 Hz, 1000–1050 Hz). All sound stimuli were digitally generated, stored as WAV files (44.1 kHz sampling rate) and presented using Cogent (http://www.vislab.ucl.ac.uk/) running under Matlab 6.5.1 (The Mathworks Inc. USA), and delivered using MR-compatible headphones (MR Confon, Germany) at 75 dB SPL. Calibrations and sound intensity measurements were made for each individual stimulus with an MR-compatible condenser microphone (Bruel and Kjaer 4188) and a sound level meter (Bruel and Kjaer 2238 Mediator).

2.3   Data Preprocessing

Anatomical and functional images were analyzed using BrainVoyager software (Brain Innovation, The Netherlands). Functional activations were analyzed in voxel space without spatial smoothing. The first five EPI images were discarded from each run to allow for magnetization saturation, and slice scan times were corrected using a cubic spline interpolation procedure. Possible head motion was corrected by realigning individual images to the first volume using a Levenberg-Marquardt algorithm (three translation, three rotation parameters). The resulting time series were temporally high-pass filtered (cut-off of seven cycles per time course) to reduce linear and non-linear drifts. Functional data in native space were co-registered to the individual structural images warped into Talairach space with spatial re-sampling to isotropic $3\,\text{mm}^3$ voxels. The transitions between sound stimuli (A to B or B to A) were used as temporal markers (triggers) for each single-trial response. For each subject and run, two separate series of parametric activation maps (for transitions A to B and B to A) were generated by least-square fitting a double-gamma function to the BOLD signal within the interval of time points covering the time between one time point before and eight time points after the sound frequency transition, after subtracting the local mean signal level.

2.4   Proposed Methodology

The proposed methodology for the analysis of fMRI data related to human brain responses to tonal stimuli with different frequencies is composed of several steps. After preprocessing, a dimensionality reduction is performed with clustering methods. For each cluster a prototype is extracted as the mean of the voxels belonging to it. Then a feature selection and a classification task are performed to identify the cluster prototypes which better classify the stimulus. The methodology was tested in a nested cross validation procedure [7] with 6 external and 3 internal folds. The dataset was first divided into 6 folds in a stratified manner, in order to ensure the same percentage of points of each class in each fold. Then, for all the possible combinations of 5 on 6 folds, used as training, the following steps are repeated: a complete linkage based hierarchical clustering method [10], with spatial constraints, allowing only adjacent voxels to be merged, was executed. The Pearson correlation was used as a measure of similarity between the voxel patterns. In this case, the dissimilarity was expressed as $d(x, y) = 1 - abs(cor(x, y))$. As always happens when using clustering algorithms, a crucial point is the selection of the number of clusters, k [6,11]. Here we performed tests by using different values of $k \in \{50, 100, 150, 200, 300, 400\}$. Once the brain parcellation is obtained, for each cluster a prototype is extracted as the mean of all voxels belonging to it. This is particularly useful, since brain activity spreads over an area that can be greater than the size of a single voxel. Furthermore, using the values of groups of adjacent voxels, instead of considering each voxel independently, can increase the method stability and its predictive capabilities [4]. The 5 folds are further divided into other 3 folds, 2 used for training and one for testing, to select the optimal parameters for the random forest classifier [5] and to identify the optimal set of features. The random forest was also used to obtain a ranking of the feature importance with respect to the classification task by considering the mean decrease accuracy index [3]. Then, a new random forest was trained by using only the most relevant features. In this case, the accuracy of the model depends on three parameters: the number of decision trees in the forest, the number of features used in each step by the trees and the number of features selected as the top relevant ones. Here, the number of features used in each step is $m = \sqrt{d}$ where d is the total number of features in the data set. The number of trees (n_{trees}) and the number of features (n_{feat}) to select after the first execution of the random forest were chosen by cross-validation in the ranges: $n_{trees} \in \{10^3, 10^4, 10^5, 5 \times 10^5\}$, $n_{feat} \in \{5, 10, 15, 20, 25, 30, 35, 40, 45\}$. The classifier accuracy is then tested on the external test set. The final accuracy of the classifier was the mean of the accuracy of six runs given by the six external folds of the nested cross validation.

3 Results

In this work a random forest was trained to classify the BOLD-fmri reponse to different shifts in frequency of tone stimulus. The used frequencies were 500 Hz and 1000 Hz with an increase shift or decrease shift named big and small, respectively. We first performed a dimensionality reduction by using a complete linkage hierarchical algorithm. We compared the mean accuracy of the classifier out of six runs of a nested cross validation across the four different experimental setups and the different values of k used for the clustering. As we can see in Fig. 2, the mean accuracy increases when the number of clusters increases independently from the frequency shift. Moreover, the results obtained on the 500 small dataset are significantly better than the others. To verify the significance of these results, a permutation test was performed. For each dataset, for the number of clusters and for the optimal parameters of the random forest classifier, 1000 random permutations of the labels were generated and the model re-trained with these labels. A p-value was calculated by counting how often the accuracy on the original labels was higher than that of the random labels. As we can see in Table 1, all the results obtained a significant p-value < 0.05.

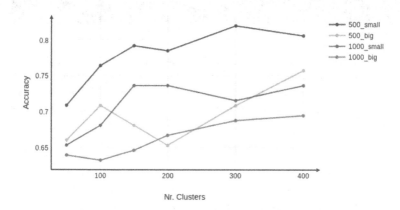

Fig. 2. Mean classification accuracy for each dataset and for different cluster size.

Table 1. Permutation test results for each dataset for the optimal results.

	500small	500big	1000small	1000big
K	300	400	150	400
Selected	16	32	63	10
Mean accuracy	82%	76%	74%	69%
PValue	0.000	0.002	0.008	0.031

Then, we further investigated how many times each voxel in the mask was selected as being part of one of the relevant clusters for classification across the six folds of the cross-validation. The map with the counting for each voxel is shown in Fig. 3. We also investigated which were the most frequently selected areas for each data set across the six cross validation executions to verify which were the most stable features. In particular, the Right Superior Temporal Gyrus was selected 6/6 times in the 500 small data set, the Right Claustrum was selected 6/6 times in the 500 big data set, the Left Traverse Temporal Gyrus was selected 6/6 times in the 1000 small data set and the Right Middle Temporal Gyrus, the Superior Temporal Gyrus and the Postcentral Gyrus were selected 6/6 in the 1000 big data set.

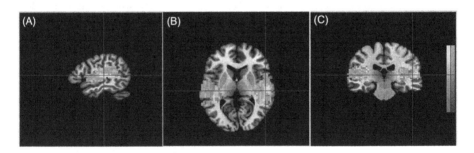

Fig. 3. Maps of the relevant voxels of the 500small data set obtained with the six experimental setups. The map shows how many times (across the six folds of the cross validation) each voxel was selected as part of a relevant cluster for classification. Yellow means higher values. The most frequent selected region is the RH Superior Temporal Gyrus. The maps are shown in the sagittal (A), horizontal (B) and coronal (C) planes. (Color figure online)

4   Conclusion

The aim of this work was to investigate the capability to classify the response of BOLD signal in response to tone stimulus with different shifts in frequency. We analysed the BOLD-response to tone stimuli (with a shift between 500 to 510 and 525 Hz and 1000 to 1010 and 1025 Hz) in the bilateral auditory cortex of three health subjects. We first performed a dimensionality reduction based on voxel clustering in order to identify regions of the brain with similar behaviour. Then, a feature selection and a classification tasks were performed to identify the cluster prototypes that better classify the stimulus. Our results suggest that distinguishing shifts in small frequency is easier than distinguishing shifts in big frequencies and multiple parts of the auditory cortex are involved in the process, corroborating the hypothesis that just-noticeable frequency differences could be derived by integrating information over a large and distributed population of auditory neurons. Of course, this is a pilot study that needs extensive analysis and experimentation in order to validate the results.

References

1. Bitterman, Y., Mukamel, R., Malach, R., Fried, I., Nelken, I.: Ultra-fine frequency tuning revealed in single neurons of human auditory cortex. Nature **451**(7175), 197 (2008)
2. Bolstad, B.M., Irizarry, R.A., Åstrand, M., Speed, T.P.: A comparison of normalization methods for high density oligonucleotide array data based on variance and bias. Bioinformatics **19**(2), 185–193 (2003)
3. Breiman, L.: Random forests. Mach. Learn. **45**(1), 5–32 (2001)
4. Fratello, M., et al.: Multi-view ensemble classification of brain connectivity images for neurodegeneration type discrimination. Neuroinformatics **15**(2), 199–213 (2017)
5. Fratello, M., Tagliaferri, R.: Decision Trees and Random Forests. Elsevier, Amsterdam (2018). https://doi.org/10.1016/b978-0-12-809633-8.20337-3
6. Kodinariya, T.M., Makwana, P.R.: Review on determining number of cluster in k-means clustering. Int. J. **1**(6), 90–95 (2013)
7. Raschka, S.: Model evaluation, model selection, and algorithm selection in machine learning (2018)
8. Saenz, M., Langers, D.R.: Tonotopic mapping of human auditory cortex. Hear. Res. **307**, 42–52 (2014)
9. Scheffler, K., Seifritz, E.: Low-impact noise acquisition magnetic resonance imaging. US Patent 7,112,965, 26 Sept 2006
10. Serra, A., Tagliaferri, R.: Unsupervised learning: clustering. In: Reference Module in Life Sciences, Elsevier (2018). https://doi.org/10.1016/b978-0-12-809633-8.20487-1
11. Sugar, C.A., James, G.M.: Finding the number of clusters in a dataset: an information-theoretic approach. J. Am. Stat. Assoc. **98**(463), 750–763 (2003)

Neural Models for Brain Networks Connectivity Analysis

Razvan Kusztos, Giovanna Maria Dimitri$^{(\boxtimes)}$ (iD), and Pietro Lió

Computer Laboratory, University of Cambridge, Cambridge, UK
gmd43@cam.ac.uk

Abstract. Functional MRI (fMRI) attracts huge interest for the machine learning community nowadays. In this work we propose a novel data augmentation procedure through analysing the inherent noise in fMRI. We then use the novel augmented dataset for the classification of subjects by age and gender, showing a significant improvement in the accuracy performance of Recurrent Neural Networks. We test the new data augmentation procedure in the fMRI dataset belonging to one international consortium of neuroimaging data for healthy controls: the Human Connectome Projects (HCP).

From the analysis of this dataset, we also show how the differences in acquisition habits and preprocessing pipelines require the development of representation learning tools. In the present paper we apply autoencoder deep learning architectures and we present their uses in resting state fMRI, using the novel data augmentation technique proposed.

This research field, appears to be unexpectedly undeveloped so far, and could potentially open new important and interesting directions for future analysis.

Keywords: fMRI · Deep learning · Data augmentation

1 Introduction

Data-centric approaches to science are becoming the norm in biomedical fields. Open-science initiatives, the extensive documentation of data protocols, and the unprecedented efforts for data curation are making machine learning permeate fields that were previously inaccessible. These fresh approaches inspired the field of *omics* for biomedical sciences, building comprehensive databases that capture biochemical processes. Together with *genomics*, studying the structure and function of the DNA, came *connectomics*, with the ample task of understanding the brain and its processes. The early attempts to understand the function of the human brain using *Electroencephalograms* (EEG, 1930s), were superseded by the *Magnetic Resonance* (MR, 1960s) revolution, which allowed the capturing of stunning *in vivo* structural imagery. This became a basis for studying the relationship between two forms of brain connectivity: *functional* – the coherent activations of brain regions, and *structural* – the neural fibre infrastructure.

© Springer Nature Switzerland AG 2020
M. Raposo et al. (Eds.): CIBB 2018, LNBI 11925, pp. 212–226, 2020.
https://doi.org/10.1007/978-3-030-34585-3_19

The field crystallised around the introduction of *Functional MRI* (fMRI, 1990s), which enables the observation of function, without the limiting *spatial resolution* of the EEG.

In recent years, a plethora of consortia began planning extensive studies aiming for uniform data acquisition and analysis. Projects like the *Human Connectome Project* (HCP), the *1000 Functional Connectome Projects*, the *developing Human Connectome Project* (dHCP), and many others started to collect data according to pre-specified protocols, at various *modalities* (EEG, MEG, fMRI, dMRI, physiological data, etc.) and made them freely and easily available to the research community.

The *open-science* approach gives researchers access to a constant flux of medical data and techniques. Functional MRIs can be found at various resolutions, when subjects are either at rest or during task, capturing different physiological features of the subjects (such as sex, age, handedness, BMI), socio-economic background, disease status. Apart from these sources of variation, together with individual irregularities or the noise error from the apparatus, new researchers are faced with understanding an incessantly increasing number of tools.

Advances in machine learning, particularly *deep learning* have left their mark on the field, with new techniques and ideas being continuously developed. New research in *representation learning* has brought forward inspiring *neural network* models, such as *autoencoders* or the feature extraction from other *deep networks*. These techniques have quickly become state-of-the art in *computer vision, signal processing* and *natural language processing*, and their ability to capture latent factors is often mesmerising [1]. In the following sections we briefly provide some background for fMRI techniques. We proceed to introducing data augmentation techniques, as well as discussing data and methodologies used.

2   Scientific Background

2.1   fMRI: functional Magnetic Resonance Imaging

fMRI is a relatively young non-invasive brain imaging technique, but in less than three decades managed to revive the fields of cognitive and social neurosciences and the study of mental illnesses. The first use of dynamic magnetic resonance imaging (MRI) for detecting functional behaviour of the brain was described by *Belliveau et al.* [2]. The idea is to observe metabolic changes during neural activations, based on changes in brain's magnetic properties. The current standard technique is to capture the *blood oxygen level dependent contrast* (BOLD). A plethora of methods for acquiring, preprocessing and analysing fMRI data has been developed. A summary of the main fMRI processing steps is presented in Fig. 1.

The impact of fMRI can be appreciated in many different fields of computational neuroscience. It provides new insights into the organisation of the brain during disease and enables perspectives on the functional integration and segregation of cortical regions amongst others. Among other important applications

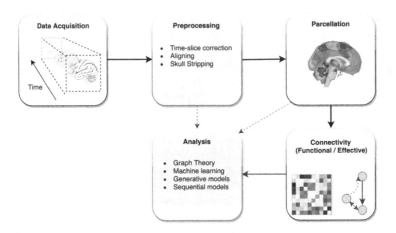

Fig. 1. In a first step data are acquired, then in a second step they are preprocessed using physiological corrections as well as image processing. Subsequently the brain is parcelled followed by analysis. Deep learning could alleviate the need for connectivity analysis by skipping straight from parcellation to the analysis step.

we can for example list the study of plasticity [3], default mode network studies [4] and mental illness studies [5].

2.2   Data Augmentation Techniques

The goal of data augmentation is to build synthetic datasets, artificially increasing the amount of data and allowing the training complex neural networks architectures, when real label data is limited [6]. This allowed in our experiments the better learning of intra-class variances in the data. These techniques are particularly challenging for timeseries, and consequently in fMRI research. Some example techniques are Window Slicing (WS) technique, and Window Warping (WW) [6]. The first one, originally used in the computer vision applications, was then applied to the timeseries domain for the first time in [7]. In this case, a timeseries is divided into slices, which are assumed to belong to the same class as the parent timeseries. Classification is then performed using the slices. Alternatively, with WW, slices of the original timeseries are picked and then stretched or squeezed [6]. In fMRI studies, the WS technique has been used in the context of training RNN, but we hypothesise that this could lead to training data leakage. In our paper, we suggest that augmentation could be done by injecting noise into the data and we further present some results of these techniques.

3   Materials and Methods

International consortia were established, aiming to make large and homogeneous datasets publicly available and push forward our understanding of the human brain. Among the notable projects, we will focus and present our results using the Human Connectome Project (HCP) (www.humanconnectome.org). This is a comprehensive database, containing imaging at various modalities, as well as metadata representing the perfect benchmark for testing our data augmentation techniques.

3.1   The Datasets: HCP

HCP consists of multiple subprojects, each dealing with separate issues of human brain mapping, such as *Young Adult HCP*, *Lifetime HCP* or *Connectomes Related to Human Disease* (www.humanconnectome.org).

Part of this initiative is the WU-Minn-Oxford consortium which collected not only resting-state fMRI, but also task-based fMRI, structural MRI, EEG, MEG of no less than 1200 subjects [8]. Data are easily accessible through user friendly interfaces (https://db.humanconnectome.org), contributing to its popularity.

We summarize in Table 1 the HCP dataset's main characteristics: In our experiments we focused on the parcelled connectome (see Parcellation in Table 1). The reason for this is that the timeseries representation of parcelled fMRI scan has a similar the same data structure but a smaller dimensionality. In particular we will use our data augmentation in order to improve classification of metadata from the same cohort, showing potentialities and performance accuracy of the augmented dataset with respect to the original one. We considered the age and gender labes for the HCP cohort, as this gives access to the most complete cohort. We made the age component discrete, considering the following classification categories: under 26, 26–30, 31–36 and 36 or over.

Table 1. Description of the HCP dataset. The *Timesteps* represent the number of volumes captured during the scan. *Volume* represents the spatial resolution of the scan: how many *voxels* along each dimension the dataset makes available.

Property	HCP
Focus	Healthy Controls (HC)
Subjects	1003, from a single site
Timesteps	4800
Volume	$91 \times 109 \times 91$
Preprocessing	MELODIC, as per [9]
Parcellation	ICA followed by PCA, keeping the first 25, 50, 100, 200, 300 or components

3.2   Methods: Data Augmentation Techinque

The starting observation that lead us to the implementation of our own data augmentation technique, was the success of linear SVM (Support Vector Machine) to differentiate the dense connectomes, obtaining consistently good accuracies for a variety of labels (sex, age, etc.) (see Figs. 2 and 3 for the gender and age classification accuracies results over time series and dense connectomes).

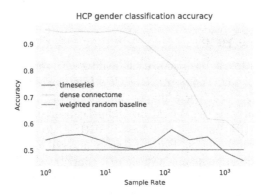

Fig. 2. Figures showing the classification accuracies (test set = 30%) using a linear SVM classifier for gender in the HCP (200 components).

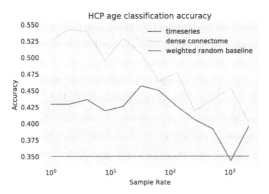

Fig. 3. Figures showing the classification accuracies (test set = 30%) using a linear SVM classifier for age in the HCP (200 components).

As we can see from the two Figs. 2 and 3 when timeseries data features were considered, performances dropped significantly. This was somehow unexpected, since by computing the correlation matrix, we are essentially discarding information. Consequently, we hypothesised that this was due to two main reasons: the inherent normalisation of the *correlation matrix* that made the classification easier for SVM-like algorithms and the high-variance noise element in the timeseries. These correlation matrices (*dense connectome*) are, in fact, normalised, whereas the timeseries are not. Among the variety of normalization techniques available before applying machine learning, for the HCP we proceeded with the min-max re-scaling. We chose to apply such type of normalization in the $[-1,1]$ interval since the features of the HCP are by default oscillating around the 0 axis. This type of re-scaling is defined as: s

- **Min-max rescaling**: Consider a hypothetical training set, X consisting of subjects as *rows* and their features as *columns*.

$$X_{minmax} = 2\frac{X - min(X)}{max(X) - min(X)} - 1 \qquad (1)$$

where min and max are computed across the rows (so we obtain a value for each subject).

The method as described above works on 1D arrays, so the 3D structure of the current data (*subjects, components, time*) allows multiple axis on which these transformations could be applied (see Fig. 4 for a visual description). In order to compare the performance of each transformation, we analysed the effect on applying them on the accuracy of the classification task using SVM linear classifiers.

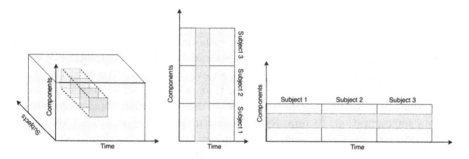

Fig. 4. The left most image shows a voxel wise application, whereas the second and third respectively show group component and group time applications.

From the results Table 2 we drew a number of conclusions. Firstly, the scaling methods do not affect the output of the SVM; nonetheless, they are likely to affect neural machine learning techniques.

Secondly, we noticed that the *group component* analysis appears to erase the information from dense connectomes. Intuitively, this operation affecting the relative values of the components' activations, so a possible explanation is that some information lies in the relative order between the activation magnitude at each time-step. Other studies in literature confirm our hypothesis. For example *Sannino et al.* [10] where the author proposes a field-of-vision graph approach.

Table 2. Table showing the results of SVM linear classifier on timeseries and dense connectome data after applying normalisation measures.

Label	Method	Timeseries			Dense Connectome		
		Minmax	Norm	Scale	Minmax	Norm	Scale
Sex	Voxel	0.51	0.52	0.53	0.80	0.80	0.81
	Time	0.52	0.53	0.53	0.80	0.81	0.81
	Component	0.50	0.48	0.48	0.81	0.80	0.82
	Group time	0.52	0.54	0.50	0.81	0.80	0.81
	Group comp	0.51	0.49	0.50	0.50	0.51	0.52
Age	Voxel	0.39	0.38	0.39	0.44	0.42	0.42
	Time	0.37	0.39	0.38	0.43	0.42	0.43
	Component	0.37	0.38	0.39	0.41	0.43	0.44
	Group time	0.40	0.43	0.40	0.42	0.42	0.43
	Group comp	0.38	0.37	0.37	0.40	0.38	0.38

3.3 Noise

As a second step to normalization we proceeded with the data augmentation procedure. As we were mentioning before, the timeseries vary greatly in terms of variance and mean and seem to be trumped by the level of noise. To analyse this situation, we propose the generation of a synthetic dataset: this consists solely of random Gaussian noise, such that each generated point (i.e. activation of component i at timestep j) matches the mean and variance of the values at the point across the real subjects. The observation that adding small quantities of noise is not damaging to the classification rate of the SVM might open possibilities for data augmentation. In representation learning such approaches have had a great success in image denoising, for example in the case of the denoising autoencoders architectures [11]. However, it is challenging to think about interpreting the result of such algorithm in fMRI. In computer vision the visual inspection is paramount, and it is certainly a strong driver for the popularity of autoencoders, GANs and other high level feature extractors. The confirmatory glance on the output of such system is not available for fMRI, where the processes are not entirely understood. Nonetheless, we can use the previously discussed lax model of noise to augment the dataset, by essentially generating from each datapoint a set of noisy variants, all of which can be assumed to have same label. In more detail, the data X is split into two subsets, X_{train} and X_{test} (in the real work we use three, including a X_{val} subset for hyper-parameter tuning). From the values of X_{train}, we infer a noise generator, as described earlier in this subsection. This generator $\mathcal{G}$ is in fact a list:

$$\mathcal{G} = [(\mu_0, \sigma_0), (\mu_1, \sigma_1), ..., (\mu_N, \sigma_N)] \tag{2}$$

where N is the number of components and the μ_j and σ_j are computed group-wise on the temporal axis. Sampling from $\mathcal{G}$ is equivalent to sampling T times (the number of timesteps) from each of the $\mathcal{N}(\mu_j, \sigma_j)$. From each subject x in X_{train}, we generate K samples $\hat{x}_i$, where K is referred to as the noise inflation factor.

$$\epsilon_i \sim \mathcal{G}, \tag{3}$$
$$\hat{x}_i = (1 - p) \times x + p \times \epsilon_i \quad i \leq K \tag{4}$$

This way we obtain a dataset K times bigger. p allows the control over how much data is replaced by noise. In the experiments below, it was set to 0.1. Performing this operation after the train-test split is essential to ensure correct machine learning practices. Otherwise, we might end up in a situation where $\hat{x}_i$ is in the training set and $\hat{x}_j$ is in the testing set.

3.4 Modelling Different Labels

An immediate extension to this idea was splitting the data into groups according to their true labels and constructing the synthetic datasets to match the distribution of each group. Comparing the ability of the linear SVM to classify gender or age in true and synthetic datasets might indicate the importance of the *means* and *standard deviations* in performing classification tasks.

In Fig. 5, the values at the 0 mark correspond to the real dataset, whereas those at the 1 mark describe the synthetic dataset. The values in between do not have an application *per se* (the *test* labels had to be seen to know which *kind of noise to add*). It is, however, a validation of the effect mixing has: simultaneously reducing the correlation and the *Mahalanobis distance* between the test point and their label's distribution. As this model of noise seem to be overfitting (by allowing nearly perfect classification rates) we also analysed a more lax model, which considers the values of each component to be drawn from the same Gaussian distribution (see Fig. 5). Both models can be seen to remove the correlation between the data, which is the expected behaviour.

4   Experimental Results

In the last few years machine learning and neural networks development, has seen an incredible success due to the possibility of their employments in many different fields. Due to the increased computational power available the standard neural networks architectures have acquired depth making it possible to scale to highly dimensional data such as images, that would have been impossible to tackle with previous techniques. For a complete reference regarding deep learning and their different architectures please refer to [12] and [13]. In our experiments we made use of two types of architectures: the LSTM and Seq2Seq autoencoder architectures. Such architectures are designe for processing sequential data, such as time series [13]. LSTM were specifically created and thought to be able to manage the *vanishing gradient* problem, when numerical errors happening during the gradient descent calculation problem [13]. An example of an LSTM cell is

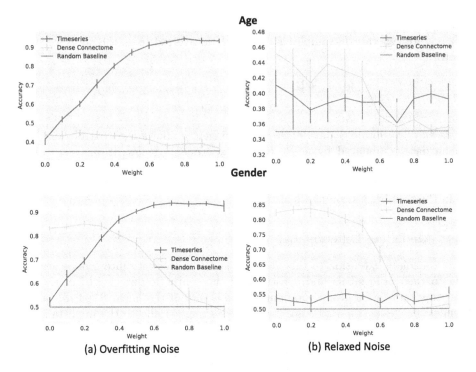

Fig. 5. Figure showing the two models of noise described above. The first columns shows an overfitting model of noise, which is too good a discriminator between classes and therefore not a faithful model. The second column shows a more relaxed version, which does not affect the classification rate through the timeseries. We can see that the classification accuracies for the timeseries are not greatly affected and for lower mixing coefficients the dense connectome proves to be robust as well. This inspired the data augmentation that we implemented.

presented in Fig. 6. Another important concept in deep learning is the concept of a generative model [12,13]. The premise behind the study of the generative models is the desire to understand the putative process that created the dataset. Autoencoders are an example of a tool developed an used for this purpose. The autoencoder framework prioritises, as suggested by its name, the process of

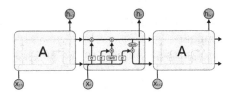

Fig. 6. An LSTM cell presented from an unrolled perspective. Designed by Chris Olah: http://colah.github.io/posts/2015-08-Understanding-LSTMs

encoding the information in a compact representation. This is achieved by using two constructs: an encoder $f_\theta(.)$ and a decoder, $g_\theta(.)$. The encoder computes the representation $h = f_\theta(x)$ for a data point x, whereas the decoder maps back from the representation space to the input space [12,13]. A graphical representation of an autoencoder is given in Fig. 7.

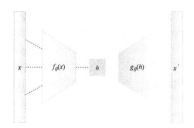

Fig. 7. Graphical representation of an autoencoder framework

Here we present the classification accuracies obtained by different neural models, applied to the original timeseries and the augmented ones, showing the improvements brought by the augmented dataset.

We can use the previously discussed model of noise to augment the dataset, by essentially generating from each datapoint a set of *noisy* variants, all of which can be assumed to have same label.

4.1 LSTM: Long Short Term Memory

A principal driver of this research was the disagreement between the sequential nature of the fMRI scan and the lack of recurrent neural networks in its study. In fact, the out-of-the-box LSTM obtains low accuracies on the timeseries data, mirroring the results we saw by using the linear SVM. In Table 3, we present the results for SVM, LSTM and the bidirectional LSTMs trained on the correlation, the timeseries and the augmented timeseries dataset [14]. The best performing architecture was the bidirectional LSTMs, and the augmented dataset, presented an important and significant improvement in accuracy. The LSTM architecture used for the experiments is the vanilla *Keras* [15] implementation, with *tanh* activation functions. The normalisation preprocessing is also dubbed by *Batch Normalisation* [16]. The optimisation routine is *Adam* [17], with a learning rate of 10^{-5}. The train-test-validation split is 60-30-10.

As observed through cross-validation, the bidirectional LSTM (bLSTM) was the most successful architecture, mimicking the findings of natural language processing [18].

The normalisation techniques were found to be beneficial to the bLSTM, increasing the accuracy by a significant margin, confirming earlier intuitions throughout the work.

Table 3. Table showing the mean accuracies for gender classification and their errors, after 5–10 repeated trials in each case. Although not applied on the timeseries, the correlation value is provided as a signal to the discrepancies between classification on timeseries and dense connectomes)

Method	Preprocessing	Accuracy
Baseline	–	0.5
Linear SVM	–	0.52 (0.01)
Linear SVM	Correlation	0.83 (0.02)
LSTM	–	0.49 (0.02)
bLSTM	–	**0.57 (0.02)**
bLSTM	Group scale	0.61 (0.02)
bLSTM	Group minmax	0.63 (0.01)
bLSTM	**Scale + augment**	**0.74 (0.03)**

An important discovery is the improvement brought by the data augmentation procedures described in Section. In Fig. 8, I present a more detailed look of this last model, indicating that increasing the augmentation factor is correlated with the accuracy increase.

This is very encouraging for researching recurrent neural models for fMRI data even further. Since the lack of data was observed to be a limiting factor, perhaps these techniques will gain traction as more and more data is being collected.

Furthermore, the observation that this LSTM model is performing better than alternatives might inspire that stronger data augmentation methods could be developed through the use of sequential autoencoder generation. An initial attempt at this is presented in the next subsection, which describes preliminary results obtained with a Seq2Seq Autoencoder.

4.2   Seq2seq: Sequence to Sequence Autoencoders

The success of training LSTMs on fMRI data can open possibilities for more structured machine learning approaches. As an example, we provide some preliminary results on representation learning using a sequence to sequence autoencoder (Seq2Seq).

The Sequence to Sequence Autoencoder architecture is an application of the *Encoder-Decoder* architecture in an unsupervised learning setup. A popular use case is the work of *Srivastava et al.* [19] for learning video representations.

The use of such structured autoencoders is crucial as an initial first step towards *end-to-end* learning in fMRI. Through using sequential operations, we limit the number of parameters from $\mathcal{O}(N^2)$ to $\mathcal{O}(NT)$ where N is the number of components (voxels, parcels, etc.) and T the number of timesteps in the analysis. Since the training times and the amount of data necessary for such machine

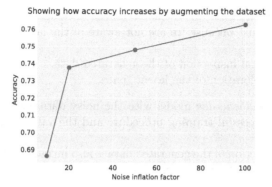

Fig. 8. This figure shows the increase in accuracy we can gain by augmenting the dataset. The noise inflation factor shows how many more sample we can create.

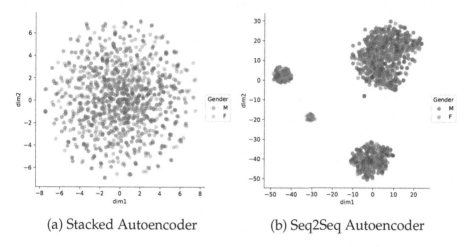

(a) Stacked Autoencoder   (b) Seq2Seq Autoencoder

Fig. 9. Two scatter plots obtained using the latent dimensions for the full dataset obtained by two autoencoders, followed by PCA and t-SNE. The interesting aspect is the presence of a clustering structure in the rightmost image.

learning algorithms to reach convergence are limiting, it is important to examine their feasibility on reduced views on such datasets.

In fact, the results for this type of autoencoder were not found to be in any way better from other simple implementation (such as a the simple autoencoder described before). However, any evaluation on this topic would be inconclusive.

The autoencoder is used in a purely unsupervised fashion, therefore ignoring all the information presented from the labels. The only a few available tools for assessing their success are:

– The reconstruction loss, whose absolute value is not informative, but we can judge through its dynamic whether the system is over-fitting or learning anything.

- Classification tasks using the available labels, although we can discard good representations just because we are not aware of the latent factors they end up capturing.
- Through the visual inspection of the generated data.
- Through the exploration of the latent space.

The *Sequence autoencoder* model with the noisy data augmentation on the HCP leads to a successful training procedure and the validation loss decreases during training.

The visual inspection of the generated data is also inconclusive. When observing pictures or more structured data, we are capable of assessing whether the dependencies between the features are "correct". In the case of fMRI, however, our knowledge of the *inherent structure of the data* is limited to the neuroscience field knowledge rather than being innate. This suggests that the representations learned by autoencoders should be used to draw conclusions, which are then validated by the wealth of experimental research.

A potential way to achieve this is through the exploration of the latent space. Because of the limited success of both simple autoencoders and Seq2Seq autoencoders in classification tasks, analysing the latent space is a last resort to querying the effectiveness of these methods.

In Fig. 9, I present in parallel the two latent representations. For the purpose of these illustrations, I am simply using the *Principle Component Analysis* (PCA) algorithm, followed by an application of *t-SNE* which has recently become popular for multivariate data visualisation purposes [20]. A remarkable aspect is that the sequential autoencoder (Fig. 9) is capable of finding a clustering structure in the data. We are unfortunately unable to detect what that relation is. This is in part due to the impossibility to capture all the factors that can affect one's brain in *metadata* files. Additionally, the source of the separation could be a signal of different error types, different machines, different parameters in preprocessing techniques and so on.

5   Conclusions

We describe a novel method for performing data augmentation on fMRI time-series, using statistical properties of the dataset. This allows increasing the accuracy of the LSTM on toy classification tasks, of which sex classification is detailed carefully. Furthermore, augmenting the dataset, we are also capable of training for the first time a sequence-to-sequence (Seq2Seq) autoencoder. Albeit with limited results, the definitive cluster structuring obtained through data encoding is inspiring and could be a starting point for developing these models further.

The application of sequential machine learning techniques to high dimensional and noisy datasets is crucial to advancing the field of bioinformatics. In the study of fMRI, combining international efforts for collecting more data, data augmentation techniques and more expressive models could lead to an increased understanding of the human brain. Using simple recurrent neural networks is only scratching the surface. Further work could include applying more expressive generative models, such as generative adversarial networks and variational autoencoders, as well as using graph convolutional networks and WaveNet-like models of convolution.

Acknowledgements. G.M.D. is funded by the Engineering and Physical Sciences Research Council (EPSRC) with International Doctoral Scholarship [number 1649557]. P.L. would like to acknowledge funding from the European Union's Horizon 2020 research and innovation programme PROPAGAGEING under grant agreement No. 634821.

References

1. Mikolov, T., Karafiát, M., Burget, L., Černockỳ, J., Khudanpur, S.: Recurrent neural network based language model. In: Eleventh Annual Conference of the International Speech Communication Association (2010)
2. Belliveau, J.W., et al.: Functional mapping of the human visual cortex by magnetic resonance imaging. Science **254**(5032), 716–719 (1991)
3. Burton, H., Snyder, A.Z., Conturo, T.E., Akbudak, E., Ollinger, J.M., Raichle, M.E.: Adaptive changes in early and late blind: a fMRI study of braille reading. J. Neurophysiol. **87**(1), 589–607 (2002)
4. Greicius, M.D., Supekar, K., Menon, V., Dougherty, R.F.: Resting-state functional connectivity reflects structural connectivity in the default mode network. Cereb. Cortex **19**(1), 72–78 (2009)
5. Wang, L., et al.: Changes in hippocampal connectivity in the early stages of Alzheimer's disease: evidence from resting state fMRI. Neuroimage **31**(2), 496–504 (2006)
6. Le Guennec, A., Malinowski, S., Tavenard, R.: Data augmentation for time series classification using convolutional neural networks. In: ECML/PKDD Workshop on Advanced Analytics and Learning on Temporal Data (2016)
7. Cui, Z., Chen, W., Chen, Y.: Multi-scale convolutional neural networks for time series classification. arXiv preprint arXiv:1603.06995 (2016)
8. Van Essen, D.C., Smith, S.M., Barch, D.M., Behrens, T.E.J., Yacoub, E., Ugurbil, K., Wu-Minn HCP Consortium, et al.: The Wu-Minn human connectome project: an overview. Neuroimage **80**, 62–79 (2013)
9. Glasser, M.F., et al.: The minimal preprocessing pipelines for the human connectome project. Neuroimage **80**, 105–124 (2013)
10. Sannino, S., Stramaglia, S., Lacasa, L., Marinazzo, D.: Visibility graphs for fMRI data: multiplex temporal graphs and their modulations across resting-state networks. Netw. Neurosci. **1**(3), 208–221 (2017)
11. Geng, X.-F., Xu, J.-H.: Application of autoencoder in depression diagnosis. DEStech Trans. Comput. Sci. Eng. (CSMA) (2017)
12. LeCun, Y., Bengio, Y., Hinton, G.: Deep learning. Nature **521**(7553), 436 (2015)
13. Goodfellow, I., Bengio, Y., Courville, A.: Deep Learning. MIT Press, Cambridge (2016)
14. Hochreiter, S., Schmidhuber, J.: Long short-term memory. Neural Comput. **9**(8), 1735–1780 (1997)
15. Chollet, F., et al.: Keras (2015)

16. Ioffe, S., Szegedy, C.: Batch normalization: accelerating deep network training by reducing internal covariate shift. arXiv preprint arXiv:1502.03167 (2015)
17. Kingma, D.P., Ba, J.: Adam: a method for stochastic optimization. arXiv preprint arXiv:1412.6980 (2014)
18. Irsoy, O., Cardie, C.: Opinion mining with deep recurrent neural networks. In: Proceedings of the 2014 Conference on Empirical Methods in Natural Language Processing (EMNLP), pp. 720–728 (2014)
19. Srivastava, N., Mansimov, E., Salakhudinov, R.: Unsupervised learning of video representations using LSTMs. In: International Conference on Machine Learning, pp. 843–852 (2015)
20. van der Maaten, L., Hinton, G.: Visualizing data using t-SNE. J. Mach. Learn. Res. 9(Nov), 2579–2605 (2008)

Exposing and Characterizing Subpopulations of Distinctly Regulated Genes by K-Plane Regression

Fabrizio Frasca[1]([⊠]) [iD], Matteo Matteucci[1] [iD], Marco J. Morelli[2,3] [iD],
and Marco Masseroli[1] [iD]

[1] Dipartimento di Elettronica, Informazione e Bioingegneria, Politecnico di Milano,
20133 Milan, Italy
`fabrizio.frasca@mail.polimi.it`,
{`matteo.matteucci,marco.masseroli`}`@polimi.it`
[2] Center for Genomic Science of IIT@SEMM, Istituto Italiano di Tecnologia (IIT),
20139 Milan, Italy
[3] Center for Translational Genomics and Bioinformatics,
IRCCS San Raffaele Scientific Institute, Via Olgettina 58, 20132 Milan, Italy
`morelli.marco@hsr.it`

Abstract. Understanding the roles and interplays of histone marks and transcription factors in the regulation of gene expression is of great interest in the development of non-invasive and personalized therapies. Computational studies at genome-wide scale represent a powerful explorative framework, allowing to draw general conclusions. However, a genome-wide approach only identifies generic regulative motifs, and possible multi-functional or co-regulative interactions may remain concealed. In this work, we hypothesize the presence of a number of distinct subpopulations of transcriptional regulative patterns within the set of protein coding genes that explain the statistical redundancy observed at a genome-wide level. We propose the application of a K-Plane Regression algorithm to partition the set of protein coding genes into clusters with specific shared regulative mechanisms. Our approach is completely data-driven and computes clusters of genes significantly better fitted by specific linear models, in contrast to single regressions. These clusters are characterized by distinct and sharper histonic input patterns, and different mean expression values.

Keywords: Gene expression · Epigenetic transcriptional regulation · K-Plane Regression

1 Introduction

In both complex and simple organisms, the regulation of gene expression is crucial in allowing cellular differentiation and response to environmental stimuli.

This work was supported by the Advanced ERC Grant "Data-Driven Genomic Computing (GeCo)" project (2016–2021), funded by the European Research Council.

M. Raposo et al. (Eds.): CIBB 2018, LNBI 11925, pp. 227–238, 2020.
https://doi.org/10.1007/978-3-030-34585-3_20

Among the many layers of gene regulation, those occurring at the stage of the initiation of transcription are considered to be the most flexible and effective [1]. Transcriptional regulation typically acts at the epigenetic level, i.e., without any modification of the underlying sequence, rather with a combination of binding of regulating molecules (transcription factors, or TF) to the DNA, and changing the structure of chromatin through the addition or removal of chemical residues on histone molecules (histone marks, HM).

Many studies have revealed the implications of epigenetic regulations: for example, their oncogenic role played in cancer etiology by gene expression alterations [2]. At the same time, epigenetic interactions can be targeted by non-invasive promising therapeutic possibilities, leveraging on their intrinsic reversibility [3]. These premises have contributed to the birth of the field of 'targeted cancer therapy', where epigenetic approaches could be used to treat cancer in a personalized manner, by kick-starting specific immune responses, or bringing back gene expression to physiological levels. Understanding the fundamental mechanisms by which histone marks and transcription factors operate to regulate the expression of specific genes is then of paramount interest as it is the necessary prerequisite in order to design effective and precise "epigenetic" drugs.

Next-generation sequencing (NGS) technologies nowadays routinely allow genome-wide measurements of gene expression and epigenetic signals in the cell lines or tissues of interest [4]. Large datasets are publicly available in repositories generated by multinational projects, such as ENCODE [5] and Roadmap Epigenomics [6]. One can now leverage on these large collections of data to quantitatively model the processes of interest and understand the specific roles, interplays and effects of epigenetic transcriptional regulators.

In particular, several statistical models have been conceived to study the association between gene-related epigenetic signals and messenger RNA (mRNA) abundance at a genome-wide scale [7]. Within this context, the problem is usually framed as a regression or classification task, where all the protein-coding genes are *samples*, signals from HMs and/or binding of TFs are *input features*, and the aim is to predict the *response value*, i.e., either mRNA levels (regression), or activities of genes (binary classification).

The relevance of genome-wide modeling resides in its omnicomprensive, explorative angle, as general conclusions can be drawn about the role and interplay of TFs and HMs. Interestingly, if at this level of resolution such features have been shown to be predictive for mRNA abundance [7], they have also been observed to exhibit certain *statistical redundancy* within themselves. In [8], the genome-wide resolution level itself has been addressed as the main cause for such observed redundancy in the regression task, as variations in the relative predictive power of TFs and HMs are observed at the finer resolution of groups of ontology-classified biological processes.

The work in [8] resorts, however, to manually curated external sources of information; in contrast, in the case of investigative analyses it is useful to let conclusions directly arise from data, including the least possible prior knowledge. This is, for instance, the case of targeted cancer therapy and personalized medicine, where the main objects of research are the possibly unknown alterations in epigenetic patterns and anomalies in their regulative effects: here, it is essential for statistical modeling to be data-driven.

In this paper, we address the problem of exposing, in a completely data-driven and interpretable manner, subpopulations of protein coding genes sharing specific dynamics of epigenetic transcriptional regulation. To this aim, we employ a version of the K-plane Regression algorithm to compute clusters where specific linear regression models are fitted to learn the internally shared regulative dynamics. With this approach, a one-to-one association between linear models and gene clusters follows and interpretative analyses are supported at best: regulative patterns can be investigated both at a gene-specific level and, statistically, at a gene-cluster level, and the regulative behavior can be a posteriori matched with the most-represented biological processes within a group.

The remainder of this paper is organized as follows. Section 2 introduces related scientific works. In Sect. 3, we discuss used data and techniques: we describe the considered epigenetic features, data sources and pre-processing steps, as well as the specific versions of K-Plane Regression algorithm we employ in our pipelines. In Sect. 4 we present the results: the models obtained are analyzed in their predictive accuracy and regulative patterns they captured. The conclusions are addressed in Sect. 5.

2   Scientific Background

So far, only few attempts have tried overcome the statistical redundancy observed in [8] in a data-driven manner: in [9] a mixture of Bayesian linear elastic nets revealed to better fit transcriptional regulation w.r.t. a single regression model and assign a distinct predictive relevance to epigenetic features. With this modeling approach, the statistical redundancy addressed in [8] is further specified in terms of feature *multi-functionality*: the epigenetic-transcriptional association is better modeled with an ensemble of models where HMs and TFs may assume different roles. However, even though in [9] the models accounting to the mixture are distinctly defined, genes in the dataset are only softly clustered, as the expression for a gene is the weighted sum of the outputs of all models.

As the 'soft' approach hinders the interpretation of the results, in this work we hypothesize the presence, within the set of protein coding genes, of a number of distinct, heterogeneous subpopulations with different transcriptional regulative behaviors, and let these clusters emerge from data by performing a hard partitioning of the whole gene set where specific linear regression models are fitted to learn the regulative dynamics of each gene sub-group.

Our problem is therefore similar to a *piecewise linear affine model fitting*, usually approached with *hinging hyperplane* [10] or *bounded error* [11] methods. However, our main goal is not to fit a supposed non-linear dynamic with piecewise linear functions, but rather learning different linear models in a scenario where dynamics are likely to be overlapped, discontinuous, and partially lying on sub-dimensional manifolds.

A more suited approach is then represented by *K-Plane Regression* [12], firstly introduced in the general context of fitting possibly discontinuous functions with an ensemble of linear models. The method is based on a clustering approach: it finds a fixed number (K) of hyperplanes such that each point in the training set is close to one of the hyperplanes, and all points in a partition are as close as possible in the input feature space. Resulting hyperplanes are found by minimizing the following objective function with an Expectation-Maximization (EM) algorithm:

$$E(\Theta) = \sum_{k=0}^{K-1} \sum_{i \in \Theta(k)} (t_i - \tilde{\boldsymbol{w}}_k^T \tilde{\boldsymbol{x}}_i)^2 + \gamma \|\boldsymbol{x}_i - \boldsymbol{\mu}_k\|_2^2, \tag{1}$$

where K is the pre-defined number of clusters, Θ defines the partitioning over the dataset – with $\Theta(k)$ the set of samples in cluster k, $\tilde{\boldsymbol{x}}_i$ and t_i are, respectively, the input feature and the target value for sample $i \in \Theta(k)$, $\boldsymbol{w}_k$ is the weight vector of the least square solution for those points, $\boldsymbol{\mu}$ terms refer to centroids in the feature space and γ is a user-defined parameter deciding the relative weight of the two additive terms in the objective function. The 'tilde' notation is used to indicate the inclusion of the bias term in the regression.

Given the capability of K-Plane Regression to fit discontinuous functions and the increased flexibility offered by a clustering approach, we built upon this last piece of work to solve our problem of modeling epigenetic transcriptional regulation with a hard ensemble of linear regression models.

3   Materials and Methods

This work aims at modeling epigenetic transcriptional regulation with a hard ensemble of linear regression models, each one explaining mRNA levels as a function of epigenetic signals for a specific gene sub-group, i.e., a cluster of genes.

3.1   Biological Setting and Data

All considered measurements are over the chronic myeloid leukemia K562 immortalized cell line (human blood tissue), and only involve protein coding genes. GENCODE v10 reference annotation for the hg19 assembly was used to retrieve their transcription start sites (TSSs). The Roadmap Epigenomics Mapping Consortium's (REMC) [6] repository was chosen as the only data source in this work.

Genes are epigenetically characterized by data in the form of processed ChIP-Seq peaks for all the $m = 12$ histone modifications measured in K562 cell line within the REMC project (no TF was accounted for). Peaks for the considered HMs were retrieved in the formats of 'bed narrowPeak' or 'bed broadPeak', according to the nature of the considered mark, i.e., TF-like (sharp) marks or broad marks. These choices are summarized in Table 1. The epigenetic status of the generic gene g is represented as an m-dimensional input vector $\boldsymbol{x}_g$, whose

Table 1. HMs and their chosen data format.

Marks	Format
H3K4me1, H3K4me2, H3K4me3, H3K9ac, H3K9me1, H3K27ac, H2A.Z	narrowPeak
H3K9me3, H3K27me3, H3K36me3, H3K79me2, H4K20me1	broadPeak

elements contain the maximum peak enrichment value attained within a symmetric window region of 10 kbases centered on g's TSS, thus summarizing the g-related status of a specific monitored HM. In accordance to [7], signals closer to the genes' TSSs (roughly, within promoters) are the most informative to predict gene expression. These vectors, considered together for all our $n = 19{,}794$ genes, form an input matrix X, with dimensions $n \times m$.

For the transcriptional characterization of genes, we consider mRNA quantifications, measured with RNA-sequencing. The transcriptional status of gene g is encoded by $t_g = \sqrt{\log(1 + \tau_g)}$, where τ_g is the original mRNA quantification, log represents the natural logarithm and the application of two sub-linear, monotonically increasing functions aims at reducing the heteroskedasticity in regression residuals. Finally, consider t_g as the $(g + 1)$-th element in n-dimensional target vector T collecting the transcriptional statuses of all the genes.

Together, X and T form our dataset $D = \langle X, T \rangle$, which is going to be partitioned by K-Plane Regression algorithm.

3.2 K-Plane Regression

K-plane Regression is designed to minimize the following objective function:

$$E(\Theta) = \sum_{k=0}^{K-1} \sum_{i \in \Theta(k)} (t_i - \tilde{\boldsymbol{w}}_k^T \tilde{\boldsymbol{x}}_i)^2, \tag{2}$$

where we dropped the second additive term $\gamma \|\boldsymbol{x}_i - \boldsymbol{\mu}_k\|_2^2$ originally present in Eq. 1, which enforces points belonging to the same partition to be close to each other. Such a 'closeness' term was explicitly introduced in [12] to avoid EM finding sub-optimal solutions, and to force the obtained partitions not to contain points from disjoint regions of the input feature space. In our application, where the objective is the disambiguation of overlapped expression dynamics from heterogeneous gene subpopulations, not only the l2-distance measure might not be suitable over historic inputs, but also partitions spanning disjoint feature space regions are not necessarily to be avoided. Concerning the issue of possible sub-optimality, we just resorted to a simple multiple re-initialization technique.

The K-Plane Regression procedure is run R times; each time it starts by a random partition and optimizes Eq. 2 as described in [12], i.e., by iteratively alternating a Maximization step – hyperplanes to clusters fitting – and an Expectation step – gene-cluster reassignments. A tolerance parameter τ is used to verify numerical convergence of the optimization procedure. Initializations are designed to construct a completely random partitioning made up of equally-sized clusters. In the end, among the R solutions obtained, the one attaining the best objective value is returned.

A draft of our K-Plane Regression pipeline is reported below.

```
procedure ClusterHyperplanes(D, K, τ, R)
    S ← empty solution dictionary
    O ← array of R elements
    for r = 0, 1, ..., R − 1                          ▷ perform R runs
        Θ₀ ← RandomInit(|D|, K)                       ▷ randomly initialize
        s, o ← K-PlaneRegression(D, K, Θ₀, τ)         ▷ run K-Plane
        S[r] ← s
        Oᵣ ← o
    best ← arg maxᵣ Oᵣ                                ▷ choose best
    return S[best]
```

```
procedure K-PlaneRegression(D, K, Θ, τ)
    X, T ← input matrix and target vector from D
    oᵖʳᵉᵛ ← ∞
    oᶜᵘʳʳ ← E(Θ)
    while (oᵖʳᵉᵛ − oᶜᵘʳʳ > τ)                        ▷ loop EM until convergence
        M ← empty model dictionary
        for (k = 0, 1, ..., K − 1)                    ▷ Maximization
            M[k] ← OLS solution over ⟨X[Θ(k)], T[Θ(k)]⟩
        Θ ← empty cluster assignment
        for (i = 0, 1, ..., |D| − 1})                 ▷ Expectation
            Θ[i] ← arg minₖ |M[k](Xᵢ) − Tᵢ|
        oᵖʳᵉᵛ ← oᶜᵘʳʳ
        oᶜᵘʳʳ ← E(Θ)
    return Θ, oᶜᵘʳʳ
```

```
procedure RandomInit(n, K)
    σ ← random permutation of [0, 1, ..., n − 1]
    Θ ← empty cluster assignment
    for (k = 0, 1, ..., K − 1)
        Θ[k] ← kᵗʰ subpart of σ
    return Θ
```

4 Experimental Results

In our experiments, the procedure ClusterHyperplanes has been run with the following parameters: $\mathcal{D} = D, K \in [2 \ldots 6], \tau = 0.1, R = 30$. The parameter K ranges in $[2 \ldots 6]$ as the best value is hardly guessable a priori and might depend on the nature of the specific problem. Better solutions, in terms of cost functions, have been observed for larger values of K: Fig. 1 depicts the trend of the convergence objective value as a function of this parameter. In the following, results for the value $K = 4$ are discussed: this choice is less prone to overfit spurious correlations, still yielding a good value of the objective function. In other words, it represents a convenient trade-off between goodness of fit and biological interpretability, given the current knowledge about HM (co-)activity.

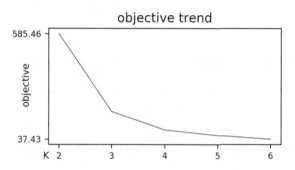

Fig. 1. Values of best solutions (objective) from re-initialized K-Plane Regression as a function of number of clusters (K).

Let $\Theta = \{\vartheta_0, \ldots, \vartheta_{K-1}\}$ be our obtained solution, with $K = 4$ and ϑ_k representing the $(k+1)$-th cluster of genes computed by the algorithm. In correspondence with this partitioning, an ensemble of cluster-wise linear models can be considered as $M = \{\mu_0, \ldots, \mu_{K-1}\}$, where μ_k represents the hyperplane being the least square solution over genes in ϑ_k. Our solution Θ is compared with $\Theta_{gw} = \{\vartheta_{gw}\}, \vartheta_{gw} = \{0, 1, \ldots, n-1\}$, the degenerate partitioning made of a single cluster indexing the whole dataset D. This solution corresponds to setting $K = 1$, i.e., to the use of a single linear model fitted over the entire dataset D. In the following, we call this model the genome-wide model, labeled as μ_{gw}.

4.1   Enhanced (Cluster-wise) Fitting

Our K-Plane Regression managed to cluster genes with common regulative behaviors, as the obtained model ensemble effectively enhanced data fitting.

Not only the objective value associated with Θ_{gw} is much larger than that associated with our solution Θ (3892.08 vs. 339.83), but also fitting is better at the level of all the computed clusters. The regression scores computed specifically over clusters in Θ, for both cluster-wise and genome-wide models, are reported in Table 2, in terms of residual sum of squares (RSS) and coefficients of determination (R^2). The i-th row of the table contains scores for models μ_i and μ_{gw} over Cluster ϑ_i – subscripts '$_{cw}$' and '$_{gw}$', respectively.

In Table 2, the effectiveness of our approach is confirmed by the fact that clusters are always better fitted by cluster-wise models than by μ_{gw}. Moreover, the

Table 2. Cluster specific figures of merit. For cluster k, RSS_{cw} and RSS_{gw} are the residual sum of squares of μ_k, μ_{gw} over ϑ_k, while R^2_{cw} and R^2_{gw} refer to the coefficients of determination of μ_k, μ_{gw} over ϑ_k.

Cluster (cardinality)	RSS_{cw}	RSS_{gw}	R^2_{cw}	R^2_{gw}
$0(2,717)$	79.22	1393.00	0.80	-2.50
$1(7,547)$	82.05	1514.44	0.54	-7.57
$2(5,045)$	85.64	714.70	0.84	-0.37
$3(4,485)$	92.90	269.92	0.92	0.76

specific linear models are very good in explaining the epigenetic transcriptional regulation of a large part of the genes (see to column "R^2_{cw}"). In three clusters out of four, the R^2 scores from μ_{gw} are negative, implying that the fitting over the genes of the clusters is worse than the constant mean model.

The intuition that μ_{gw} is likely to only capture the regulative mechanisms of genes with "intermediate" regulative behaviour, such as those in Cluster 3, is supported by what observed in Fig. 2, where cluster-specific residuals (y-axis) from cluster-wise and genome-wide models are plotted against target values (x-axis).

Residuals from the genome-wide model are generally more disperse and heteroskedastic, except for Cluster 3 – the only one where μ_{gw} attains positive R^2 – where they are similar to those from the cluster-wise model μ_3, fitting genes very well. The overall R^2 of 0.66 attained by μ_{gw} on the whole dataset D is, consequently, an intermediate value resulting from putting together mildly-modeled genes (Cluster 3) with the remaining ones, where the genome-wide model seems to be rather inadequate.

Hard hyperplanes clustering has revealed the criticality of single genome-wide regression by exposing subsets of genes under-fitted by μ_{gw}. Clearly, in a realistic setting, such as targeted cancer therapy, it would be completely unacceptable to fit only 23% of protein coding genes (Cluster 3), as conceptually wrong conclusions might be drawn about the epigenetic regulative behaviors of the remaining genes.

4.2   Cluster Characterization

The effectiveness of hyperplanes clustering also emerges by observing how the obtained clusters are distinct in terms of the input patterns and mean expression value for the genes they contain. In this sub-section we leverage on the enhanced interpretability coming from a hard gene partitioning to characterize the computed clusters.

For the generic cluster-wise model μ_i, let $\tilde{\boldsymbol{w}}_i$ be its weight vector, made of the learnt intercept and regression coefficients ($m+1$ elements). For gene g in ϑ_i, let $\boldsymbol{\psi}_g$ be its weighted input vector, obtained by an element-wise multiplication between its input vector $\tilde{\boldsymbol{x}}_g$ and $\tilde{\boldsymbol{w}}_i$ – the input vector is 1-edged to account

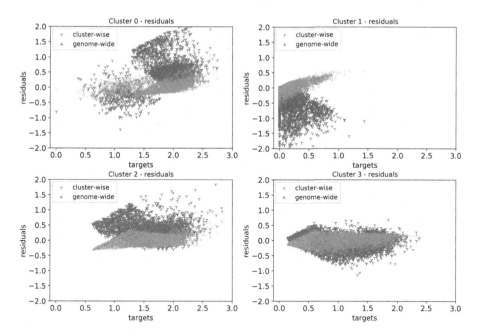

Fig. 2. Cluster specific residuals from cluster-wise (orange) and genome-wide (blue) models. (Color figure online)

for bias. The weighted input vectors are an effective means to quickly assess the importance of single features in determining the predicted response value, as $y_g = \sum_{j=0}^{m}(\psi_{g_j})$. The weighted input vectors for all the genes in a cluster generate feature-wise boxplots which illustrate the frequency distributions of cluster-specific histone contributions and their associated dispersion.

Figure 3 depicts the patterns obtained by considering the feature-wise medians of the weighted input vectors, specifically for each cluster, along with the 25th and 75th percentiles of their distributions. In the patterns, intercepts are in orange, whilst HMs are green if associated with positive regression weight (computed *activators*) and red otherwise (computed *repressors*), with semi-transparent rendering for weights not passing a statistical F-test with significance $\alpha = 0.01$. In this way, fictitious correlations are pruned, resulting in simpler and more robust patterns.

Cluster 1 is the most populated one and comprises genes with a negligible histonic activity and usually null expression: these genes are likely to never be activated, being instead repressed at a chromatin level, like for example developmental genes. The flat input and the low intercept are consistent with the related expression distribution (0.0 RPKM median value). In such a scenario, a lower signal-to-noise-ratio is the probable cause of the mild attained R^2 score in this cluster (see Table 2).

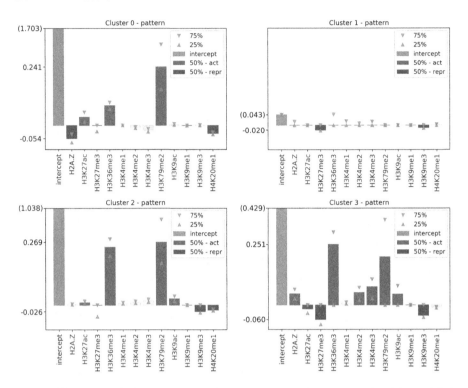

Fig. 3. Cluster specific input patterns; in green *computed activators* (positive weight), in red *computed repressors* (negative weight). (Color figure online)

Clusters 2 and 0 are made, respectively, of low and high expressed genes (RPKM medians 12.73 and 32.23). This characterization is confirmed by the intercepts and the predominant roles assumed by activator H3K79me2, and H3K36me3 specifically in Cluster 2, with their large variations explaining higher expression levels. The two clusters show different relative regulative relevance from repressors H2A.Z and H3K9me3, and activators H3K27ac and H3K9ac.

Cluster 3 embraces null to low transcriptional activity (RPKM median 2.32) and is characterized by a more complex input pattern: more relevant than in other clusters are H3K27me3, H3K4me2 and H3K4me3. The simultaneous presence of activating and repressive marks (H2A.Z and H3K27ac, respectively) recapitulate the characteristic of bivalent promoters, whose genes could be poised for fast activation when needed. Moreover, single HMs might counterbalance one another and/or co-work to induce particular effects.

It is interesting to notice the similarity between the pattern of Cluster 3 with that of the genome-wide one, as shown side-by-side in Fig. 4. This is a further confirmation the algorithm has managed to expose the sub-group of genes possessing the largest leverage in bending one single regression hyperplane. As emerging from the cluster characterization above, the remaining population has instead been set apart in a well differentiated manner: genes lacking the single

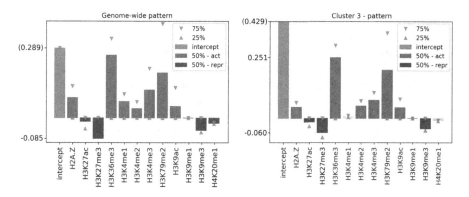

Fig. 4. Genome-wide input pattern (left) vs. Cluster 3 input pattern (right).

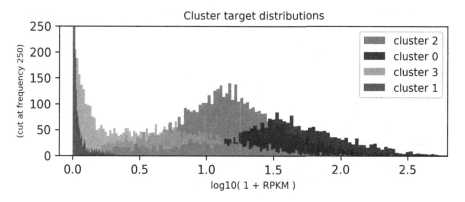

Fig. 5. Cluster target distributions (base-10 log of RPKMs).

genome-wide fit have been naturally stratified according to their expression value in groups with distinct characteristic input patterns.

Such discussed stratification for transcriptional activity is finally recapitulated in Fig. 5, which reports, for each cluster, the distribution of the expression response value in base-10 logarithm of RPKM.

5   Conclusions

We proposed the application of a randomly re-initialized version of K-Plane Regression to expose subpopulations of protein coding genes commonly regulated at an epigenetic-histonic level. The proposed approach has revealed how a single regression model only captures the fit of a sub-group of genes with null to low expression and how poor scores from μ_{gw} on the remaining genes are due to unfitting rather than linear under-fitting. The hard gene partitioning produced by our method allowed instead a statistical characterization of the computed clusters in terms of input contribution patterns, revealing how clusters stratify for higher and higher expression levels, with histone marks assuming specific roles of different relevance.

Future developments will involve the biological characterization of the found gene clusters: grouped genes will be analyzed to find possibly enriched biological processes they are involved into. Also, further investigations will target the optimal choice of the pre-determined number of clusters, i.e., K: we will analyze cluster-specific target distributions and patterns as a function of parameter K to understand the partitioning dynamic behavior of the algorithm.

References

1. Maston, G., Evans, S., Green, M.R.: Transcriptional regulatory elements in the human genome. Annu. Rev. Genomics Hum. Genet. **7**, 29–59 (2006)
2. Vaquerizas, J., Kummerfeld, S., Teichmann, S., Luscombe, N.: A census of human transcription factors: function, expression and evolution. Nat. Rev. Genet. **10**, 252–263 (2009)
3. Bannister, A.J., Kouzarides, T.: Regulation of chromatin by histone modifications. Cell Res. **21**(3), 381–395 (2011)
4. Levy, S.E., Myers, R.M.: Advancements in next-generation sequencing. Annu. Rev. Genomics Hum. Genet. **17**, 95–115 (2016)
5. ENCODE Project Consortium: An integrated encyclopedia of DNA elements in the human genome. Nature **489**(7414), 57–74 (2012)
6. Kundaje, A., et al.: Integrative analysis of 111 reference human epigenomes. Nature **518**, 317–330 (2015)
7. Cheng, C., et al.: A statistical framework for modeling gene expression using chromatin features and application to modENCODE datasets. Genome Biol. **12**, R15 (2011)
8. Budden, D., Hurley, D., Cursons, J., Markham, J., Davis, M., Crampin, E.: Predicting expression: the complementary power of histone modification and transcription factor binding data. Epigenetics Chromatin **7**, 36 (2014)
9. do Rego, T.G., Roider, H.G., de Carvalho, F.A.T., Costa, I.G.: Inferring epigenetic and transcriptional regulation during blood cell development with a mixture of sparse linear models. Bioinformatics **28**(18), 2297–2303 (2012)
10. Breiman, L.: Hinging hyperplanes for regression, classification, and function approximation. IEEE Trans. Inf. Theory **39**, 999–1013 (1993)
11. Amaldi, E., Mattavelli, M.: The MIN PFS problem and piecewise linear model estimation. Discrete Appl. Math. **118**, 115–143 (2002)
12. Manwani, N., Sastry, P.: K-plane regression. Inf. Sci. **292**, 39–56 (2015)

Network Propagation-Based Semi-supervised Identification of Genes Associated with Autism Spectrum Disorder

Hugo F. M. C. Martiniano[1,2]($\boxtimes$)(ID), Muhammad Asif[1,2](ID),
Astrid Moura Vicente[1,2](ID), and Luís Correia[1](ID)

[1] Faculdade de Ciências, BioISI - Biosystems and Integrative Sciences Institute,
Universidade de Lisboa, Campo Grande, 1749-016 Lisbon, Portugal
`hfmartiniano@ciencias.ulisba.pt`, `muhasif123@gmail.com`
[2] Instituto Nacional de Saúde Doutor Ricardo Jorge,
Avenida Padre Cruz, 1649-016 Lisbon, Portugal

Abstract. Autism Spectrum Disorder (ASD) is an etiologically and clinically heterogeneous neurodevelopmental disorder with more than 800 putative risk genes. This heterogeneity, coupled with the low penetrance of most ASD-associated mutations presents a challenge in identifying the relevant genetic determinants of ASD. We developed a machine learning semi-supervised gene scoring and classification method based on network propagation using a variant of the random walk with restart algorithm to identify and rank genes according to their association to know ASD-related genes. The method combines information from protein-protein interactions and positive (disease-related) and negative (disease-unrelated) genes. Our results indicate that the proposed method can classify held-out known disease genes in a cross-validation setting with good performance (area under the receiver operating curve ∼0.85, area under the precision-recall curve ∼0.8 and Matthews correlation coefficient 0.57). We found a set of top-ranking novel candidate genes identified by the method to be significantly enriched for pathways related to synaptic transmission and ion transport and specific neurotransmitter-associated pathways previously shown to be associated with ASD. Most of the novel candidate genes were found to be targeted by *denovo* single nucleotide variants in ASD patients.

Keywords: Semi-supervised learning · Network propagation · Autism Spectrum Disorder · Machine learning · Protein-protein interactions

1 Introduction

Autism Spectrum Disorder (ASD) is an etiologically and clinically heterogeneous neurodevelopmental disorder, with an approximate prevalence of 1% of the population [1]. It is characterized by early-onset of difficulties in social interaction

© Springer Nature Switzerland AG 2020
M. Raposo et al. (Eds.): CIBB 2018, LNBI 11925, pp. 239–248, 2020.
https://doi.org/10.1007/978-3-030-34585-3_21

and communication, and repetitive, restricted behaviors, interests, or activities, frequently associated with co-morbidities like intellectual disability, epilepsy and language disabilities [2].

Twin studies comparing the concordance in identical and fraternal twins provide evidence for a string contribution of genetic factors to ASD risk variance [3]. In the past decade, an intense effort to identify the genetic determinants of ASD has been undertaken by several international consortia. This produced several large-scale genomic datasets using high-throughput techniques, including Single Nucleotide Polymorphisms (SNPs) from genome-wide association studies, whole exome sequencing and whole genome sequencing [3–5].

However, despite the enormous amount of data generated over the past decade [3], a clear genetic cause can only be identified in 10% to 20% of affected individuals, and diagnostic still relies on clinical observation, rather than ASD-specific etiology. Genetic causes identified in ASD patients include large chromosomal abnormalities, Copy Number Variants (CNVs) and single-gene mutations, either transmitted or *de novo*. More than putative 800 ASD-risk genes have been identified [3], but most CNVs and Single Nucleotide Variants (SNVs) are characterized by incomplete penetrance [3,6].

An increasing body of evidence suggests that ASD risk variance is the result of a combination of common and rare variants, acting on specific biological pathways (for example, neuronal development and axonal guidance, synaptic function, and chromatin remodeling) [7], with various transmission modes, and possibly combined with gene-environment interactions [3].

This heterogeneity presents a challenge for translational approaches and therefore clinical application lags behind the research knowledge [6].

One of the consequences of the genetic heterogeneity of ASD is that the sample sizes necessary for establishing statistically significant genotype-phenotype associations are very large (of the order of tens of thousands of individuals).

Given this scenario, one possible alternative to large-scale genomic studies is to target genes involved in known ASD-related biological pathways to gain insight into the specific processes disrupted in each individual patient or group of patients.

One promising approach for the identification of novel ASD-related genes and pathways is through the application of machine learning techniques.

2 Scientific Background

Krishnan *et al.* [8] used Support Vector Machines (SVM) to discriminate ASD-related from ASD-unrelated genes based on information from a human brain-specific gene interaction network. Despite obtaining a good (AUC = 0.8), they used an arbitrary candidate gene-weighing scheme.

Asif *et al.* [9] achieved an AUC of 0.8 for the identification of ASD-related genes using a Random Forest trained on gene semantic similarity measures calculated from Gene Ontology.

In this paper we describe a semi-supervised machine learning approach to identify and rank potentially relevant disease risk genes, based on prior information from publicly available databases. We present the results of the application of this method to the discovery of novel ASD risk genes.

We used a modification of the semi-supervised classification method proposed by Zhou et al. [10], which is closely related to Network Propagation (NP). NP is a family of methods that use the flow of information through network connections as a means to establish relationships between nodes. Several NP variants are widely used to identify genes and genetic modules that underlie a process of interest (see for instance [11] for a review). The main concept behind the technique, when applied to gene-gene interaction networks, is that genes underlying the same phenotype tend to interact. NP has been has been applied to the discovery of significantly connected gene modules associated with ASD [12].

To the best of our knowledge, reported applications of NP (for example [12]), are done within a positive-unlabeled learning framework. In this work we use a different approach, where the algorithm learns from both positive-labeled (disease-associated) and negative-labeled (non disease-associated) genes. Although it is not currently possible to define a true set of negative genes for ASD the inclusion of a negative gene set adds more information for the method to exploit, as genes thought to be unrelated to the disease (and their neighbors) are down-weighed.

3 Materials and Methods

3.1 Semi-supervised Learning Algorithm

We use a modified version of the semi-supervised classification method first described by Zhou et al. [10]. The algorithm uses as inputs a network $G(\mathcal{V}, \mathcal{E})$, where the nodes, $\mathcal{V}$, represent genes, and the edges, $\mathcal{E}$, represent gene-gene interactions, and two gene sets, $\mathcal{P}$ and $\mathcal{N}$, containing disease-related and non disease-related genes, respectively.

Using information from the predefined gene sets, we build an initial score vector (f^0) for all genes in the network, where the element f_i^0 corresponding to gene g_i is:

$$f_i^0 = \begin{cases} \dfrac{1}{|\mathcal{P}|} & g_i \in \mathcal{P} \\[2mm] -\dfrac{1}{|\mathcal{N}|} & g_i \in \mathcal{N} \\[2mm] 0 & g_i \notin \mathcal{P} \wedge g_i \notin \mathcal{N} \end{cases} \tag{1}$$

The initial scores are then propagated through the network, using the iterative formulation of the random walk with restart (RWR) algorithm [11]:

$$f^{t+1} = (1 - \lambda)\mathbf{W}f^t + \lambda f^0 \tag{2}$$

Where f^t is the vector of gene scores for step t, λ is the restart coefficient ($\lambda = 0$ corresponds to a random walk without restart) and $\mathbf{W}$ is a weight matrix derived from the normalization of the network adjacency matrix: $\mathbf{W} = \mathbf{D}^{\frac{1}{2}}\mathbf{A}\mathbf{D}^{\frac{1}{2}}$, where $\mathbf{A}$ is the adjacency matrix and $\mathbf{D}$ is the diagonal node degree matrix $\mathbf{D} = \mathrm{diag}(d_1, d_2, \ldots, d_j, \ldots, d_n)$, where $d_j = \sum_i A_{ij}$.

In the case of weighed networks, the A_{ij} matrix element is the weight of the edge connecting the i and j nodes, and the D_{ii} matrix element contains the sum of the weights of all edges connected to node i.

The iterative update (Eq. 2) is performed until convergence of f^t, that is until it verifies $||f^t - f^{t-1}||^2 < 1 \times 10^{-6}$.

In the context of the random walk with restart framework, this is equivalent to a random walk starting with equal probabilities from the positive and negative nodes. After convergence, the probabilities of the visits from negative nodes are subtracted from the probabilities of visits from positives nodes, yielding the final score vector.

Performance was evaluated using 10-fold cross-validation, with λ varying from 0.1 to 0.9 in 0.1 steps. As performance evaluation metric we used the area under the receiver operating curve (AUROC), the area under the precision-recall curve (AUPRC), the Matthews correlation coefficient (MCC), and report mean values over all folds for each λ parameter.

3.2   Gene Sets

The Simons Foundation for Autism Research (SFARI)[1] curates an authoritative list of ASD-associated genes. The Human Gene Module of the SFARI database lists 1007 genes, grouped into several categories, according to the level of evidence linking them to ASD (Table 1). The genes in the *Syndromic* category are implicated in syndromic forms of autism, in which subpopulations of patients with a specific genetic syndrome, such as Angelman syndrome or fragile X syndrome, present symptoms of autism. These can also be in one of the other categories, or just classified as *Syndromic*.

Table 1. Categories of Autism candidate risk genes from SFARI.

Category	Description	Number of genes
S	*Syndromic*	143
1	*High confidence*	25
2	*Strong candidate*	58
3	*Suggestive evidence*	176
4	*Minimal evidence*	405
5	*Hypothesized but untested*	157
6	*Evidence does not support a role*	21
–	*Uncategorized*	88

---

[1] https://gene.sfari.org/database/human-gene/, accessed 1 May 2018.

For this application we labeled as positive class the genes from SFARI categories 1, 2, 3 and 4, as well as those classified just as syndromic. We labeled as negatives the non mental genes used by Krishnan *et* al. [8], from which we removed the ones overlapping with genes from the positive class. All other genes in the network, including those from SFARI categories 5 to 6, are unlabeled and therefore are assigned a score of 0, according to Eq. 1.

For both labeled gene sets we converted the gene identifiers to the latest Hugo Gene Nomenclature Consortium (HGNC) symbols, discarding all genes for which there was no correspondence. This resulted in two sets with 739 positive and 1132 negative genes, which correspond to the $\mathcal{P}$ and $\mathcal{N}$ sets mentioned above.

3.3 Biological Network

As input network we used the STRING [13] (version 10.5) protein-protein interaction (PPI) database. The STRING database contains both experimental data and interactions inferred from text-mining the scientific literature, with an associated confidence score.

From the human subset of PPI interactions we converted all protein identifiers to the respective HGNC symbol using data obtained from the Ensembl BioMart [14], discarding those involving symbols for which conversion was not possible. Redundant interactions were removed, keeping the interaction with the highest confidence score.

We built a network with the entire set of edges, assigning the confidence score to edge weights. Only the largest connected component was selected and all other nodes were deleted. The final String-derived network contains 18003 genes and 5007158 edges.

3.4 Enrichment Analysis

To characterize the identified ASD-related gene sets we used enrichment analysis (hypergeometric test) for pathways from the Reactome database, using the *Enrichr* web Application Programming Interface[2]. All reported p-values are adjusted for multiple testing with the Benjamini-Hochberg correction method.

4 Results and Discussion

4.1 Performance Evaluation

To evaluate the capability of the method to identify disease-related genes we performed 10-fold cross-validation on the gene sets defined above. Generally, we observe that the method exhibits a good capability to classify genes in the held-out folds.

---

[2] http://amp.pharm.mssm.edu/Enrichr, accessed 1 Jun. 2018.

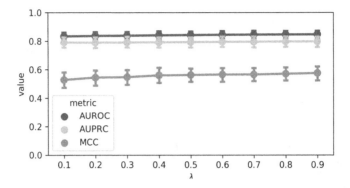

Fig. 1. Mean AUROC, AUPRC and MCC values from 10-fold cross validation for various values of lambda. Error bars represent 95% confidence intervals.

None of the evaluation metrics are very sensitive to the restart parameter (λ), as can be seen in Fig. 1. Maximum values for all metrics are obtained for $\lambda = 0.9$, with an AUROC values 0.85, an AUPRC of 0.8 and a MCC of 0.57.

For comparison we also tried applying the procedure to networks produced by setting edge cutoffs based on the interaction confidence, for values, 400 and 700, corresponding to medium and high confidence interactions, respectively (results not shown). Cross-validated mean AUROC values for these networks, with or without confidence scores as weights are substantially ($\sim$5% and $\sim$15%, respectively) lower than for the full weighed network.

Treating the networks as weighed or binarizing interactions after applying cutoffs yields AUROC values differing by less than 1%. This indicates that the use of weighed networks has little impact on the performance of the method. In the following we analyze the scores obtained with $\lambda = 0.9$.

4.2 Selection and Characterization of Top Ranking Genes

The final score vector (f^∞) contains a score for each gene in the network, corresponding to the degree of association with the disease.

To assess the relevance of the top-ranking genes identified by the method we selected, from the cross-validation procedure, a threshold corresponding to the score of 95th percentile. Genes with scores above the threshold are considered as strong candidates for association to ASD.

The composition of the set of top candidate genes is displayed in Fig. 2. As expected, the top ranking genes are constituted by genes in the SFARI categories 1 to 4 or Syndromic, which are defined beforehand as belonging to the positive class. In addition to those, the method identifies six genes (*SYN3, COP1, CBLN1, HTR2A, GABRB1, CLSTN3*) from SFARI category 5 (*Hypothesized but untested*).

Eight genes (*DLGAP3, GRM1, PPFIA1, SLC24A2, GRIK3, PHF8, PTPRT, CACNA1B*) added to the SFARI database but, as of now, not yet assigned to any category (*Unc* in Fig. 2), are also present in the candidate gene sets.

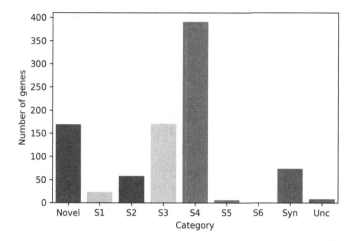

Fig. 2. Composition of the selected gene set. S1 to S6 correspond to SFARI gene categories 1 to 6, *Syn* denotes SFARI genes classified as Syndromic, *Unc* denotes genes in the SFARI gene list which are not yet categorized, *Novel* denotes genes not present in SFARI.

We consider the remaining 177 top-ranking genes as putative new candidate risk genes (*Novel* in Fig. 2). Of these, only some are directly implicated in human pathologies. The Online Mendelian Inheritance in Man (OMIM) database lists the *DLG3* gene as associated with *X-linked mental retardation-90* (MRX90, OMIM:300850), *CDH15* is associated to autosomal dominant mental retardation-3 (MRD3, OMIM:612580), *CHRNB2*, *CACNB4* and *CHRNA2* are associated with several types of nocturnal frontal lobe epilepsy and *CACNB4* is associated with type 5 episodic ataxia (EA5, OMIM:613855).

It is noteworthy that the method selects only one gene (*GRM8*) from SFARI category 6 (*Evidence does not support a role*), of which 21 are present in the network with initial weights set to zero. We take this as another indication of the accuracy of the method, as these genes have been shown not to be implicated with ASD and are not ranked as such.

For the subset of novel candidate genes we performed enrichment analysis to find significantly overrepresented biological pathways from the Reactome database. We report only highly significantly enriched pathways.

Results for pathway enrichment analysis (Table 2) show a significant enrichment in pathways related to synaptic transmission and calcium ion transport, with several other neurotransmitter-related pathways also being identified as significant.

The novel candidate risk gene set is enriched in genes involved in specific biological processes related to glutamate binding and AMPA receptor activity, both of which have been linked to ASD [3].

We also note a significant enrichment in genes related to the *cardiac conduction* pathway. These are genes of the *Voltage-dependent calcium channel*

Table 2. Subset of significantly enriched biological pathways (multiple correction adjusted p-values $< 1^{-10}$) for novel candidate risk genes.

Term name	Adjusted p-value
Transmission across Chemical Synapses (R-HSA-112315)	1.69×10^{-40}
Neuronal System (R-HSA-112316)	1.02×10^{-35}
Neurotransmitter Receptor Binding And Downstream Transmission In The Postsynaptic Cell (R-HSA-112314)	3.31×10^{-21}
Phase 2 - plateau phase (R-HSA-5576893)	4.42×10^{-19}
Phase 0 - rapid depolarisation (R-HSA-5576892)	4.42×10^{-19}
Phase 1 - inactivation of fast Na+ channels (R-HSA-5576894)	5.92×10^{-19}
Dopamine Neurotransmitter Release Cycle (R-HSA-212676)	5.59×10^{-16}
Neurotransmitter Release Cycle (R-HSA-112310)	8.85×10^{-15}
Glutamate Binding, Activation of AMPA Receptors and Synaptic Plasticity (R-HSA-399721)	1.51×10^{-12}
Trafficking of AMPA receptors (R-HSA-399719)	1.51×10^{-12}
Cardiac conduction (R-HSA-5576891)	1.39×10^{-11}
Developmental Biology (R-HSA-1266738)	1.52×10^{-11}
Serotonin Neurotransmitter Release Cycle (R-HSA-181429)	1.79×10^{-11}
Norepinephrine Neurotransmitter Release Cycle (R-HSA-181430)	1.79×10^{-11}

(*CACN*) family. Calcium signaling disregulation is reported as being involved in ASD [3] and the genes in this pathway overlap with those in the AMPA-receptor-related pathways.

For confirmation we searched for *denovo* variants from genome or exome sequencing of ASD patients obtained from the Simons Simplex Collection (SSC) subset of denovo-db (version 1.6.1)[3]. Of the 177 candidate genes identified, 70% (123) were found to harbor at least one *denovo* variant, with more than half of these (88) having a variant in at least two samples.

5   Conclusion

We present a semi-supervised machine learning method for systematic discovery of novel candidate disease risk genes for ASD. The method consists on network propagation using a variant of the random walk with restart algorithm, combining information from biological networks and prior information on both positive (disease-related) and negative (disease-unrelated) genes.

Our results indicate that the method can classify held-out known disease genes in a cross-validation setting with very good performance (AUROC ~ 0.85, AUPRC ~ 0.8, MCC ~ 0.57).

---

[3] http://denovo-db.gs.washington.edu/denovo-db/, accessed 1 February 2019.

The identified putative novel candidate genes for association with ASD were analyzed for significant enrichment in gene sets from pathways from the Reactome database, and found to be significantly enriched for pathways related to synaptic transmission and ion transport and specific neurotransmitter-associated pathways previously shown to be associated with ASD.

A large percentage of the identified novel genes were found to be targeted by *denovo* variants in patients from the Simons Simplex Collection (SSC) dataset.

We expect that the outcomes of this work will contribute to the diagnosis and general understanding of the molecular mechanisms underlying ASD. Future developments will include the use of the derived gene ranking to prioritize genetic variants for gene and variant prioritization in the analysis of genomics data from ASD patients.

Acknowledgments. The authors would like to acknowledge the support by the UID/MULTI/04046/2019 centre grant from FCT, Portugal (to BioISI). A.M. is recipient of a fellowship from BioSys PhD programme (Ref SFRH/BD52485/2014) from FCT (Portugal). This work used the EGI infrastructure with the support of NCG-INGRID-PT (Portugal) and BIFI (Spain).

References

1. Elsabbagh, M., et al.: Global prevalence of autism and other pervasive developmental disorders. Autism Res. **5**(3), 160–179 (2012). https://doi.org/10.1002/aur. 239
2. American Psychiatric Association. Diagnostic and Statistical Manual of Mental Disorders, 5th edn., Washington, DC (2013). https://doi.org/10.1176/appi.books. 9780890425596
3. Chaste, P., Roeder, K., Devlin, B.: The Yin and Yang of autism genetics: how rare De Novo and common variations affect liability. Ann. Rev. Genomics Hum. Genet. **18**, 167–187 (2017). https://doi.org/10.1146/annurev-genom-083115-022647
4. Pinto, D., et al.: Functional impact of global rare copy number variation in autism spectrum disorders. Nature **466**(7304), 368 (2010). https://doi.org/10. 1038/nature09146
5. Pinto, D., et al.: Convergence of genes and cellular pathways dysregulated in autism spectrum disorders. Am. J. Hum. Genet. **94**(5), 677–694 (2014). https://doi.org/ 10.1016/j.ajhg.2014.03.018
6. Vorstman, J., Parr, J., Moreno-De-Luca, D., Anney, R., Nurnberger Jr., J., et al.: Autism genetics: opportunities and challenges for clinical translation. Nat. Rev. Genet. **18**(6), 362–376 (2017). https://doi.org/10.1038/nrg.2017.4
7. Ansel, A., Rosenzweig Joshua, P., Zisman, P.D., Melamed, M., Gesundheit, B.: Variation in gene expression in autism spectrum disorders: an extensive review of transcriptomic studies. Front. Neurosci. **10**, 601 (2017). https://doi.org/10.3389/ fnins.2016.00601
8. Krishnan, A., et al.: Genome-wide prediction and functional characterization of the genetic basis of autism spectrum disorder. Nat. Neurosci. **19**(11), 1454–1462 (2016). https://doi.org/10.1038/nn.4353
9. Asif, M., et al.: Identifying disease genes using machine learning and gene functional similarities, assessed through Gene Ontology. PloS one **13**(12), e0208626 (2018). https://doi.org/10.1371/journal.pone.0208626

10. Zhou, D., Bousquet, O., Lal, T.N., Weston, J., Schölkopf, B.: Learning with local and global consistency. In: Advances in Neural Information Processing Systems, pp. 321–328 (2004)
11. Cowen, L., Ideker, T., Raphael, B.J., Sharan, R.: Network propagation: a universal amplifier of genetic associations. Nat. Rev. Genet. **18**(9), 551–562 (2017). https://doi.org/10.1038/nrg.2017.38
12. Mosca, E., et al.: Network diffusion-based prioritization of autism risk genes identifies significantly connected gene modules. Front. Genet. **8**, 129 (2017). https://doi.org/10.3389/fgene.2017.00129
13. Szklarczyk, D., et al.: The STRING database in 2017: quality-controlled protein-protein association networks, made broadly accessible. Nucleic Acids Res. **45**(Database issue), D362–D368 (2017). https://doi.org/10.1093/nar/gkw937
14. Smedley, D., Haider, S., et al.: The BioMart community portal: an innovative alternative to large, centralized data repositories. Nucleic Acids Res. **43**(1), W589–W598 (2015). https://doi.org/10.1093/nar/gkv350

Designing and Evaluating Deep Learning Models for Cancer Detection on Gene Expression Data

Arif Canakoglu$^{(\boxtimes)}$, Luca Nanni$^{(\boxtimes)}$, Artur Sokolovsky ,
and Stefano Ceri

Dipartimento di Elettronica, Informazione e Bioingegneria,
Politecnico di Milano, Milan, Italy
{arif.canakoglu,luca.nanni,artur.sokolovsky,stefano.ceri}@polimi.it

Abstract. Transcription profiling enables researchers to understand the activity of the genes in various experimental conditions; in human genomics, abnormal gene expression is typically correlated with clinical conditions. An important application is the detection of genes which are most involved in the development of tumors, by contrasting normal and tumor cells of the same patient. Several statistical and machine learning techniques have been applied to cancer detection; more recently, deep learning methods have been attempted, but they have typically failed in meeting the same performance as classical algorithms. In this paper, we design a set of deep learning methods that can achieve similar performance as the best machine learning methods thanks to the use of external information or of data augmentation; we demonstrate this result by comparing the performance of new methods against several baselines.

Keywords: Deep learning · RNA-seq · Machine learning · Cancer detection

1 Introduction

Next Generation Sequencing technologies enabled in the last years the creation of constantly growing whole-transcriptome datasets, allowing the researchers to understand the underlying mechanisms of gene expression and their relationship with sample phenotype. Large gene expression databases like The Cancer Genome Atlas [20] or the Genotype-Tissue Expression project [13] became the basis on which a lot of computational works have defined novel methodologies to extract information from gene expression datasets.

Machine Learning (ML) is the study of statistical techniques which enable computers to extract relevant patterns from data and use this information to solve specific tasks without the need of manually specify the set of instructions. Deep Learning (DL) is a specific form of ML in which algorithms are trained to

---

A. Canakoglu and L. Nanni—Co-primary authors.

© Springer Nature Switzerland AG 2020
M. Raposo et al. (Eds.): CIBB 2018, LNBI 11925, pp. 249–261, 2020.
https://doi.org/10.1007/978-3-030-34585-3_22

learn an increasingly abstract hierarchy of *feature representations* of the original data, where the first layers represent *low-level* (i.e., concrete) features and last layers represent *high-level* (i.e., abstract) features.

DL models have the ability to learn expressive representations of the data without the need of specific pre-processing steps [11]. After training, the learned data representation is usually able to embed very complex non-linear relationships between the sample features, which can then be used to perform the classification task. The power of this approach has been widely demonstrated in many fields like computer vision, speech recognition, and natural language processing. Most of DL models achieve this by relying on Neural Network architectures [16], which usually require more parameters to be trained than classic ML algorithms. This reflects in an increased need of training data, and for this reason, only very recently researchers have started to apply DL architectures to biological tasks.

Three main challenges in the training and testing of DL models with gene expression data are (a) the lack of training samples, (b) the unbalanced populations of the different classes, and (c) high dimensionality of the problem. In this paper, we show a set of specific deep learning methods that, thanks to their ability to adapt to the problem, can achieve similar performances as the best machine learning methods.

2   Scientific Background

Gene expression levels, measuring the transcription activity, are widely used to predict abnormal gene activities, in particular for distinguishing between healthy and tumor cells [22]. This is a classical problem that has been addressed by a variety of ML methods, e.g., see [6,10,18]. There is a long story of learning algorithms applied to gene expression datasets: relevant applications count identifying gene expression signatures specific to a cancer type or sub-type [17,23], differentiating between tumor characteristics like grade or stage [14] and predicting clinical outcomes [21].

More recently, new artificial intelligence methods have emerged, and some attempts have been made of using DL methods for cancer detection and classification [5,12].

As discussed above, the three main challenges using DL models with gene expression data are (a) the lack of training samples, as the open datasets available in The Cancer Genome Atlas [20] are distributed across 33 tumor types, resulting in very few samples for most of them (b) the unbalanced populations of the different classes, as very few normal cells are available in comparison with cancer cells, and (c) high dimensionality of the problem, as the number of features (coding and noncoding genes) may be of the order of 100 K depending on the technology used.

The first challenge could possibly be alleviated by a suitable generation of synthetic data; however, such practice is not easy in the case of gene expression [3]. The second challenge can be approached by sampling techniques, but this is not recommended in the general lack of training data. Finally, the huge dimensions of the search space can be dealt with some pre-filtering technique on the genes [8], but this might cause the omission of relevant information.

To address these challenges, in this work we describe three innovative DL methods and compare them to standard ML methods and a baseline DL method; our attempt is to match the quality of classical machine learning for binary classification, in the 99% accuracy range, thereby also testing their applicability to more difficult cancer classification problems. Innovation addresses two orthogonal perspectives: the data dimension and the provisioning of additional information. Along the first perspective, we can use *feature engineering* to simplify the training task or *data augmentation* to increase the information used for training. Along the second one, we can use external information for training, either in the form of *biological knowledge* or of *other compatible datasets* to improve model training.

Table 1. Four neural network methods described in this work

	No information	External information
Feature engineering	*Feed Forward Network*	*Ontology-Guided CNN*
Data augmentation	*Ladder Network*	*Transfer Learning*

We will next describe the datasets used for gene expression, then apply the baseline models in order to classify normal and tumor cells, then present each of three interesting cases of baseline augmentations (Table 1), and finally, we will compare their performance. All the considered models take as input gene expressions and produce as output the normal/tumor labels for three datasets corresponding to specific tumors.

3   Materials and Methods

3.1   Datasets

We used RNA-seq data from the TCGA public dataset [20]: we downloaded Illumina HiSeq 2000 log2 scaled data matrix from the Xena Browser[1] in December 2017. Due to the lack of samples and the imbalance between normal and tumor samples in most cancer types, we considered the three most represented cancer types by origin tissue, which are: 1. **Breast**: Breast invasive carcinoma (BRCA); 2. **Lung**: Lung adenocarcinoma (LUAD) and Lung squamous cell carcinoma (LUSC); 3. **Kidney**: Kidney Chromophobe (KICH), Kidney renal clear cell carcinoma (KIRC) and Kidney renal papillary cell carcinoma (KIRP). In Table 2, we reported the number of available data and we also performed a principal components analysis (PCA) on the three selected datasets (Fig. 1) displaying the first two components and the relationship with the sample label (normal or tumor).

---

[1] http://xena.ucsc.edu.

Table 2. TCGA sample counts for each tissue

Tissue	Total	Normal	Tumor	Normal (%)
Breast	1218	114	1104	9.36%
Kidney	1020	129	891	12.65%
Lung	1129	110	1019	9.74%

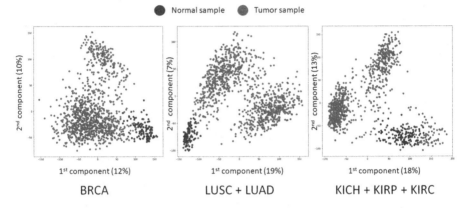

Fig. 1. First two components of Principal Component Analysis (with their percentage of explained variance) of the three cohort datasets considered in this study.

3.2 Baselines

As baseline for classical machine learning, we trained a *Support Vector Machine with linear kernel (SVM)*, known as a high-performance model for this task [6]. We also evaluated a classical *5-nearest neighbours classifier* and a *10-tree random forest*, which were used for clinical outcome prediction with gene expression data. The PCA of Fig. 1 shows that the problem is intrinsically separable; hence, improving over the l-SVN baseline is very hard.

As baseline for deep learning, we trained a classic *Feed Forward Network* (FFN). As FFN is used as baseline, the network architecture is as simple as possible: for each considered dataset, we selected the top 5000 most variant genes and performed a Min-Max normalization of the expression values. The architecture is composed by 2 hidden layers that contain 100 and 20 neurons respectively, with the ReLu activation function. We used binary cross entropy as loss function to be minimized:

$$\mathcal{L}(y, \hat{y}) = -\frac{1}{N} \sum_{i=1}^{N} [y_i \log(\hat{y}_i) + (1 - y_i) \log(1 - \hat{y}_i)] \tag{1}$$

We anticipate from the discussion that the performance of machine learning baselines is generally superior to FFN, used as deep learning baseline. We study new deep learning models with the objective of improving over the relative baseline and achieve comparable results with the machine learning models.

3.3   Ladder Network

A ladder network [15] combines both supervised and unsupervised parts in a single deep neural network; the training of both parts is simultaneous, without using layer-wise pre-training.

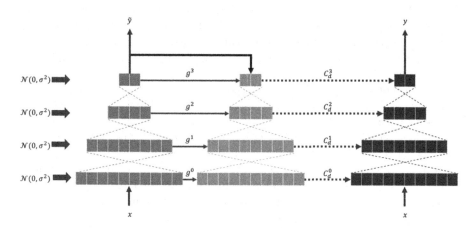

Fig. 2. Ladder networks are constituted by the combination of three components. The first one (on the left) is an encoder, corrupted with Gaussian noise ($\mathcal{N}(0, \sigma^2)$); the second one (in the middle) is a decoder, with a denoising functions (g^n); the third one is an encoder without corruption. Denoising cost (C_d^l) is used to calculate the loss function of the system.

In the Fig. 2, ladder network has three paths; two of them are feed forward paths, one standard (on the left) and one corrupted (on the right). In the middle, a denoising decoding path connects two forward passes. The unsupervised part, which contains corrupted feed forward pass and decoding pass, works as Denoising Auto Encoder (DAE) [19], in which the Gaussian noise is added to each hidden layer of corrupted forward pass. The supervised part is the feed forward network on the right; its weights are batch normalized by the unsupervised part, by using a denoising cost specific to each layer.

The loss function of the network is weighted sum of the supervised and unsupervised parts. The former is one is cross entropy cost on the top of the standard forward pass(y) and cost functions of each layer (C_d^l). The unsupervised cost is denoising square error costs weighted by a hyper parameter at each layer of the decoder.

We tuned the network by using different parameters as in [7], the most relevant ones are the number of layers (single layer or 2, 3, 5, 7 and 10 hidden layers) and the training feed size (10, 20, 30, 40, 60, 80 and 120 labeled data). We selected 5 hidden layers with 5000, 1500, 500, 250, 10 neuron sizes in the forward paths and we feed 120 labelled data which contains an equal number of elements from each class to the supervised path.

3.4   Ontology-Driven Convolutional Neural Networks

The concept of neighborhood in the context of biology can be applied to an extremely vast set of entities and it is usually associated with the concept of *interaction*. Biological entities can be thought to be *near* if they share a common behaviour, or present a similar pattern, or are semantically correlated.

Convolutional Neural Networks (CNN) exploit the spatial relationships between the input features to derive high-level representations which are then provided as inputs of a classical feed-forward network (FFN). The convolutional layer uses a set of kernels transforming the input features by aggregation, locally applied to near features.

The concept of neighborhood in the context of biology can be applied to an extremely vast set of entities, which are *near* if they share a common behavior, or present a similar pattern, or are semantically correlated. We enriched the prediction capability of a CNN by guiding the convolution layers using distance relations between genes derived from prior biological knowledge.

In order to calculate neighbour genes, we need to first define the distances between them. A distance matrix $\mathbf{D}$ is a symmetric matrix in which the diagonal elements are zero and the other non-negative elements contain the distance d_{xy} between genes g_x and g_y . In order to compute distances, we used the Genomic and Proteomic Data Warehouse (GPDW) [4], which contains gene datasets(e.g., Entrez gene, Ensembl gene, ...), their annotations from the Gene Ontology [2] (all three aspects: cellular component, molecular function and biological process), and also the ontological relationships between them. The minimum distance information extracted from GPDW for each gene pair $<g_x, g_y>$ is:

$$d\left(g_x, g_y\right) = \begin{cases} \min\limits_{\substack{o_i \in \Omega_x, o_j \in \Omega_y, \\ o_a \in \mathrm{oe}(\Omega_x) \cap \mathrm{oe}(\Omega_y)}} \mathrm{od}\left(o_i, o_a\right) + \mathrm{od}\left(o_j, o_a\right) & \text{if } \begin{array}{l} x \neq y \wedge \\ \mathrm{oe}\left(\Omega_x\right) \cap \mathrm{oe}\left(\Omega_y\right) \neq \emptyset \end{array} \\ 0 & \text{if } x = y \\ +\infty & \text{otherwise} \end{cases}$$

where Ω_x, Ω_y are the sets of ontological terms directly annotated respectively to gene g_x and g_y, oe is a semantic expansion function towards its hypernyms and od is a function giving the minimum distance between two ontology concepts calculated as the number of the edges between them. The k-neighbours of a gene g_x are the genes g_k for which d_{xk} is in the set of the k lowest distances between x and all the genes with $i \neq k$.

We then used this distance measure to derive the sets of genes on which apply the convolution. For each gene, we derived the nearest 4 genes and applied a 1-dimensional convolutional filter to each set of 5 genes, using a stride of 5. In this way, we derive an aggregated representation of the neighborhoods of genes, which we then input to a FFN having 2 hidden layers of 200 and 50 neurons (characterized all by a ReLu activation function). In Fig. 3 a schematic representation of the pipeline is presented.

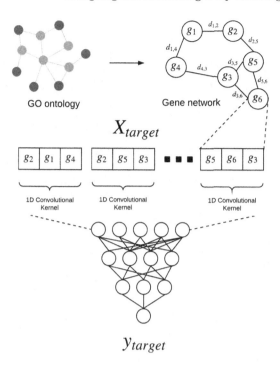

Fig. 3. Schematic representation of the ontology-driven convolutional neural network pipeline. As a first thing, we extract from the GO ontology a gene interaction network. Then, for each gene we extract its k neighbours in the graph and we perform a 1D convolution step on the feature vector composed by the gene and its neighbours. The resulting values are finally provided to a Feed Forward Neural Network

3.5   Transfer Learning Using a Combined Set of Tumors

The lack of data when concentrating on a single tumor for classification produces often unreliable and low performing models. To deal with this issue, we designed a transfer learning procedure that makes use of external information provided by a generic classification problem on the whole set of tumors.

Transfer learning is widely used in deep learning setups both academically and in industry. It provides a relatively simple method for enriching datasets submitted to the learning task. However, the model designer has to be particularly careful in associating datasets which share common characteristics and distributions. When instead the general model is trained with a dataset without a strong affinity to the target one, the hybrid model performances are degraded. In addition to this, if the target dataset has strong linearity properties with respect to the label, then the addition of the general model will yield to an over-complicated hybrid model, which will also get a performance degradation.

Transfer learning occurs as follows: Suppose that we want to detect the presence of cancer for tumor t. We first train a general network model M_g on all the tumor types except t. We then train a hybrid model M_h composed by the

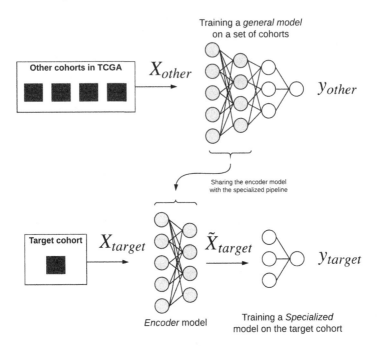

Fig. 4. Schematic representation of the transfer learning pipeline. The first step is to train a model on a set of TCGA cohorts with enough training samples of both classes. Then we share the encoding part of the model with a smaller neural network, which is trained on the few samples of the target cohort.

concatenation of M_g and a small neural network M_s, where the weights of M_g are kept unchanged during the training. One can see this procedure as a function composition $M_h = M_g \circ M_s$ where the various models are functions $M(x) = \tilde{x}$ which extract relevant features from the input vector except for the final model, which produces as output the class probability $y = \tilde{x} = M(x)$ (see Fig. 4).

For our application the general model is constituted by a FFN with four hidden layers respectively with 500, 200, 100 and 50 neurons for a total of 2625901 trainable parameters. Between all the layers the ReLu activation function is used with the exception of the output, which applies a Sigmoid function. The small network used for the fine tuning of the hybrid network on the specific tumor type has two hidden layers of 50 an 10 neurons. Like in the FFN case, we pre-filtered the genes to be used for training and testing taking only the top 5000 variant ones. This speeds up the training and enables us to increase the performance of the model.

For this method we are going also to show the performance evaluation on an additional TCGA dataset characterized by non-trivial data distribution and low sample number: **Bladder Urothelial Carcinoma** (BLCA), whose PCA plot is shown in the Fig. 5a.

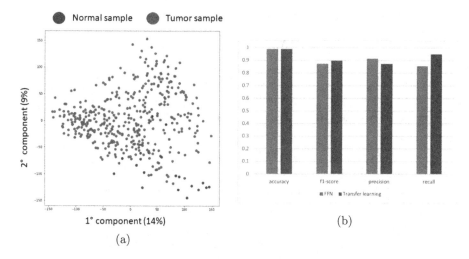

Fig. 5. (a) PCA graph for the BLCA dataset. (b) Performance measures of transfer learning on the BLCA dataset and comparison with FFN

4 Experimental Results

All the methods were evaluated using 5 times repeated 5-fold cross validation strategy and taking the mean of the various metrics as aggregated performance score.

When a classification task has a strong class imbalance, the choice of the correct evaluation metrics is particularly relevant [9]. In order to evaluate correctly the models, we must assess both the specificity and the sensitivity of the models. Therefore, together with the widely spread *accuracy* measure, we compute for each fold also the *precision, recall* and *f1-score*.

Each fold splits the data in training and test set and a portion of the training data (25%) is taken as validation set. All the models were programmed in Python using the Keras library.

Table 3 compares the baselines. A linear model like SVM achieves high accuracy in all the cohorts that we selected for the study, thanks to an intrinsic linear separability property of the cohorts that we selected, confirmed by PCA analysis (Fig. 1). KNN and Random Forest show lower performance metrics with exception for recall; this is principally due to the shape of the clusters that the two classes form in the datasets, which enables a neighbor-based approach like KNN to achieve better recognition capability for the normal samples. Random Forest shows lower recall but better precision than KNN; prediction accuracy is lower than all the other baselines. FFN achieves in general worse performance than the linear SVM, including a worse precision. This is expected because of difficulty of training the model with a small number of samples; on the other hand, it is remarkable that such a simple architecture and pre-filtering of genes can achieve comparable results with state-of-art machine learning methods.

Table 3. Performance measures for the four baseline models.

	Accuracy	F_1 score	Precision	Recall
Linear SVM				
BRCA	0.994913	0.973241	0.964699	0.982609
LUAD+LUSC	0.996986	0.984847	0.9739	0.996364
KIRC+KIRP+KICH	0.996668	0.986865	0.988303	0.986031
KNN				
BRCA	0.991622	0.957749	0.921784	0.998261
LUAD+LUSC	0.991849	0.960519	0.925095	1
KIRC+KIRP+KICH	0.996866	0.987742	0.983696	0.992185
Random Forest				
BRCA	0.987843	0.932008	0.965448	0.903083
LUAD+LUSC	0.992561	0.960789	0.972691	0.950909
KIRC+KIRP+KICH	0.992943	0.970947	0.99202	0.952062
Feed Forward Network				
BRCA	0.988511	0.945621	0.910416	0.991304
LUAD+LUSC	0.995039	0.97576	0.957081	0.996364
KIRC+KIRP+KICH	0.996079	0.984266	0.989459	0.979815

Table 4 compares our three proposed methods. Ladder Network achieves better recall than both linear SVM and FFN, but does not manage to compete at the level of accuracy. Note that this model does not use any gene pre-filtering, as it extracts the relevant inner features from the *whole* set of genes, characterizing itself as a good knowledge extraction method for gene expression. Ladder network has equal number of samples for the supervised learning path, this leads to better recall and lightly worse precision.

The ontology-guided CNN approach achieves better performance than FFN in all the metrics with the exception of the LUNG dataset. In addition, it is able to overcome machine learning methods in the Kidney dataset and almost same performance in the other tissues, while it is always superior in recall. We checked if the performance improvement could be caused by the convolutional layer by itself by testing the classification accuracy with a randomly generated distance matrix, but we achieved worse performance. Therefore, we conclude that the GO ontological network is able to guide the convolution to a better inner representation of the features.

The transfer learning approach provides accuracy results comparable with FFN; on the BRCA datasets, it improves in all the measures w.r.t. the machine learning baselines, including linear SVM. We also applied transfer learning to a different cancer type, **Bladder Urothelial Carcinoma** (BLCA), for which very few normal datasets are available. In Table 5, we show better performance with respect to the FFN baseline for what concerns recall (10% improvement)

Table 4. Performance measures for three deep learning models.

	Accuracy	F$_1$ score	Precision	Recall
Ladder Network				
BRCA	0.988998	0.944120	0.899841	0.992982
LUAD+LUSC	0.992737	0.964003	0.932088	0.998182
KIRC+KIRP+KICH	0.995294	0.981595	0.971168	0.992248
Ontology-Guided CNN				
BRCA	0.993436	0.965749	0.950609	0.982609
LUAD+LUSC	0.99292	0.965749	0.935173	1
KIRC+KIRP+KICH	0.998039	0.992305	0.992593	0.992308
Transfer Learning				
BRCA	0.990963	0.953823	0.920312	0.990909
LUAD+LUSC	0.993974	0.970087	0.948913	0.992727
KIRC+KIRP+KICH	0.995683	0.982932	0.982227	0.984431

Table 5. Transfer learning results

BLCA dataset	Accuracy	F$_1$ score	Precision	Recall
Linear SVM	0.992016	0.899937	0.948667	0.876667
Feed Forward Network	0.991069	0.872635	0.914000	0.856667
Transfer Learning	0.989668	0.897123	0.871524	0.950000

and F$_1$ score; transfer learning proved to be a valid tool when applied to cancer classification in the case of severe lack of samples.

5   Conclusions

To the best of our knowledge, this work presents the first systematic evaluation of machine and deep learning methods for the cancer classification problem on gene expression data, improving earlier work by [1].

In this paper we have presented, evaluated and compared four different deep learning models for cancer detection using gene expression data. We have also compared our results with three state-of-the-art machine learning models (SVM, KNN and Random Forest). We have used as benchmark three tissue datasets coming from the TCGA repository: Breast, Kidney and Lung. We have also used a challenging dataset (BLCA) from the point of view of cardinality and data distribution to test the capabilities of a transfer learning approach.

We generally demonstrated that three deep learning architectures can compete with machine learning methods even in the presence of few training samples, population unbalancing and high problem dimensionality. We have demonstrated that structuring the feature space using gene relationships given by the GO ontology and then using convolution to extract high level representations of the input is useful for improving the detection recall and accuracy. We have also proposed a framework of transfer learning which enables to increase the recall in the three tissues while keeping comparable accuracy and precision.

Our results show that, for the considered datasets, deep learning models provide results comparable with classical machine learning methods, but are not always able to improve their performance, which is highly affected by small fluctuations (for example, an accuracy decrease of 0.01 corresponds to the misclassification of 12.18 samples on average, which is rather significant); this is due to the small cardinality and high dimensionality of datasets. With the constant growth of the sizes of genomic databases, we expect these methods to become fully applicable.

Availability. The source of all methods are available in https://github.com/DEIB-GECO/cancer_classification.

Acknowledgment. This work is supported by the ERC Advanced Grant 693174 GeCo (Data-Driven Genomic Computing) and by the Amazon Machine Learning Research Award on Data-driven Machine and Deep Learning for Genomics.

References

1. Agrawal, S., Agrawal, J.: Neural network techniques for cancer prediction: a survey. Procedia Comput. Sci. **60**, 769–774 (2015). https://doi.org/10.1016/j.procs.2015.08.234
2. Ashburner, M., et al.: Gene ontology: tool for the unification of biology. Nature Genet. **25**, 25 (2000)
3. Blagus, R., Lusa, L.: Smote for high-dimensional class-imbalanced data. BMC Bioinform. **14**(1), 106 (2013). https://doi.org/10.1186/1471-2105-14-106
4. Canakoglu, A., et al.: Integrative warehousing of biomolecular information to support complex multi-topic queries for biomedical knowledge discovery. In: BIBE, pp. 1–4. IEEE (2013)
5. Danaee, P., Ghaeini, R., Hendrix, D.A.: A deep learning approach for cancer detection and relevant gene identification. In: Pacific Symposium on Biocomputing 2017, pp. 219–229. World Scientific (2017)
6. Furey, T.S., et al.: Support vector machine classification and validation of cancer tissue samples using microarray expression data. Bioinformatics **16**(10), 906–914 (2000)
7. Golcuk, G., Tuncel, M.A., Canakoglu, A.: Exploiting ladder networks for gene expression classification. In: Rojas, I., Ortuño, F. (eds.) IWBBIO 2018. LNCS, vol. 10813, pp. 270–278. Springer, Cham (2018). https://doi.org/10.1007/978-3-319-78723-7_23
8. Guyon, I., et al.: Gene selection for cancer classification using support vector machines. Mach. Learn. **46**(1), 389–422 (2002). https://doi.org/10.1023/A:1012487302797
9. He, H., Garcia, E.A.: Learning from imbalanced data. IEEE Trans. Knowl. Data Eng. **21**(9), 1263–1284 (2009). https://doi.org/10.1109/TKDE.2008.239

10. Hijazi, H., Chan, C.: A classification framework applied to cancer gene expression profiles. J. Healthc. Eng. **4**(2), 255–284 (2013). https://doi.org/10.1260/2040-2295.4.2.255
11. LeCun, Y., et al.: Deep learning. Nature **521**, 436 (2015)
12. Liu, J., Wang, X., Cheng, Y., Zhang, L.: Tumor gene expression data classification via sample expansion-based deep learning. Oncotarget **8**(65), 109646–109660 (2017). https://doi.org/10.18632/oncotarget.22762
13. Lonsdale, J., et al.: The genotype-tissue expression (GTEx) project. Nat. Genet. **45**(6), 580 (2013)
14. Rahimi, A., Gönen, M.: Discriminating early-and late-stage cancers using multiple kernel learning on gene sets. Bioinformatics **34**(13), i412–i421 (2018)
15. Rasmus, A., et al.: Semi-supervised learning with ladder networks. In: Advances in Neural Information Processing Systems, pp. 3546–3554 (2015)
16. Schmidhuber, J.: Deep learning in neural networks: an overview. Neural Netw. **61**, 85–117 (2015)
17. Shen, R., Olshen, A.B., Ladanyi, M.: Integrative clustering of multiple genomic data types using a joint latent variable model with application to breast and lung cancer subtype analysis. Bioinformatics **25**(22), 2906–2912 (2009). https://doi.org/10.1093/bioinformatics/btp543
18. Statnikov, A., et al.: A comprehensive comparison of random forests and support vector machines for microarray-based cancer classification. BMC Bioinform. **9**(1), 319 (2008)
19. Vincent, P., et al.: Stacked denoising autoencoders: learning useful representations in a deep network with a local denoising criterion. J. Mach. Learn. Res. **11**(Dec), 3371–3408 (2010)
20. Weinstein, J.N., et al.: The cancer genome atlas pan-cancer analysis project. Nat. Genet. **45**(10), 1113–1120 (2013)
21. Yousefi, S., et al.: Predicting clinical outcomes from large scale cancer genomic profiles with deep survival models. Sci. Rep. **7**(1), 11707 (2017)
22. Zhang, L., et al.: Gene expression profiles in normal and cancer cells. Science **276**(5316), 1268–1272 (1997). https://doi.org/10.1126/science.276.5316.1268
23. Zuyderduyn, S.D., et al.: A machine learning approach to finding gene expression signatures of the early developmental stages of squamous cell lung carcinoma. Cancer Res. **66**(8 Supplement), 431–432 (2006). http://cancerres.aacrjournals.org/content/66/8_Supplement/431.4

Analysis of Extremely Obese Individuals Using Deep Learning Stacked Autoencoders and Genome-Wide Genetic Data

Casimiro A. Curbelo Montañez$^{(\boxtimes)}$ [iD], Paul Fergus [iD],
Carl Chalmers [iD], and Jade Hind

Department of Computer Science, Liverpool John Moores University,
Liverpool L3 3AF, UK
c.a.curbelomontanez@ljmu.ac.uk

Abstract. Genetic predisposition has been identified as one of the components contributing to the obesity epidemic in modern societies. The aetiology of polygenic obesity is multifactorial, which indicates that lifestyle and environmental factors may influence multiples genes to aggravate this disorder. Several low-risk single nucleotide polymorphisms (SNPs) have been associated with BMI. However, identified loci only explain a small proportion of the variation observed for this phenotype. The linear nature of genome wide association studies (GWAS) used to identify associations between genetic variants and the phenotype have had limited success in explaining the heritability variation of BMI and shown low predictive capacity in classification studies. GWAS ignores the epistatic interactions that less significant variants have on the phenotypic outcome. In this paper we utilise a novel deep learning-based methodology to reduce the high dimensional space in GWAS and find epistatic interactions between SNPs for classification purposes. SNPs were filtered based on the effects associations have with BMI. Since Bonferroni adjustment for multiple testing is highly conservative, an important proportion of SNPs involved in SNP-SNP interactions are ignored. Therefore, only SNPs with p-values $<1 \times 10^{-2}$ were considered for subsequent epistasis analysis using stacked autoencoders (SAE). This allows the nonlinearity present in SNP-SNP interactions to be discovered through progressively smaller hidden layer units and to initialise a multi-layer feedforward artificial neural network (ANN) classifier. The classifier is fine-tuned to classify extremely obese and non-obese individuals. The performance of classifications using progressively smaller compressed layers was compared and the results reported. The best results were obtained with 2,000 compressed units (SE = 0.949153, SP = 0.933014, Gini = 0.949936, Logloss = 0.1956, AUC = 0.97497 and MSE = 0.054057). Using 50 compressed units it was possible to achieve (SE = 0.785311, SP = 0.799043, Gini = 0.703566, Logloss = 0.476864, AUC = 0.85178 and MSE = 0.156315).

Keywords: Classification · Deep learning · Dimensionality reduction · Epistasis · GWAS · Polygenic obesity · SNPs · Stacked autoencoder

© Springer Nature Switzerland AG 2020
M. Raposo et al. (Eds.): CIBB 2018, LNBI 11925, pp. 262–276, 2020.
https://doi.org/10.1007/978-3-030-34585-3_23

1 Introduction

Obesity prevalence has increased over several decades and has now reached epidemic proportions [1]. This has had a significant impact on morbidity and mortality. Individuals suffering from obesity are at higher risk of developing numerous non-communicable diseases (NCDs), including type 2 diabetes mellitus, cardiovascular disease, and certain types of cancer [2].

The aetiology of common or polygenic obesity is multi-factorial, indicating that combinations of lifestyle and environmental factors may interact with multiple genes, causing this disorder. This is further supported by twin, adoption, and family studies which found that variation in body mass index (BMI) was largely due to heritable genetic differences, with heritability (the proportion of the variability of a trait that is attributable to genetic factors) estimates in adults ranging between 40% and 70% [3]. Hence, it is believed that obesity risk is higher among those individuals genetically predisposed to weight gain in obesogenic environments where gene-environment interactions occur.

The advent of high-throughput technologies has enabled hypothesis free approaches based on single-locus analysis such as genome-wide association studies (GWAS) [4]. In GWAS, Single Nucleotide Polymorphism (SNP) [5] are independently tested for association with a phenotype of interest, omitting, therefore, the existence of interactions between loci. Currently, several low-risk common genetic variants have been associated with BMI, including variants within the FTO or MC4R genes. However, these identified loci only explain a small proportion ($\sim$1.4–2.7%) of the variation observed in BMI [6]. The linear nature of the framework employed by GWAS may explain the limited success of these studies when explaining heritability variation of BMI. This is now regarded as a significant limitation in GWAS, particularly when studying complex disorders that rely on an understanding of gene-gene and gene-environment interactions [7]. An important challenge in the analysis of high-throughput genetic data is the development of computational and statistical methods to identify epistasis interactions. Investigating all combinations between SNPs in genome-wide studies is computationally very costly since the number of tests and time necessary to perform the exhaustive search increases exponentially with the order of interactions considered, limiting scalability. Therefore, epistatic analysis has primarily been restricted to two locus interactions.

In this paper a deep learning (DL) stacked autoencoder (SAE) is used to deal with nonlinearity present in SNP-SNP interactions and to initialise a multi-layer feedforward artificial neural network (ANN) classifier. The classifier is fine-tuned to classify extremely obese and non-obese individuals. Better approximations to non-linear functions can be generated by using DL models than those with a shallow structure. The combination of SAE and a multi-layer feedforward ANN is used to model the epistatic effects of a subset of filtered genetic variants (p-values $< 1 \times 10^{-2}$) based on their association with BMI after quality control (QC) procedures, demonstrating the potential of DL for GWAS.

The remainder of this paper is organized as follows. Section 2 describes the Materials and Methods used in the study. The results are presented in Sect. 3 and discussed in Sect. 4 before the paper is concluded and future work is presented in Sect. 5.

2 Materials and Methods

2.1 Study Participants

Case and control data utilised in this study was requested from the database of Genotypes and Phenotypes (dbGaP) [8]. Participants were extracted from different study cohorts in the Geisinger MyCode project. A subset of 1,231 primary patients of a Geisinger Clinic with non-urgent visits to the clinic were used as control group. Conversely, 962 unique samples with a mean BMI of 49.17 (±8.83 SD), part of a cohort of primary Caucasian patients from the Geisinger Clinic with extreme obesity who have undergone bariatric surgery were considered as control group.

A total of 2,193 participants of which 917 are males and 1,236 are females form the case/control dataset in this study. Each participant contains 594,034 markers. Furthermore, 99.5% of the participants belong to a white ethnical background (Caucasians).

2.2 Genetic Analysis

Quality Control

To conduct association analyses, only those individuals reported to be white Caucasian were selected to reduce potential bias due to population stratification. Quality control, an imperative step prior to any GWAS analysis [9], was also applied to identify potentially problematic individuals and SNPs. The first step was to filter out and remove data samples with discordant sex. Related or duplicated samples were removed using Identity by Descent (IBD) coefficient estimates (IBD > 0.185). To obtain high-quality data for the association analysis, the data set was pruned with the following criteria: sample call rate >99%; SNP call rate >99%; and a threshold for Hardy–Weinberg equilibrium of 0.0001 in control cohort. Due to limited power of rare variants in an association study, we only kept SNPs with minor allele frequencies >0.05. After QC, 1,997 individuals (879 cases and 1,118 controls) and 240,950 genetic variants remained for subsequent analysis.

Association Analysis

Statistical association testing between individual SNPs and obesity was conducted under an additive model using logistic regression [10] in PLINK (v1.9) [11].

Given a, a linear increase in risk for each copy of the a allele is assumed. For example, if the risk is γ for Aa/aA, then there is a risk of 2γ for aa. Let i be the individuals (i = 1, 2, …, n), Y_i the phenotype for individual i and X_i the genotype of individual i at a particular SNP. Let $Y \in \{0, 1\}$ be a binary phenotype for case/control status and $X \in \{0, 1, 2\}$ be a genotype at the typed locus, where 0, 1 and 2 represent homozygous major allele AA, heterozygous allele Aa, and homozygous minor allele aa

respectively. Logistic regression modelling of the expected value of a phenotype Y_i, given a genotype X_i can be defined as a linear predictor function:

$$logit(E(Y_i|X_i)) \sim \beta_0 + \beta_1 X_i \tag{1}$$

Correction for multiple testing is commonly adopted in association analysis, using adjustments such as Bonferroni correction [12]. However, it has been suggested that adjustments, such as Bonferroni, are too conservative and may result in missing significant associations when performing multiple association tests. Consequently, we did not adjust for multiple testing but rather considered the results of all association tests with p-values lower than 1×10^{-2}. This approach has been adopted in other previous studies of epistasis in obesity [13] and T2D studies [14] where p-value $< 1 \times 10^{-1}$ and p-value $< 1 \times 10^{-3}$ were considered respectively.

Utilising logistic regression, while not ideal, enables the number of SNPs with insignificant marginal effects to be reduced to meet the computational needs required for epistatic analysis and machine learning tasks.

2.3   Multi-layer Feedforward Artificial Neural Network

A multi-layer feedforward ANN or Multilayer Perceptron (MLP) is implemented based on the formal definitions in [15], to conduct binary classification. Labelled training samples $(x^{(i)}, y^{(i)})$ from case-control genetic data are considered for a supervised learning problem. A complex non-linear hypothesis $h_{W,b}(x)$ is defined using a feed forward ANN, with parameters W, b fitted to our data. The network first calculates the output matching the input (feed-forward stage). Then, the backpropagation algorithm [16] is used to calculate error propagating to previous layers, and finally, the weights of the network are adjusted.

Taking a set of labelled samples $\{x_1, x_2, ...x_n\}$ and a bias unit b as input, single computational units or neurons output

$$h_{W,b}(x) = f(W^T x) = f\left(\sum_{i=1}^{n} W_i x_i + b\right) \tag{2}$$

where $f : \mathbb{R} \rightarrow \mathbb{R}$ represents the activation function and W the weight. Activation functions, such as the sigmoid function, hyperbolic tangent (tanh) and rectifier linear unit (ReLU) are common activation functions used in many neural network configurations. However, rectifier functions have shown faster learning compared to sigmoid or tanh [17]. In the experiments conducted in this thesis, the selection of activation functions is determined using random search optimization methods to simplify model configuration [18].

Input, hidden and output layers make up the network structure where l represent the total number of layers densely connected, L_1 the input layer and L_{nl} the output layer. Several parameters constitute the neural network. The parameter $W_{ij}^{(l)}$ denotes the weight for the connection between the jth neuron in layer l, and the ith neuron in layer $l + 1$. The bias unit $b_i^{(l)}$, associated with neuron i in layer $l + 1$, is introduced to counteract the problem associated with input patterns that are zero. The number of nodes in layer l is denoted by s_l without taking $b_i^{(l)}$ into consideration. Additionally, the activation or output value of node i in layer l of the network, denoted as $a_i^{(l)}$, is equal to the activation function of the total weighted sum of inputs (including the bias term), represented as $f(z_i^{(l)})$. Given that the values from the inputs are denoted by $a^{(1)} = x$ and the activation for layer l is $a^{(l)}$, the activation output layer l plus the intercept term $+1$ ($a^{(l+1)}$) can be computed. Thus, a compact vectorised form of $z_i^{(l)}$ and $a_i^{(l)}$ is given by

$$z^{(l+1)} = W^{(l)}a^{(l)} + b^{(l)} \qquad (3)$$

$$a^{(l+1)} = f(z^{(l+1)}) \qquad (4)$$

Equations (3–4) can be used to compute the output of the network, successively calculating all the activations in layer L_2, then L_3 and so on up to the output layer Ln_l. Learning using the proposed FNN-based model (MLP) is performed by adjusting the connection weight values to minimise the prediction error on training data.

The neural network hypothesis is defined as $h_{W,b}(x)$ based on a given set of fixed parameters W, b. The neural network is trained using training samples $(x^{(i)}, y^{(i)})$ where $y^{(i)} \in \mathbb{R}^2$. The parameter x is a vector of input features representing individuals while outputs for the two class labels (obese or non-obese in this study) are represented using y. Therefore, two neurons will compose the output layer as shown in Fig. 1. The figure represents the network architecture utilised in this paper for classification analysis.

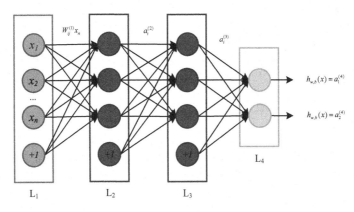

Fig. 1. MLP network with an input layer L_1, two hidden layers L_2 and L_3 and an output layer L_4 with two output units.

Given a training set $\{(x^{(1)}, y^{(1)}),\ldots, (x^{(m)}, y^{(m)})\}$ of m samples, the neural network is trained using gradient descent optimisation and the overall cost function is defined as

$$J(W,b) = \left[\frac{1}{m} \sum_{i=1}^{m} J(W,b,x^{(i)},y^{(i)}) \right] + \frac{\lambda}{2} \sum_{l=1}^{n_l-1} \sum_{i=1}^{s_l} \sum_{j=1}^{s_l+1} (W_{ji}^{(l)})^2$$

$$= \left[\frac{1}{m} \sum_{i=1}^{m} (\frac{1}{2} ||h_{W,b}(x^{(i)}) - y^{(i)}||^2) \right] + \frac{\lambda}{2} \sum_{l=1}^{n_l-1} \sum_{i=1}^{s_l} \sum_{j=1}^{s_l+1} (W_{ji}^{(l)})^2 \tag{5}$$

where the first term is the average sum of squared errors and the second, a weight decay or regularization term that helps prevent overfitting by reducing the magnitude of the weights. The relative importance of the two expressions is controlled with the weight decay parameter λ. Therefore, backpropagation is used to efficiently compute the partial derivatives [15] of the cost function for a single sample with respect to any weight or bias in the network, $J(W, b; x, y)$. Finally, gradient descent is used to reduce our cost function $J(W, b)$ before training the neural network used in this study for classification purposes.

2.4   Autoencoders

Based on the previous definition of a multilayer feedforward ANN, autoencoders (AE) and SAE are used in this study to learn the epistatic relationships between filtered SNPs in the rules and produce a significant smaller input feature space to initialise the weights of a multi-layer feedforward ANN (MLP) classifier. This latent information is representative of the epistatic interactions that occur between SNPs.

Autoencoders belong to the unsupervised learning class of algorithms and can be used to pre-train neural networks [19]. A basic AE is a three-layer neural network that applies backpropagation to learn an output $\hat{x}$ that is similar to the input x. Hence, an AE tries to learn a function $h_{W,b}(x) \approx x$, given a set of unlabelled training samples $\{x^{(1)}, x^{(2)}, x^{(3)}, \ldots\}$, where $x^{(i)} \in \mathbb{R}^n$. An example of a single layer AE is illustrated in Fig. 2, where the first and the third layers are the input and the reconstruction or output layer with 5 units, respectively. The second layer or hidden layer aims to generate the deep features by minimizing the error between the input vector and the reconstruction vector. Thus, an AE is a neural network with a single hidden layer composed by two parts, an encoder and a decoder.

The output of the encoder z is a reduced representation of x used by the decoder to reconstruct the original input x. An AE with a code dimension lower than the input dimension is termed undercomplete [20]. This forces the AE to capture the most prominent features of the training data. First, the encode phase maps input data into a feature vector z so that, for each sample $x^{(i)}$ from the input set $\{x^{(1)}, x^{(2)}, x^{(3)}, \ldots\}$, we have

$$z^{(i)} = f(W^{(1)}x^{(i)} + b^{(1)}) \tag{6}$$

while in the decode phase, the decoder reconstructs the input x, producing a reconstructed space $\hat{x}$ defined by

$$\widehat{x}^{(i)} = f(W^{(2)}z^{(i)} + b^{(2)}) \tag{7}$$

where $W^{(1)}$ and $W^{(2)}$ represent the input-to-hidden and the hidden-to-output weights respectively, $b^{(1)}$ and $b^{(2)}$ represent the bias of hidden and output neurons, whereas $f(\cdot)$ denotes the activation function.

Parameters $W^{(1)}$, $W^{(2)}$, $b^{(1)}$ and $b^{(2)}$ in the AE can be learned by minimising the reconstruction error

$$J(W, b; x, \widehat{x}) = \frac{1}{2} \left\| h_{W,b}(x) - \widehat{x} \right\|^2 \tag{8}$$

This is a measurement of discrepancy between input x and reconstructed $\hat{x}$ with respect to a single sample. For a training set of m samples, the cost function of an autoencoder is as shown below:

$$
\begin{aligned}
J(W, b) &= \left[\frac{1}{m} \sum_{i=1}^{m} J\left(W, b; x^{(i)}, \widehat{x}^{(i)}\right) \right] + \frac{\lambda}{2} \sum_{l=1}^{n_l-1} \sum_{i=1}^{s_l} \sum_{j=1}^{s_l+1} \left(W_{ji}^{(l)}\right)^2 \\
&= \left[\frac{1}{m} \sum_{i=1}^{m} \left(\frac{1}{2} \left\| h_{W,b}\left(x^{(i)}\right) - \widehat{x}^{(i)} \right\|^2 \right) \right] + \frac{\lambda}{2} \sum_{l=1}^{n_l-1} \sum_{i=1}^{s_l} \sum_{j=1}^{s_l+1} \left(W_{ji}^{(l)}\right)^2
\end{aligned}
\tag{9}
$$

where m denotes the overall training set size and the square error is used as the reconstruction error for each training sample. The second term remains as explained in the previous section and represents a weight decay term introduced to decrease the magnitude of the weights and aid to prevent overfitting. Equation (9) can be minimised using stochastic gradient descent.

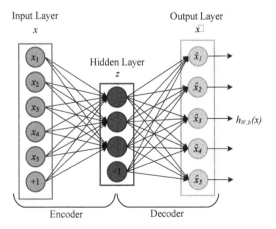

Fig. 2. Autoencoder diagram.

The AE will learn any structure present in the data. Basic AEs typically learn a low-dimensional representation as similarly performed by principal component analysis (PCA). The hidden layer is forced to summarise the data, to compress it. After training an AE, the output layer (reconstruction) and its parameters are discarded, and the learned reduced features remain in the hidden layer which can then be used for classification or as the input of an extended network to extract deeper features. The strength of AEs lies in this type of reconstruction-oriented training that only uses information in the hidden layer which represents learned features from the input. Therefore, the learned non-linear transformation, defined by weights and biases, describes a feature extraction step.

By stacking a sequence of AEs layer by layer, an SAE can be constructed [21]. Once a single layer AE has been trained, a second AE can be trained using the hidden layer from the first AE as shown in Fig. 3. By repeating this procedure, it is possible to create SAEs of arbitrary depth. The first single layer AE maps inputs into the first hidden vector. Once the first layer AE is trained, its reconstruction layer is removed, and the hidden layer becomes the input layer of the next AE.

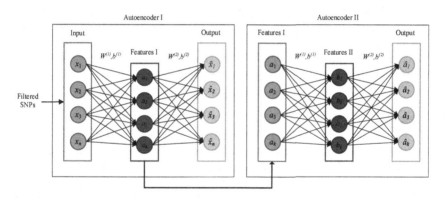

Fig. 3. Example of SAE formed by two single AEs.

In this paper, AEs are stacked to enable greedy layer-wise learning where the l_{th} hidden layer is used as input to the $l + 1$ hidden layer in the stack. The results produced by the SAE are utilised to pre-train the weights for the proposed MLP (See Fig. 1), rather than randomly initialising the weights to small values to classify extreme cases of obesity and normal individuals. Greedy layer-wise pre-training helps the model initialise the parameters near to a good global minimum and transform the problem space to a better form of optimisation [21]. By adopting this approach, it is expected to achieve smoother convergence and higher overall performance in the classification task [22].

An SAE with 2,000, 1,000, 500 and 50 hidden neurons in each hidden layer was considered in the experiments conducted in this paper (See Fig. 4). Consequently, selected layers are used as input features for classification using an MLP. In Fig. 4 an instance of the SAE proposed in this study connected with an MLP is depicted. The

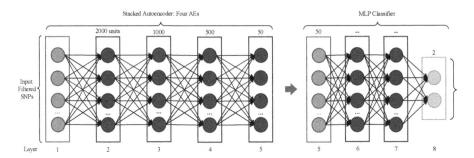

Fig. 4. Instance of proposed SAE connected with an MLP.

classification scheme represented is composed of 8 layers: one input layer, four hidden layers from AEs, two hidden layers from the MLP, and an output layer. This represents the network configuration where SNPs are compressed progressively from 2,465 to 50 neurons.

The goal of the proposed SAE architecture is to extract a mapping that decodes the input (set of SNPs) as closely as possible without losing significant SNP-SNP patterns. The encoder decreases the dimensionality of the original data (SNP set) stack by stack, leading to a reduction in noise while preserving important information patterns [21]. Consequently, AEs are used in this study to gradually extract deep features representative of obesity epistasis.

3   Results

Only SNPs with p-values lower than 1×10^{-2} were considered for machine learning analysis. First, a SAE is used to learn the deep features of a subset of 2,465 SNPs (p-value $< 1 \times 10^{-2}$) in an unsupervised manner, to capture information about important SNPs and the cumulative epistatic interactions between them. This task is conducted layer wise by stacking four single layer AEs with 2000-1000-500-50 hidden units, where the original 2,465 SNPs are compressed into progressively smaller hidden layers (See Fig. 4). The final SAE hidden layer is then used to initialise a deep learning network, based on a multi-layer feedforward ANN. The classifier is trained with stochastic gradient descent using back-propagation and fine-tuned to classify case-control instances in the validation and test sets. Consequently, four classification experiments were conducted.

Model performance in this paper is measured using a range of numerical and graphical approaches [23]. The performance of the classifiers is assessed through sensitivity (SE), specificity (SP), gini, logloss, area under the curve (AUC) and mean squared error (MSE) as performed in [24, 25]. Classifiers with good predictive capacity possess SE, SP, gini and AUC values close to 1 but logloss and MSE values close to 0. Additionally, a three-way data split procedure is utilised (training, validation and test) - 60% for training, 20% for validation and 20% for model testing. The dataset is thus partitioned based on 3:1:1 ratio, using a 60/20/20 split as recommended in [26].

3.1   Hyperparameters Selection

To maximise the predictive capacity of the classifiers, the network architecture and the regularization parameters were tuned. To achieve this, random search was utilised and a maximum of 200 models were generated to obtain the best parameters. The adaptive learning rate ADADELTA [27] was used for stochastic gradient descent optimisation, with parameters *rho* and *epsilon* set to 0.99 and 1×10^{-8} respectively, to balance the global and local search efficiencies.

More specific tuning parameters were considered for each model in the training phase to obtain optimal results. In the first classifier trained with 2,000 compressed hidden units as input, two hidden layers with 10 neurons were used. RectifierWithDropout was used as the activation function throughout the network. To prevent overfitting and to add stability and improve generalisation, Lasso (L1) and Ridge (L2) regularisation values were set to 4.7×10^{-5} and 2.0×10^{-5} respectively. In the second classifier (1,000 compressed hidden units as input), two hidden layers with 20 neurons each, a RectifierWithDropout activation function, and L1 = 8.5×10^{-5} and L2 = 6.0×10^{-6} regularisation values were considered this time for training purposes. The third classifier with 500 compressed hidden units as input, was trained considering two hidden layers with 20 neurons each, a TanhWithDropout activation function, and L1 = 6.8×10^{-5} and L2 = 5.1×10^{-5} regularisation values. The final classifier considered in this study, with the lowest number of compressed hidden units as input (50 compressed units), was trained with two hidden layers of 50 neurons each, a RectifierWithDropout activation function, and L1 = 4.2×10^{-5} and L2 = 6.1×10^{-5} regularisation values. Based on empirical analysis, these configurations produced the best results.

3.2   Classifier Performance

To measure the performance, each MLP classifier was initialised using the different compressed units obtained using the SAE defined in the study. Performance metrics for the validation set are provided in Table 1 while Table 2 shows the performance metrics on the test data when the trained models are used.

The first layer composed of 2,000 compressed hidden units was used to initialize and fine-tune a classifier model. An optimised F1 threshold value of 0.4977 was used to extract the validation set metrics as indicated in Table 1. Successive layers of the SAE were used to initialise and fine-tune the remaining models with 1000, 500 and 50 hidden compressed units as input respectively. On this occasion, metrics were obtained using optimised F1 threshold values 0.6188, 0.4978 and 0.2701 for each of the models respectively.

Table 2 shows the performance metrics obtained using the test set. Optimised F1 threshold values 0.5363, 0.3356, 0.3899 and 0.4615 were used to obtain these metrics by training the models with 2,000, 1,000, 500 and 50 compressed input units respectively.

Early stopping was adopted to avoid overfitting. Model building stops when the logloss on the validation set does not improve by at least 1% for 2 consecutive scoring epochs (stopping rounds). As shown in Fig. 5 the AUC plots provide useful information about early divergence between the training and validation curves, highlighting if overfitting occurs.

Table 1. Performance metrics for validation set.

Compression	SE	SP	Gini	Logloss	AUC	MSE
2,000	0.920213	0.938326	0.960798	0.181744	0.9804	0.054684
2,000-1,000	0.840426	0.938326	0.903365	0.288873	0.95168	0.084837
2,000-1,000-500	0.867021	0.889868	0.882815	0.31456	0.94141	0.09635
2,000-1,000-500-50	0.920213	0.577093	0.697629	0.477655	0.84881	0.159281

Table 2. Performance metrics for test set.

Compression	SE	SP	Gini	Logloss	AUC	MSE
2000	0.949153	0.933014	0.949936	0.1956	0.97497	0.054057
2000-1000	0.915254	0.875598	0.910253	0.294813	0.95513	0.087531
2000-1000-500	0.909605	0.875598	0.900468	0.285151	0.95023	0.087162
2000-1000-500-50	0.785311	0.799043	0.703566	0.476864	0.85178	0.156315

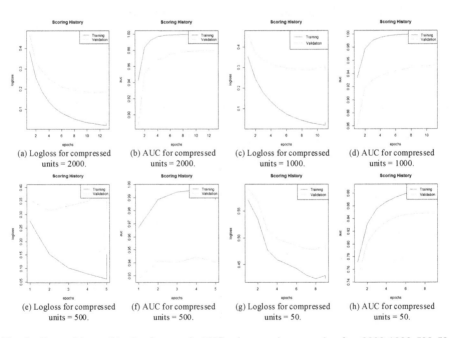

(a) Logloss for compressed units = 2000.   (b) AUC for compressed units = 2000.   (c) Logloss for compressed units = 1000.   (d) AUC for compressed units = 1000.

(e) Logloss for compressed units = 500.   (f) AUC for compressed units = 500.   (g) Logloss for compressed units = 50.   (h) AUC for compressed units = 50.

Fig. 5. From (a) to (h), Logloss and AUC plots against epochs for 2000-1000-500-50 compressed units.

Model Selection

The cut-off values for the false and true positive rates in the test set are depicted by the ROC curves in Fig. 6. The ROC curves show a gradual deterioration in the performance of the classifiers as the initial 2,465 features (SNPs) are progressively compressed down to 50 hidden units in the SAE. Despite the observable deterioration, the results remain high with 50 compressed hidden units.

4 Discussion

After QC, 1,997 individuals (879 cases and 1,118 controls) and 240,950 genetic variants remained for subsequent analysis. Logistic regression under an additive genetic model was performed to assess the association between SNPs and binary disease status. However, in logistic regression analysis, SNPs are independently tested for association with the phenotype under investigation, omitting epistatic interactions between these genetic variants. Hence, we performed unsupervised feature extraction in a set of 2,465 SNPs (p-value $< 1 \times 10^{-2}$) stacking four single layer AEs with 2000-1000-500-50 hidden units. To evaluate the effectiveness of our SAE extracted features, we used a supervised learning model to classify extremely obese samples from normal control samples. Four MLPs were trained with the compressed hidden units considered in the SAE.

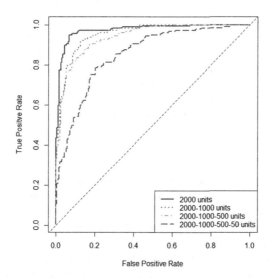

Fig. 6. ROC curves for the test set using trained models with the different compressed units.

Overall, performance metrics obtained using the test set showed slight better improvement in classification accuracy than metrics obtained in the validation set. The results were only higher in the validation set when 2,000 features were used as can be observed in Table 1. The best result using the test set was obtained using 2,000 features (SE = 0.949153, SP = 0.933014, Gini = 0.949936, Logloss = 0.1956, AUC = 0.97497 and MSE = 0.054057). Conversely, the lowest performance in the test set was achieved when the features were compressed to 50 hidden units (SE = 0.785311, SP = 0.799043, Gini = 0.703566, Logloss = 0.476864, AUC = 0.85178 and MSE = 0.156315). In fact, performance metrics started deteriorating when the layers of the SAE were gradually compressed (See Tables 1 and 2). It can be noted that the specificity in the validation set (SP = 0.577093) when 50 features were used was noticeably lower than the value obtained with the same number of compressed neurons in the test set (SP = 0.799043).

Figure 5 shows that there is no significant indications of overfitting between the training and validation datasets. On the other hand, Fig. 6 shows a gradual deterioration in performance when the features are compressed into smaller number of hidden units although the performance is still high even with 50 units. Although the results were higher in the classifiers with the higher number of hidden units and less stacked layers, we managed to compress the initial 2,456 SNPs to 50 hidden units and still get over 85% AUC with relatively low overfitting as shown in Fig. 5. This demonstrates the potential of using our deep learning methodology to abstract large, complex and unstructured data into latent representations capable of capturing the epistatic effect between SNPs in GWAS. Using deep learning stacked autoencoders to initialise the multi-layer feedforward ANN classifier outperformed the results obtained in our previous study [25] using the same dataset. In [25], SNPs were filtered considering suggestive and Bonferroni levels of significance and then used for classification analysis using a multi-layer feedforward ANN. These SNPs were selected based on highly conservative threshold designed to mitigate type I errors but, in our experiments, we demonstrated that significant SNPs had insufficient discriminative capacity to discern between obese and non-obese individuals. Hence, assessing the impact of single variants on the phenotype under study using traditional statistic is insufficient, since epistatic interactions between SNPs are omitted. Our results in this study show an improvement especially in the performance metrics Gini, Logloss and MSE (more reliable models) in comparison with the values obtained in our previous study. Utilising deep learning SAE is a better alternative than using statistical approaches such as logistic regression in GWAS for classification tasks, with potential to help explaining the missing heritability phenomena in polygenic obesity.

This paper presents a novel approach with emphasis in the feature extraction and classification phases, using latent information extracted from high-dimensional genomic data for the identification of individuals with higher predisposition to obesity. However, compressing the features using SAEs makes it difficult to identify which of the 2,465 SNPs contributes to the compress hidden units. This is a well-known problem in neural networks where model interpretation is difficult to achieve [28]. This limitation fosters the need to create robust methods for the interpretation of deep learning networks.

5 Conclusion

In this paper, a novel approach to investigate complex interactions between genetic variants in polygenic obesity, utilising cases and controls from Geisinger MyCode project has been presented. We combined common genetic tools and techniques for QC and association analysis with deep learning to capture relevant information and the epistatic interactions between SNPs. Quality control, association analysis, multi-layer feedforward ANN classifier and deep learning SAE constitute the components of our proposed methodology. Overall, the results highlight the benefits of using deep learning stacked autoencoders to detect epistatic interactions between SNPs in genomic data and how these can be used to model MLPs to classify obese and non-obese observations from the eMERGE MyCode dataset. This contributes to the computational biology and bioinformatics field and provides new insights into the use of deep learning algorithms when analysing GWAS that warrants further investigation.

Although the utilization of deep learning SAE and multilayer feedforward ANN (MLP) has been previously considered in many areas of research, we claim that this is the first time that it has been applied to study epistasis in GWAS of polygenic obesity. This framework has been successfully employed for preterm birth classification in African-American woman [24].

Despite we have presented encouraging results, the study needs further research. Mapping SNPs inputs to hidden nodes in the SAE is still a recognized limitation as deep learning approaches act as black box where models become difficult to interpret. In future work, we aim to create a robust method that helps interpreting the outcome of the compressed units by combining association rule mining (ARM) [29] and the strength of SAEs.

References

1. James, W.P.T.: WHO recognition of the global obesity epidemic. Int. J. Obes. **32**(S7), S120–S126 (2008)
2. Borrell, L.N., Samuel, L.: Body mass index categories and mortality risk in US adults: the effect of overweight and obesity on advancing death. Am. J. Public Health **104**(3), 512–519 (2014)
3. Walley, A.J., Blakemore, A.I.F., Froguel, P.: Genetics of obesity and the prediction of risk for health. Hum. Mol. Genet. **15**(SUPPL. 2), 124–130 (2006)
4. Rao, K.R., Lal, N., Giridharan, N.V.: Genetic & epigenetic approach to human obesity. Indian J. Med. Res. **140**(November 2014), 589–603 (2015)
5. Walker, J.M.: Genetic Variation, vol. 628. Humana Press, Totowa (2010)
6. Locke, A.E., Kahali, B., Berndt, S.I., Justice, A.E., Pers, T.H., Day, F.R.: Genetic studies of body mass index yield new insights for obesity biology. Nature **518**(7538), 197–206 (2015)
7. Moore, J.H., Williams, S.M.: Epistasis and its implications for personal genetics. Am. J. Hum. Genet. **85**(3), 309–320 (2009)
8. Chen, K.Y., Janz, K.F., Zhu, W., Brychta, R.J.: Redefining the roles of sensors in objective physical activity monitoring. Med. Sci. Sports Exerc. **44**(301), S13–S23 (2012)
9. McCarthy, M.I., et al.: Genome-wide association studies for complex traits: consensus, uncertainty and challenges. Nat. Rev. Genet. **9**(5), 356–369 (2008)

10. Li, W.: Three lectures on case control genetic association analysis. Brief. Bioinform. **9**(1), 1–13 (2007)
11. Purcell, S., et al.: PLINK: a tool set for whole-genome association and population-based linkage analyses. Am. J. Hum. Genet. **81**(3), 559–575 (2007)
12. Clarke, G.M., Anderson, C.A., Pettersson, F.H., Cardon, L.R., Morris, A.P., Zondervan, K.T.: Basic statistical analysis in genetic case-control studies. Nat. Protoc. **6**(2), 121–133 (2011)
13. Lee, S., Kwon, M.-S., Park, T.: Network graph analysis of gene-gene interactions in genome-wide association study data. Genomics Inform. **10**(4), 256 (2012)
14. Gül, H., Aydin Son, Y., Açikel, C.: Discovering missing heritability and early risk prediction for type 2 diabetes: a new perspective for genome-wide association study analysis with the Nurses' Health Study and the Health Professionals' Follow-Up Study. Turk. J. Med. Sci. **44** (6), 946–954 (2014)
15. Ng, A.: Sparse Autoencoder. In: CS294A Lecture Notes, pp. 1–19 (2011)
16. Rumelhart, D.E., Hinton, G.E., Williams, R.J.: Learning representations by back-propagating errors. Nature **323**(6088), 533–536 (1986)
17. Glorot, X., Bordes, A., Bengio, Y.: Deep sparse rectifier neural networks. In: Proceedings of the 14th International Conference on Artificial Intelligence and Statistics (AISTATS), pp. 315–323 (2011)
18. Bergstra, J., Bengio, Y.: Random search for hyper-parameter optimization. J. Mach. Learn. Res. **13**, 281–305 (2012)
19. Le, Q.V.: A Tutorial on Deep Learning Part 2: Autoencoders, Convolutional Neural Networks and Recurrent Neural Networks, Mountain View, CA (2015)
20. Goodfellow, I., Bengio, Y., Courville, A.: Deep Learning, p. 1. MIT Press, Cambridge (2016)
21. Bengio, Y., Lamblin, P., Popovici, D., Larochelle, H.: Greedy layer-wise training of deep networks. In: Advances in Neural Information Processing Systems, pp. 153–160. MIT Press (2007)
22. Danaee, P., Ghaeini, R., Hendrix, D.A.: A deep learning approach for cancer detection and relevant gene identification. In: Pacific Symposium on Biocomputing, vol. 22, no. 4, pp. 219–229, January 2017
23. Salari, N., Shohaimi, S., Najafi, F., Nallappan, M., Karishnarajah, I.: A novel hybrid classification model of genetic algorithms, modified k-nearest neighbor and developed backpropagation neural network. PLoS One **9**(11), e112987 (2014)
24. Fergus, P., Curbelo, C., Abdulaimma, B., Lisboa, P., Chalmers, C., Pineles, B.: Utilising deep learning and genome wide association studies for epistatic-driven preterm birth classification in African-American women. IEEE/ACM Trans. Comput. Biol. Bioinform. (2018)
25. Curbelo, C., Fergus, P., Curbelo, A., Hussain, A., Al-Jumeily, D., Chalmers, C.: Deep learning classification of polygenic obesity using genome wide association study SNPs. In: 2018 International Joint Conference on Neural Networks (IJCNN), pp. 1–8 (2018)
26. Lever, J., Krzywinski, M., Altman, N.: Model selection and overfitting. Nat. Methods **13**(9), 703–704 (2016)
27. Zeiler, M.D.: ADADELTA: an adaptive learning rate method. arXiv:1212.5701, December 2012
28. Manning, T., Sleator, R.D., Walsh, P.: Biologically inspired intelligent decision making. Bioengineered **5**(2), 80–95 (2014)
29. Agrawal, R., Imieliński, T., Swami, A.: Mining association rules between sets of items in large databases. In: Proceedings of 1993 ACM SIGMOD International Conference on Management of Data, vol. 22, no. 2, pp. 207–216, June 1993

Predicting the Oncogenic Potential of Gene Fusions Using Convolutional Neural Networks

Marta Lovino[1][(✉)], Gianvito Urgese[1], Enrico Macii[2],
Santa di Cataldo[1], and Elisa Ficarra[1]

[1] Department of Control and Computer Engineering, Politecnico di Torino,
Corso Duca Degli Abruzzi 24, 10129 Turin, Italy
{marta.lovino,gianvito.urgese,santa.dicataldo,elisa.ficarra}@polito.it
[2] Interuniversity Department of Regional and Urban Studies and Planning,
Politecnico di Torino, Corso Duca Degli Abruzzi 24, 10129 Turin, Italy
enrico.macii@polito.it

Abstract. Predicting the oncogenic potential of a gene fusion transcript is an important and challenging task in the study of cancer development. To this date, the available approaches mostly rely on protein domain analysis to provide a probability score explaining the oncogenic potential of a gene fusion. In this paper, a Convolutional Neural Network model is proposed to discriminate gene fusions into oncogenic or non-oncogenic, exploiting only the protein sequence without protein domain information. Our proposed model obtained accuracy value close to 90% on a dataset of fused sequences.

Keywords: Gene fusions · Deep learning · Convolutional Neural Networks

1 Introduction

Nowadays, the increased availability of Next Generation Sequencing (NGS) data enables new unforeseen insights into the relation between some genetic rearrangements and cancer development. In this regard, a challenging area is represented by the study of gene fusions, a genetic aberration where two separate DNA regions (usually two distinct genes) join together into a hybrid gene. The genes retained at 5p' and 3p' of the fused sequence are conventionally called 5p' gene and 3p' gene, respectively. If the promoter region of at least one of the two genes is retained in the fusion, the erroneous sequence is transcribed at the RNA level, and the aberrated transcript can result into an abnormal protein [1].

Since the discovery of the first genetic rearrangement by Nowell and Hungerford in 1960, a large number of gene fusions have been associated to cancer development and used as cancer predictors [1]. However, gene fusions do not automatically relate to carcinogenic processes, as they can be found in large number even in non-tumoral samples [2]. In light of the above, predicting whether

M. Raposo et al. (Eds.): CIBB 2018, LNBI 11925, pp. 277–284, 2020.
https://doi.org/10.1007/978-3-030-34585-3_24

an aberrated transcript will result into a functional protein or a cancer driver is a very critical and challenging task in the study of cancer development.

To the best of our knowledge, all current approaches reconstruct the candidate fusion from original sequenced data and apply different types of machine learning methods to perform protein domain analysis. For example, the tools Oncofuse [3] and Pegasus [4] use respectively naive Bayes Network and decision tree classifiers to provide an oncogenic probability score for the fusion based on protein domain data.

2   Scientific Background

To this date, a large number of machine learning approaches have been proposed to solve different types of DNA sequence classification problems, with an increasing trend in the use of deep learning techniques [5,6]. More specifically, Convolutional Neural Networks (CNNs), a class of deep, feed-forward neural networks originally designed for image classification problems, are now exploited in many DNA sequence analysis tasks for their ability to automatically learn the features from the training data, avoiding the design of handcrafted descriptors. Among the many tasks, CNNs have been successfully applied to model the properties and functions of DNA sequences, to the prediction of single-cell DNA methylation states and microRNA targets, as well as to the recognition of splice junction sites and promoter sequence regions [5].

In this work we exploit CNN to classify candidate gene fusions into cancer driving and non-carcinogenic fusions, outputting a categorical class label instead of a probability score. Unlike previous approaches, our model exploits human reference sequences (and not original sequencing data), relying only on the fusion sequence, with no additional input about conserved or lost protein domains. By doing so, our aim is to avoid any possible bias that the prediction models leveraged by protein domain analysis may introduce into the classification task, as well as to improve the generalization capabilities and ease-of-retraining of the classifier. This is a very important trait in a continuously-evolving field of knowledge such as the study of cancer development.

To design a completely protein domain independent model, we provide the real amino acid composition of the fused protein to the network, without any other additional data interpretation.

3   Materials and Methods

As already mentioned, the purpose of our work is to discriminate between gene fusions with functional oncogenic potential (referred to as *Onco class*) and fusions that are not involved in a carcinogenic process (referred to as *NotOnco class*), without any previous information on the protein domains retained or lost in the fusion sequences. For this purpose, we exploit the ability of the CNN to recognize local spatial patterns that are significant for the classification without requiring any a priori feature description of the two classes.

Overall our dataset contains a total number 1741 reconstructed fused sequences, respectively 1005 for the *Onco class* and 736 for the *NotOnco class*. As CNNs traditionally take images as input, we apply and compare three different encoding methods to transform fusion sequences into image-like data structures. The process of data retrieval, encoding from sequence to images and CNN design and training are described in the following.

3.1 Fusion Data Retrieval

Gene fusion data were retrieved from two different sources, respectively for the *Onco* and the *NotOnco class*.

Cosmic, a catalogue for somatic mutations in cancer [7], was used for the *Onco class*. This catalogue provides per each fusion the transcript name of both 5p' and 3p' genes, as well as breakpoint information on the retained transcripts considering UTR regions. For our work we selected only the coding sequence (CDS) retained in the fusion. As this sequence translates into a protein which may or may not be involved in an oncogenic process, it is the only information that is significant for our classification task. For consistency with the *NotOnco class* data, we reconstructed a total number of 1011 fusion sequences from the GRCh37 version of the catalogue.

Data for the *NotOnco class* were reconstructed based on Babicenau et al. work on recurrent chimeric fusion RNAs in non-cancer tissues and cells [2]. In this work, SOAPfuse (a tool for gene fusion analysis) was applied on 171 non-neoplastic tissue samples from 27 different tissues, identifying 291 recurrent fusions (i.e. fusions that are detected in more than one sample) involving 238 gene pairs. Per each of these fusions authors report the breakpoint position on human reference genome hg19 of both fused genes. As no information is provided about which part of the transcript is retained in the fusion, we assumed the most common configuration, where the fused sequence is the result of the region near the promoter for the 5p' gene, and of the ending region for the 3p' gene. This assumption is biologically consistent, since a fused transcript needs a promoter region to be translated into protein, and has no impact to our classification task. As a matter of fact, we observed that 91% of the *Onco class* fused transcripts included the region near the promoter for 5p' gene and the ending region for the 3p' gene. On top of that, when selecting the proper CDS region according to the above mentioned configuration, the same CDS region may be involved in more than one trascript. Therefore, we decided to consider as *NotOnco class* all the fusion sequences resulting from all possible combinations of transcripts at 5p' gene with transcripts at 3p' gene. In order to avoid any biases, we discarded all the cases where the intron can be retained in the fusion trascripts. This led to obtain for the *NotOnco class* a total number of 741 fusions which involve 524 transcripts.

Three transcripts are present in both the *Onco class* and the *NotOnco class*.

3.2 Encoding: From Sequences to Images

Once all the fused sequences had been reconstructed, they were translated into protein sequences following the Amino Acid Translational Table. The translation process is in-frame, because transcripts were taken from the beginning of the coding sequence identifiable by the ATG triplet, (a.k.a. initiation codon). As CNN are inherently designed to take images as input, the fused amino acid sequence needs to be converted into a $N \times M \times C$ data structure, where N and M are length and width of the image and C the number of channels. For our purposes, N was set to 3000. Hence, we discarded longer sequences and padded the shorter ones using a fake amino acid. By doing so, we obtained a total number of 1741 strings (1005 for the *Onco class* and 736 for the *NotOnco class*, respectively) of 22 different letters, each corresponding to one amino acid (21 real amino acids plus the fake one).

Popular methods for string encoding are ordinal encoding and one-hot encoding, eventually with some variations.

Ordinal encoding substitutes the ith letter in a fusion with a fixed value corresponding to a unique amino acid. Hence, the resulting image will have minimal dimensions $N = 3000 \times M = 1 \times C = 1$, with memory saving advantages compared to other techniques. On the other hand, the incremental values assigned to the amino acids establish an artificial ordering which may bias the representation [8].

One-hot encoding assigns to the i_{th} letter a vector of length L, where each j_{th} element corresponds to a feature. In standard one-hot encoding features are the amino acids: hence, the i_{th} letter is encoded by a vector of all zeros, except for the jth element associated to the amino acid, which is set to 1.

In our work we explored yet another encoding solution, with features corresponding to 28 real amino acid properties (i.e. hydrophobicity, ionic, mass, polarity, etc.) taken from Bulka's work [9]. Hence, in the following we will refer to this strategy as *Bulka's encoding*. In case of on/off properties, the j_{th} element is set to 0 or to 1, based on the fact that the amino acid has or does not have that specific property. For the other ones that are not on/off (i.e. number of H bonds, isoelectric point and hydrophobicity) it is set to the normalized value of that property. For both one-hot and Bulka's encoding strategies, the size of the obtained images is $N = 3000 \times M = 1 \times C = L$.

As the CNN model will inherently assume spatial correlations between adjacent pixels, the data structure was arranged so that the amino acid features constitute the third dimension (i.e. channels) of the image.

Overall the encoding step is summarized in the first section of Fig. 1.

3.3 CNN Architecture and Training Paradigm

As shown in the second section of Fig. 1, we designed a CNN model with two convolutional layers (kernel size 5) followed by two max pooling layers (kernel size 2). To avoid overfitting, we set dropout to 0.5 and learning rate to 0.01. After flattening, we inserted a 1000-units dense layer with ReLU activation function

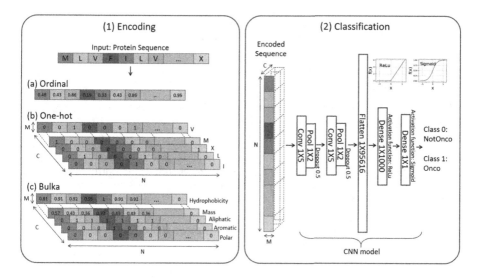

Fig. 1. Overview of the encoding and classification process.

and a final single unit dense layer with sigmoid activation function, which provides the classification output. Batch size was set to 256 and number of epochs to 50, and the network was trained by backpropagation implementing a Stochastic gradient descent optimizer.

The CNN was implemented in Keras python library under Tensorflow backend [10].

4   Experimental Results

A first set of experiments aimed at evaluating the performance of the network in the classification of completely new fused transcripts, using different types of encoding techniques. For this purpose, we created a random partition of the available dataset, where we ensured a complete independence of the training and test sets in terms of involved transcripts. More specifically, we included in the test set only fused sequences whose 5p' and 3p' genes were both not present in any of the fused sequences used for training. This configuration resulted in 1490 samples for the training set and 251 samples for the test set, respectively. With these sets, we trained and tested our CNN model with the three different encoding methods (i.e. ordinal, standard one-hot and Bulka's encoding). In order to asses the stability of the network in terms of independence from weights initialization, we trained and tested the model five times per each type of encoding. As it is visible from the trend of the loss functions during the test set, shown in Fig. 2, the network converged well within 50 epochs.

The test accuracy values obtained in the five runs per each type of encoding are shown in the form of box-plots in Fig. 3, with black boxes ranging from the

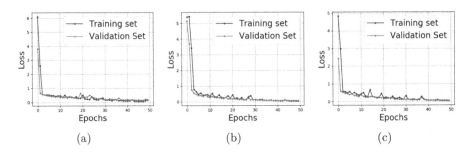

Fig. 2. Loss functions of CNN models using different encodings: (a) ordinal encoding, (b) standard one-hot encoding and (c) Bulka's encoding.

25% to the 75% percentile of the accuracy values, and red lines indicating the median accuracy value over the five runs.

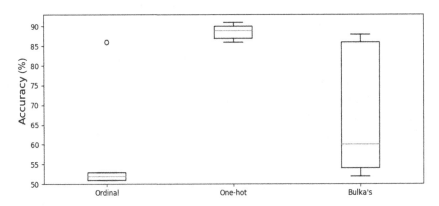

Fig. 3. Box-plot of test accuracy values over five runs, using three different types of encoding techniques.

From the plots in Fig. 3 we can make the following observations. (i) Ordinal encoding consistently achieved the lowest accuracy (around 52%, with very low variation over the five runs). (ii) Bulka's encoding obtained on average higher accuracy than ordinal encoding (median accuracy value around 60%), but at the price of a very high variability of the results (Bulka's box ranges from 52% to 88% accuracy). (iii) Standard one-hot encoding had a very good accuracy (median value 89%), coupled with reasonably low variability.

Based on our results, standard one-hot encoding provided the best compromise, in terms of classification accuracy and stability of the model. This evidence can be explained by taking into consideration the three different encoding designs. On one hand, ordinal encoding introduces a very strong bias into the representation, because it forcefully creates an alphabetical ordering of the 22 amino acids. This easily explains the low accuracy values obtained by this type of encoding. On the other hand, Bulka's encoding uses physical properties of the proteins to univocally represent each amino acid, without implying any type of ordering. Nonetheless, there is no certainty about the significance of the specific properties that were chosen for the representation, nor of their complete independence. This might explain the high instability of the classification model leveraging upon Bulka's technique. In the end, according to our results, standard one-hot encoding ensures the most unbiased data representation, and hence the highest classification accuracy. Based on this evidence, we selected this type of encoding for our next set of experiments.

Because of the limited number of transcripts involved in the dataset, one might argue that the high accuracy of the network derives from a sort of memorization of the transcript sequences, and not from a real capability of discriminating significant patterns on the input data. To prove this hypothesis wrong, we performed a second set of experiments, giving the entire transcripts of both the *NotOnco* and the *Onco class* as input to our CNN model. The rationale of this experiment is to ensure that the network does not blindly assign all the first set of transcripts to the *NotOnco class* and all the second set of transcripts to the *Onco class*, respectively.

As a result of this experiment, we obtained that only 70% of the first set of transcripts were classified as *NotOnco class* and the 78% of the second set as *Onco class*, respectively. As a fair amount of the whole transcripts were still assigned a class that is different from the one they were extracted from, we can reasonably conclude that the classification task is not driven by the transcript sequence alone.

5   Conclusion

In the end, our experiments proved that the proposed CNN approach is able to predict the oncogenicity of a gene fusion with a satisfactory level of accuracy, relying only on the fused sequence with no additional information. Tests on three different encoding methods demonstrated that standard one-hot encoding was the most suitable for the representation of the amino acid sequence.

Future works will focus on the interpretation of the features extracted by the locally connected stages of the CNN, in order to obtain a deeper understanding of the specific biological patterns that mostly influence the carcinogenic potential of a gene fusion. On top of that, we plan to increase as much as possible the number of samples used to train the CNN, with the aim of improving the generalization capabilities of our model.

References

1. Mertens, F., Johansson, B., Fioretos, T., Mitelman, F.: The emerging complexity of gene fusions in cancer. Nat. Rev. Cancer **15**(6), 371 (2015)
2. Babiceanu, M., et al.: Recurrent chimeric fusion RNAs in non-cancer tissues and cells. Nucleic Acids Res. **44**(6), 2859–2872 (2016)

3. Shugay, M., Ortiz de Mendíbil, I., Vizmanos, J.L., Novo, F.J.: Oncofuse: a computational framework for the prediction of the oncogenic potential of gene fusions. Bioinformatics **29**(20), 2539–2546 (2013)
4. Abate, F., et al.: Pegasus: a comprehensive annotation and prediction tool for detection of driver gene fusions in cancer. BMC Syst. Biol. **8**(1), 97 (2014)
5. Min, S., Lee, B., Yoon, S.: Deep learning in bioinformatics. Brief. Bioinform. **18**(5), 851–869 (2017)
6. Rizzo, R., Fiannaca, A., La Rosa, M., Urso, A.: A deep learning approach to DNA sequence classification. In: Angelini, C., Rancoita, P.M.V., Rovetta, S. (eds.) CIBB 2015. LNCS, vol. 9874, pp. 129–140. Springer, Cham (2016). https://doi.org/10.1007/978-3-319-44332-4_10
7. Forbes, S.A., et al.: COSMIC: mining complete cancer genomes in the catalogue of somatic mutations in cancer. Nucleic Acids Res. **39**(suppl–1), D945–D950 (2010)
8. Choong, A.C.H., Lee, N.K.: Evaluation of convolutionary neural networks modeling of DNA sequences using ordinal versus one-hot encoding method. bioRxiv, p. 186965 (2017)
9. Bulka, B., Freeland, S.J., et al.: An interactive visualization tool to explore the biophysical properties of amino acids and their contribution to substitution matrices. BMC Bioinform. **7**(1), 329 (2006)
10. Chollet, F., et al.: Keras (2015). https://keras.io

Unravelling Breast and Prostate Common Gene Signatures by Bayesian Network Learning

João Villa-Brito[1]([✉]), Marta B. Lopes[3,4] (ID), Alexandra M. Carvalho[3] (ID),
and Susana Vinga[2,4] (ID)

[1] Instituto Superior Técnico, Av. Rovisco Pais, 1049-001 Lisbon, Portugal
joao.v.brito@tecnico.ulisboa.pt
[2] IDMEC, Instituto Superior Técnico, Universidade de Lisboa, Lisbon, Portugal
susanavinga@tecnico.ulisboa.pt
[3] Instituto de Telecomunicações, Instituto Superior Técnico, Universidade de Lisboa,
Av. Rovisco Pais, 1049-001 Lisbon, Portugal
alexandra.carvalho@tecnico.ulisboa.pt
[4] INESC-ID, Instituto Superior Técnico, Universidade de Lisboa,
R. Alves Redol 9, 1000-029 Lisbon, Portugal

Abstract. Breast invasive carcinoma (BRCA) and prostate adenocarcinoma (PRAD) are two of the most common types of cancer in women and men, respectively. As hormone-dependent tumours, BRCA and PRAD share considerable underlying biological similarities worth being exploited. The disclosure of gene networks regulating both types of cancers would potentially allow the development of common therapies, greatly contributing to disease management and health economics. A methodology based on Bayesian network learning is proposed to unravel breast and prostate common gene signatures. BRCA and PRAD RNA-Seq data from The Cancer Genome Atlas (TCGA) measured over ∼20000 genes were used. A prior dimensionality reduction step based on sparse logistic regression with elastic net penalisation was employed to select a set of relevant genes and provide more interpretable results. The Bayesian networks obtained were validated against information from STRING, a database containing known gene interactions, showing high concordance.

Keywords: Sparse logistic regression · Gene expression · Machine learning

Supported by the EU Horizon 2020 research and innovation program (grant No. 633974 - SOUND project), and the Portuguese Foundation for Science & Technology (FCT), through projects UID/EMS/50022/2019 (IDMEC, LAETA), UID/EEA/50008/2019 (IT), UID/CEC/50021/2019 (INESC-ID), PERSEIDS (PTDC/EMS-SIS/0642/2014), PREDICT (PTDC/CCI-CIF/29877/2017), NEUROCLINOMICS2 (PTDC/EEI-SII/1937/2014), and IF/00653/2012; also partially supported by internal IT projects QBigData and RAPID.

M. Raposo et al. (Eds.): CIBB 2018, LNBI 11925, pp. 285–292, 2020.
https://doi.org/10.1007/978-3-030-34585-3_25

1 Scientific Background

Due to the computerisation of our everyday life, available data is growing tremendously across all fields of research, businesses and industry. When dealing with high-dimensional data, sparse models are able to extract knowledge from data, by identifying a smaller number of relevant variables (from a whole set of variables) explaining the data. In the context of biological data, the use of sparse graphical models is expected to disclose valuable insights on the underlying biological mechanisms.

This work searches for common gene signatures between breast invasive carcinoma (BRCA) and prostate adenocarcinoma (PRAD), two of the most common types of invasive cancer in women and men, respectively. Although arising in organs with different anatomies and physiological functions, BRCA and PRAD tumours depend on gonadal steroids for their development, as the organs they originate from, being hormone-dependent. Both cancers have considerable underlying biological similarities worth being exploited with the goal of improving patient outcomes [1]. The proposed methodology uses Bayesian network learning to identify a common gene network to both cancers, as not only the genes regulating the diseases but also the interaction between them could help better understanding the diseases, while providing guidance to cancer therapy research and disease management.

2 Materials and Methods

2.1 Dimensionality Reduction

Let $\boldsymbol{Y}$ be a random variable whose n components are independently distributed with means $\boldsymbol{\mu}$, $\boldsymbol{X}$ the $n \times p$ matrix containing the set of p explanatory variables, and $\boldsymbol{\beta} = \{\beta_1, \ldots, \beta_p\}^T$ the $p \times 1$ vector of unknown regression coefficients associated with each covariate. Then, in a generalised linear model (GLM) [2]:

$$E(\boldsymbol{Y}) = \boldsymbol{\eta} = g(\boldsymbol{\mu}) = \boldsymbol{X}\boldsymbol{\beta} = \boldsymbol{x}_i\boldsymbol{\beta}; \qquad i = 1, \ldots, n. \tag{1}$$

Specifically in this work, the independent variable is binary, thus $\boldsymbol{Y}$ is assumed to follow a Binomial distribution. Defining the probability of success, p_i as the probability of $Y_i = 1$, given the associated variables vector $\boldsymbol{x}_i$, binary logistic regression (LR) models show how the response variable depends on the set of variables:

$$\boldsymbol{\eta} = \text{logit}(p_i) = \log\left(\frac{p_i}{1 - p_i}\right) = \boldsymbol{x}_i\boldsymbol{\beta}; \qquad i = 1, \ldots, n. \tag{2}$$

The unknown regression coefficients $\boldsymbol{\beta}$ are estimated using maximum likelihood. For a n sized sample, the log-likelihood function for a binary LR is

$$\ell(\boldsymbol{\beta}) = \sum_{i=1}^{n}[y_i \log(p_i) + (1 - y_i)\log(1 - p_i)]. \tag{3}$$

The estimates $\hat{\boldsymbol{\beta}}$ obtained maximise $\ell(\boldsymbol{\beta})$. Usually, they are all non-zero, and if $p > N$ (more explanatory variables than observations), they are not unique. When addressing healthcare big data problems, it is necessary to constrain the regression problem in order to estimate interpretable models, e.g. through regularised optimisation. In other words, it is necessary to encourage sparsity. In a sparse statistical model, only a relatively small number of predictors is different from zero. The more sparse the model is, the less the number of non-zero parameters.

Ridge regularisation [3] adds an ℓ_2 constraint, $\sum_{j=1}^{p} \beta_j^2$, to the log-likelihood function, promoting solutions with small norms, i.e., close to zero, but still non-sparse. The least absolute shrinkage and selection operator (lasso) [4] is a regularisation method that enjoys the stability of ridge regression, while promoting variable selection. It works by combining the log-likelihood function with an ℓ_1 constraint, $\sum_{j=1}^{p} |\beta_j|$.

The lasso penalty is able to perform both shrinkage and variable selection. While it performs well in many circumstances, it has shown some limitations. If $p > n$ the lasso selects no more than n variables before it saturates; if there are highly correlated variables, the lasso arbitrarily selects only one, not taking into account the group as a whole (there is no clustering); and when the variables are highly correlated, in $n > p$ situations, the prediction accuracy of the lasso becomes dominated by the ridge regression.

Elastic net regularisation [5] is a technique proposed to solve the mentioned problems, as a combination of both lasso and ridge, and performing as well as the lasso whenever the lasso does the best. The regulariser is defined as $\lambda \sum_{j=1}^{p} \left\{ (1 - \alpha)\beta_j^2 + \alpha|\beta_j| \right\}$. The ℓ_1 part of the penalty helps to generate a sparse model, while the ℓ_2 part makes it possible to select more than n variables (in the $p > n$ case) and encourages clustering. The tuning constants, $\lambda \geq 0$ and $\alpha \in [0, 1]$ control the magnitude of the parameters and the relative weight of each constraint, respectively.

2.2   Bayesian Networks

Graphical models are a powerful probabilistic representation that provide interpretable models of the domain. For this reason, they have been used in a large variety applications such as genetics, oncology, computational biology, and medicine and health care. The large volume of high-dimensional biological data has motivated the use of graphical models to provide understanding into novel biological mechanisms.

Bayesian networks are the most widely known directed graphical models, however, they are typically not used with high dimensional data, with $p \gg n$, as they do not scale well with the number of variables. Nonetheless, these directed models provide us with unprecedented insights about probabilistic correlations between variables under study, in this case, gene expression values. This could be of great benefit for genomics applications, with datasets such as the human transcriptome, with $p \sim 20000$.

Considering a p-dimensional random vector $\mathbf{Z} = (Z_1, \ldots, Z_p)$, whose realisation is in $\mathbf{X}$, a Bayesian network (BN) is rigorously defined as a directed acyclic graph (DAG) $G = (V; E)$ with nodes in V, coinciding with $\mathbf{Z}$, and edges in E, representing a joint probability distribution $P(\mathbf{Z})$ in a factored way, according to the DAG structure as:

$$P(Z_1, \ldots, Z_p) = \prod_{j=1}^{p} P(Z_j | \mathrm{pa}(Z_j), \theta_j), \tag{4}$$

where $\mathrm{pa}(Z_j) = \{Z_i : Z_i \rightarrow Z_j \in E\}$ is the parent set of Z_j and θ_j encodes the parameters that define the conditional probability distribution (CPD) for Z_j. Gaussian CPDs for continuous data are considered.

Learning a BN reduces to learn its structure and parameters. Having the structure fixed, parameters are quite easy to learn. The hard task is to learn the structure itself, generally approached through score-based learning; in this case, a score is used to ascribe the network fitting to the data. Most common scoring criteria are based on maximum likelihood estimation with penalisation factors to prevent data overfitting.

Aragam et al. developed a new R package, called sparsebn [6], focused on learning the structure of sparse graphical models, especially thought for large networks. To learn a BN from data, they have used a score-based approach that relies on regularised maximum likelihood estimation. The following criterion was considered:

$$\min_{\mathbf{B} \in \mathbb{D}} \ell(\mathbf{B}; \mathbf{X}) + \rho_\lambda(\mathbf{B}), \tag{5}$$

where ℓ denotes the negative log-likelihood, ρ_λ is some regulariser, the matrix $\mathbf{B}$ is the weighted adjacency matrix of a DAG, being $\mathbb{D}$ the set of weighted adjacency matrices that represent DAGs. For continuous data, a Gaussian likelihood with ℓ_1 or minimax concave penalty is used.

The package offers methods to learn the structure of a BN, to estimate its parameters $\hat{\mathbf{B}}$, to plot that structure and, for Gaussian data, to calculate the implied covariance and precision matrices. Many methods from the literature on coordinate descent such as *warm starts, active set iterations, block updates* and *sparse data structures* were used by the authors to make the algorithms run faster, distinguishing sparsebn from existing packages, for sparse structure learning and high dimensional data.

2.3 Datasets

To unravel common gene signatures to breast and prostate cancers, two datasets were extracted from the Cancer Genome Atlas (TCGA) database [7]: BRCA and PRAD datasets, corresponding to breast and prostate, respectively. For more information on the datasets refer to https://github.com/jvillabrito/common-gene-signature.

A subset of 19810 variables was selected from the BRCA dataset, and of 19660 from the PRAD dataset, corresponding to the protein-coding genes reported from the Ensembl genome browser [8] and the Consensus CDS [9] project. The data was pre-processed as follows (see Fig. 1). The variables with zero standard deviation were excluded from both datasets, and only the 19529 common to both datasets were considered for further analysis. The variables were log-transformed and normalised to zero mean and unit variance. The final datasets are $X_{\text{brca}} \in \mathbb{R}^{n_{\text{brca}} \times p}$; $y_{\text{brca}} \in \mathbb{R}^{n_{\text{brca}}}$ and $X_{\text{prad}} \in \mathbb{R}^{n_{\text{prad}} \times p}$; $y_{\text{prad}} \in \mathbb{R}^{n_{\text{prad}}}$, with $n_{\text{brca}} = 1204$, $n_{\text{prad}} = 547$ and $p = 19529$. Matrices X are the explanatory variables (genes) matrices and vectors y are the binary response vectors, with '1' and '0' corresponding to tumour and normal tissue samples, respectively. Samples presented with metastases were not considered for the analysis.

2.4   Finding Common Gene Signatures

With the goal of obtaining more interpretable results, a dimensionality reduction step was added before learning the Bayesian networks, using logistic regression with elastic net penalisation, considering two values of α ($\alpha = 0.1$ and $\alpha = 0.01$). Two approaches were tested: *jointEN* and *sepEN*. In the first, sparse logistic regression with elastic net penalty was applied to a new dataset combining BRCA and PRAD data, BRCAPRAD ($X_{\text{brcaprad}} \in \mathbb{R}^{n_{\text{brcaprad}} \times p}$; $y_{\text{brcaprad}} \in \mathbb{R}^{n_{\text{brcaprad}}}$, with $n_{\text{brcaprad}} = 1751$ and $p = 19529$), using the glmnet R package. The λ parameter was optimised by 10-fold cross-validation. The variables selected were used for further analysis. In the second approach, two sparse logistic models were fit, one for BRCA and another for PRAD, also with 10-fold cross-validation. The variables used are the ones selected separately for each cancer.

After the dimensionality reduction block (see Fig. 1), sparsebn R package was used to learn the Bayesian networks, using the method estimate.dag. The parameter 'edge.threshold' was used to force the number of edges in the solution to be less or equal than the number of nodes. The output of the method is a solution path, rather than an unique solution, consisting of a sequence of estimates for a predetermined set of lambdas $\lambda_{max} > \lambda_1 > \cdots > \lambda_{min}$ (default grid of values are used based on a decreasing log-scale). As λ decreases, there is less regularisation, i.e. the graphs are more dense, containing more edges. The select.parameter method was then used to get the optimal value of λ, based on a trade-off between the increase in log-likelihood and the increase in complexity between solutions. Only the solution for the optimal lambda was considered for further analysis. Four Bayesian networks were learnt: two from BRCA data, one using only tumour tissue samples (*tumour*BN) and another using only normal tissue samples (*normal*BN), and the same from PRAD data. The four Bayesian Networks obtained were validated by comparing the resulting edges with STRING information [10]. The *tumour*BNs were then compared to determine the number of shared edges. The same was done for *normal*BNs. Finally, *tumour*BNs were compared against *normal*BNs, to verify whether they share common edges.

3   Results

The number of edges of the Bayesian networks obtained can be found in Table 1, for reduced data and for full dimension data. In the case of reduced data, *jointEN* and *sepEN* approaches are discriminated. BN_{brca} and BN_{prad} correspond to the Bayesian networks learnt from BRCA and PRAD data, respectively. It can be noticed that the solutions for the optimal lambdas have the number of edges close to the number of variables. A considerable overlap between BN_{brca} and BN_{prad} networks (# common edges; Table 1) was obtained, which supports the fact that both types of cancer have underlying similarities . A noticeable overlap was also obtained when comparing the pairwise gene connections identified and STRING information. The percentage of known gene interactions found when learning the BNs from tumour data is approximately twice the number of gene interactions when BNs are learnt from normal data.

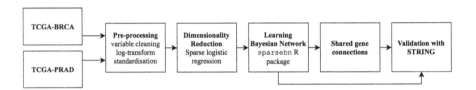

Fig. 1. Proposed solution pipeline.

Table 1. Number of edges of BNs obtained. The numbers in parentheses state the percentage of edges that represent known gene interactions, based on STRING information. (T: Tumour; NT: Non-Tumour)

α	Data	n		Approach	p	# edges		# common edges
		brca	prad			BN_{brca}	BN_{prad}	
0.1	T	1091	495	*jointEN*	546	499 (12%)	528 (11%)	57 (47%)
				sepEN	738	624 (11%)	733 (11%)	33 (45%)
	NT	113	52	*jointEN*	546	537 (7%)	534 (8%)	38 (21%)
				sepEN	738	713 (9%)	721 (7%)	33 (24%)
0.01	T	1091	495	*jointEN*	2791	2674 (16%)	2663 (13%)	268 (32%)
				sepEN	738	624 (11%)	733 (11%)	33 (45%)
	NT	113	52	*jointEN*	2791	2790 (6%)	2791 (9%)	153 (18%)
				sepEN	4370	4370 (8%)	4336 (9%)	287 (16%)
–	T	1091	495	–	19529	18411 (23%)	18553 (17%)	1589 (31%)
	NT	113	52	–	19529	19527 (9%)	19366 (10%)	1664 (16%)

Figure 2 illustrates the networks of the common edges in BN_{brca} and BN_{prad}, when BNs are learnt from tumour data, after dimensionality reduction with $\alpha = 0.01$. Besides paired genes, highly connected genes were obtained as well, called hubs, which are also reported in STRING.

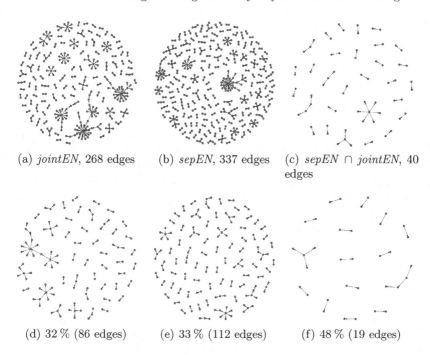

(a) *jointEN*, 268 edges   (b) *sepEN*, 337 edges   (c) *sepEN* ∩ *jointEN*, 40 edges

(d) 32 % (86 edges)   (e) 33 % (112 edges)   (f) 48 % (19 edges)

Fig. 2. Networks of common genes in BN$_{brca}$ and BN$_{prad}$ learnt form tumour data, by (a) *jointEN* and (b) *sepEN*, for $\alpha = 0.01$; (c) is the intersection of edges in (a) and (b); (d), (e), and (f) correspond to the edges from the networks in (a), (b), and (c) that are reported in STRING, respectively.

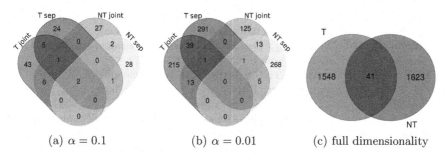

(a) $\alpha = 0.1$ (b) $\alpha = 0.01$ (c) full dimensionality

Fig. 3. Venn diagrams with common edges to breast and prostate for regularised and full approaches, BNs learnt from tumour (T) and normal (NT) tissue samples.

To infer whether the *tumour*BNs obtained are specific to BRCA and PRAD diseases or not, Venn diagrams were produced to illustrate the overlap between *tumour*BNs and *normal*BNs (see Fig. 3). For $\alpha = 0.1$, the overlap is of 9 and 3 edges for *jointEN* and *sepEN*, respectively, while for $\alpha = 0.01$ the overlap is of 14 and 6 edges. With no regularisation, 41 edges in the *tumour*BN were also found in *normal*BN. These edges are more likely related to cell machinery, and therefore of little interest, and not related to BRCA and PRAD diseases. Such little overlap might be an indicator of the specificity of the *tumor*BNs obtained to the diseases under study.

4   Conclusion

The methodology proposed was able to extract common gene signatures to both types of cancer, BRCA and PRAD, by Bayesian network learning. A considerable overlap between the gene networks identified and STRING network information was obtained, a strong indication that the networks learnt may be biologically meaningful. The present results are expected to play a role in cancer therapy research, by fostering cancer therapy research for both types of cancer. Moreover, this can be extended to multiple diseases, in the search for common gene signatures across multiple types of cancer.

References

1. Risbridger, G.P., Davis, I.D., Birrell, S.N., Tilley, W.D.: Breast and prostate cancer: more similar than different. Nat. Rev. Cancer **10**(26), 205–212 (2010)
2. McCullagh, P., Nelder, J.A.: Generalized Linear Models, 2nd edn. Chapman and Hall, London (1989)
3. Hoerl, A.E., Kennard, R.W.: Ridge regression: biased estimation for nonorthogonal problems. Technometrics **12**(1), 55–67 (1970)
4. Tibshirani, R.: Regression shrinkage and selection via the lasso. J. Roy. Stat. Soc. Ser. B (Methodol.) **58**, 267–288 (1996)
5. Zou, H., Hastie, T.: Regularization and variable selection via the elastic net. J. Roy. Stat. Soc.: Ser. B (Stat. Methodol.) **67**(2), 301–320 (2005)
6. Aragam, B., Gu, J., Zhou, Q.: Learning large-scale Bayesian networks with the sparsebn package. arXiv preprint arXiv:1703.04025 (2017)
7. The Cancer Genome Atlas - TCGA. https://cancergenome.nih.gov/
8. The Ensembl genome browser. https://www.ensembl.org/index.html. Accessed May 2017
9. The Consensus CDS (CCDS) project. Release 20. https://www.ncbi.nlm.nih.gov/projects/CCDS/CcdsBrowse.cgi. Accessed May 2017
10. Szklarczyk, D., Franceschini, A., et al.: STRING v10: protein-protein interaction networks, integrated over the tree of life. Nucleic Acids Res. **43**, D447–52 (2015)

Engineering Bio-Interfaces and Rudimentary Cells as a Way to Develop Synthetic Biology

Effect of Epigallocatechin-3-gallate on DMPC Oxidation Revealed by Infrared Spectroscopy

Filipa Pires$^{(\boxtimes)}$, Bárbara Rodrigues, Gonçalo Magalhães-Mota ,
Paulo A. Ribeiro , and Maria Raposo

CEFITEC, Department of Physics, FCT-UNL, Universidade Nova de Lisboa,
2829-516 Caparica, Portugal
af.pires@campus.fct.unl.pt, mfr@fct.unl.pt

Abstract. The daily exposure of skin cells to the sun increases the rate of production of free radicals, which threatens the healthy appearance of skin and, even more worrying, damages the structural integrity of tissues and DNA, causing inflammation and carcinogenesis. This work demonstrates the feasibility of using natural agents, in particular tea catechins, in protecting lipidic membranes from oxidative stress-induced by UV radiation exposure. For that purpose, thin cast films prepared from vesicular suspensions of dimyristoylphosphatidylcholine (DMPC) and dimyristoylphosphatidylcholine + (-)-epigallocatechin-3-gallate (DMPC + EGCG) were deposited onto calcium fluoride supports and irradiated with 254 nm UV radiation. The molecular damage after irradiation with UV light was analysed by infrared (IR) together with 2D correlation spectroscopies. Results revealed that the DMPC phospholipid polar moiety is the most vulnerable and sensitive structural target of UV radiation. To check if the presence of the EGCG molecules is protecting the lipids, the principal components analysis (PCA) mathematical method was applied, allowing to conclude that EGCG slows down the cascade of the oxidant-events in the lipid, thus protecting the polar moiety of the lipid.

Keywords: EGCG · Lipid oxidation · Infrared spectroscopy

1 Introduction

Skin is daily exposed to ultraviolet radiation (UV), a strong exogenous agent, that imbalances the pro- and antioxidant processes in the physiological system. The unprotected and prolonged exposure to UV light (290–400 nm) promotes the oxidation of the cellular components as membranes, proteins and nucleic acids [1], affecting the cellular integrity and gene expression, which in turn increases the chance of developing melanoma and nonmelanoma skin cancers.

In last decades, plenty of studies in the literature agree and sustain the idea that the daily consumption of catechins, natural molecules found in the tea plant

© Springer Nature Switzerland AG 2020
M. Raposo et al. (Eds.): CIBB 2018, LNBI 11925, pp. 295–302, 2020.
https://doi.org/10.1007/978-3-030-34585-3_26

Camelia sinensis, contributes to cancer prevention. Green tea leaves contains the following catechins: (-)-epicatechin, (-)-epigallocatechin, (-)-epicatechin-3-gallate and (-)-epigallocatechin-3-gallate (EGCG). According with *in vitro* and *in vivo* studies, EGCG has antioxidant, antibacterial and antimutagenic activity, protecting more effectively the biological tissues against the oxidative stress and reducing the risk of developing certain types of cancer [2].

2   Scientific Background

It has been found that the topical application of EGCG in SKH-1 hairless mice irradiated with ultraviolet B light (UVB) selectively stimulate the apoptosis of the malignant skin cells, decreasing the number and volume of tumors and, consequently, inhibiting the UVB-induced carcinogenesis [3]. UVB light is known to induce specific mutations in DNA (deoxyribonucleic acid) through photochemical reactions that culminates with the formation of cyclobutane pyrimidine dimers (CPDs) and pyrimidine(6-4)pyrimidone photoproducts (64PPs) at dipyrimidine sites, which if not repaired lead to carcinogenesis [4]. The topical application of catechins ($3\,mg/2.5\,cm^2$) before UVB exposure, is known to result in the inhibition of the erythema formation and prevents the penetration of UVB light across the skin, to a certain extent, leading to a reduction of the CPDs formed in the dermis and epidermis [5]. In another study, the administration of EGCG in drinking water revealed to reduce the production of inflammatory molecules and increased the levels of IL-12 in UVB exposed-mice, in way to increase the repair of UVB-induced CPDs compared to untreated mice [6].

The cellular membrane, a complex and semi-permeable barrier that protects cells from its environment and controls the influx and efflux of substances across the membrane, is another biological target of the UV radiation. Catechins have several hydroxyl groups on its structure that have the affinity to membranes, more specifically, to the lipidic membranes, being the associated interactions strong enough to alter the membrane dynamics [7]. Our preliminary results suggested that EGCG act as scavenger of reactive oxygen species thus diminishing the rate of oxidation on DMPC phospholipids [10].

In this study one intends to clarify the catechin–lipid molecular interactions using the vibrational spectra, in order to understand the protective role of EGCG against radiation. For this purpose, the IR spectra of EGCG incorporated on liposomes (artificial model that mimics the phospholipid bilayer of the cellular membranes) exposed to UV radiation over time, were systematically analysed in this work. The results showed that EGCG inserts into DMPC membrane and slows down the oxidative damage in DMPC phospholipids by protecting the phosphate and carbonyl groups of phospholipids from UV radiation. Also, the results showed that, after a certain irradiation time, the UV light attacks the gallate group of the EGCG (ring responsible to bind EGCG to the polar moiety of the DMPC lipid), which impairs its antioxidant activity on phospholipids.

3   Materials and Methods

Thin film hydration method was used to prepare dimyristoylphosphatidylcholine liposomes (DMPC, M.W. 677.93 g/mol, Avanti Polar Lipids). The first step consisted in dissolving the phospholipid in a chloroform/methanol mixture (4:1) and dried it in a gentle nitrogen stream, forming a lipidic dried film. Afterwards, this film was hydrated in an aqueous medium containing the EGCG (M.W. 458.37 g/mol, Sigma-Aldrich) to be encapsulated. Small unilamellar vesicles were obtained after sonication of the vesicle suspensions with a tip sonicator (UP50H, Hielscher Ultrasonics, GmbH, Germany), 15 times with 1 min interval between sonication cycles of 30 s. The unencapsulated EGCG was then removed by dialysis (membranes of regenerated cellulose with the cutoff size of 8–10 kDa, Spectra/Pro, Biotech USA) in a H2O Milli-Q bath for 8 h. A drop casting method was used to immobilize the vesicles onto the surface of the calcium fluoride substrates and, afterwards, these substrates were irradiated with a 254 nm UVC germicide lamp (Philips TUV PL-S 5W/2P 1CT) at an irradiance of 1.9 W/m^2. Infrared measurements were carried out with a Fourier transform infrared (FTIR) spectroscope Thermo Scientific Nicolet-model 530 (Waltham, MA, USA) in the transmittance mode, with a 4 cm^{-1} resolution and using 128 scans, in order to assess the damage induced by UV exposure. The 2D Shige free software for two-dimensional correlation spectroscopy [8], was employed to study radiation induced-damage on the cast films of DMPC and DMPC+EGCG by using FTIR spectra. Principal components analysis method (PCA) was used to highlight the similarities and differences between the infrared spectra of DMPC and DMPC+EGCG samples. The infrared region between 1180–1300, 1600–1800 and 2800–2950 cm^{-1}, corresponding to the polar moiety (phosphate and carbonyl groups) and to the lipids hydrocarbon tails, respectively, were baseline corrected and processed by PCA.

4   Experimental Results

4.1   Analysis of Vibrational IR Spectra of DMPC and DMPC+EGCG Liposomes

The molecular interactions established between EGCG and DMPC phospholipids, as well as, the structural changes in lipid membrane resulting from UV irradiation were assessed by infared spectroscopy. Fourier transform infrared spectra of DMPC and DMPC+EGCG liposomes deposited on CaF_2 solid supports and irradiated with 254 nm UV light over time are shown in Fig. 1A and B, respectively. The spectra present two distinct regions, one related with the polar head group (region below 1900 cm^{-1}) and another associated with the C-H stretching vibrations of the hydrocarbons tails of the lipids and water molecules vibrations (region 3800–2600 cm^{-1}), respectively. Focusing on the low wavenumber region of the spectra, five strong bands are present. The most intense is centred at 1741 cm^{-1} and is associated to the stretching mode of the carbonyl group (C=O). The peaks centred at 1230 cm^{-1} and 1089 cm^{-1} are assigned to

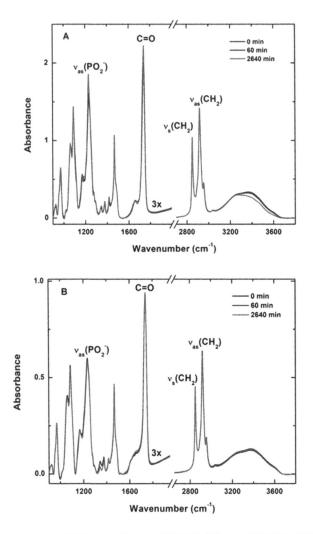

Fig. 1. Infrared spectra of thin cast films of DMPC (A) and DMPC+EGCG (B) liposomes deposited on CaF_2 solid supports before and after UV radiation exposure.

the asymmetrical and symmetrical $PO_2{}^-$ stretching modes. The others bands in this region, observed at 1060 cm^{-1} and 970 cm^{-1}, correspond to the C-O-P-O-C stretching mode and to the antisymmetric stretching of the $^+N - CH_3$, respectively. The highest wavenumber region of the spectra contains bands associated to the IR modes of the asymmetrical stretching of the CH_2 near 2918 cm^{-1}, the symmetrical stretching of the CH_2 near 2851 cm^{-1} and the O-H stretching of the hydroxyl groups and the water molecules trapped inside of intact liposomes near 3375 cm^{-1}.

The similarities between the DMPC and DMPC+EGCG spectra are notorious, since EGCG produces strong IR bands in the same wavenumber regions of the lipid. In addition, it should be taken into account that the experiments were carried out in a non-controlled atmosphere environment, fact that can hamper the observation of some characteristic peaks of the EGCG, namely those centred at $1519\,\mathrm{cm}^{-1}$ and $1695\,\mathrm{cm}^{-1}$. In fact, these peaks assigned to stretching of C=C of the aromatic ring and to the stretching of C=O of gallic acid that links the trihydroxybenzoate group and chroman group, respectively, can be masked by the water vapour and carbon dioxide.

Even more, a general analysis of the spectra of Fig. 1A, excluding the higher wavenumber region, it appears that radiation little affects the samples. In order to better analyse the spectra data and to set information on the damage caused by irradiation on these samples, 2D-correlation spectroscopy was applied to the measured spectra data. The 2D-correlation spectroscopy is a mathematical technique which allows to detect the very small changes in the measured signals when a sample is subjected to the external perturbation, which in this case, is the UV irradiation. Synchronous 2D correlation maps associated to the measured infrared spectra of DMPC and DMPC+EGCG samples are represented in Fig. 2, in which the points in the diagonal (autopeaks) indicate the vibrations affected by the UV radiation. In the case of DMPC samples, Fig. 2a), the autopeaks indicate that the groups that exhibit vibrations at 1232 and $1740\,\mathrm{cm}^{-1}$, i.e. the polar moiety of the lipid (phosphate and carbonyl groups, respectively), are being affected by the UV irradiation. Moreover, the most intense autopeak is observed at $3364\,\mathrm{cm}^{-1}$, Fig. 2(b), suggesting that some liposomes were being broken and the water retained inside of them lost after UV irradiation.

For the case of DMPC+EGCG samples, the synchronous correlation map, Fig. 2(c, d), shows four less-intense autopeaks at 1232, 1344, 1740 and $3371\,\mathrm{cm}^{-1}$. This indicates that the same molecular groups of the DMPC are also suffering changes after UV irradiation, but in a less extent. Moreover, the presence of the autopeak located at $1344\,\mathrm{cm}^{-1}$, indicates that the structure of EGCG is being affected by the UV light, in particular, the O-C=O bond of the gallate group. The degradation of the gallate moiety with UV light, breaks the EGCG-lipid interaction and, consequently, hampers the protective role of EGCG on the phosphate and carbonyl groups of the DMPC lipid.

To make easy data processing, principal component analysis (PCA), a statistical method that convert a set of data of correlated variables into a set of linearly uncorrelated variables, [9] was applied to extract relevant and useful information from the IR spectral data. Figure 3 shows the biplot of the first two principal components, having a cumulative variance of 99.79%, resulting from the 99.22% of the first component and the 0.58% of the second component. The data clearly shows that EGCG slows down the cascade of the oxidant-events responsible for lipid damage, since all the points are close together in the biplot plane, illustrating the similarly of the samples. The PCA well-captured the structural changes in DMPC liposomes induced by UV irradiation, as seen by the clear separation of the points with the increase of the irradiation time.

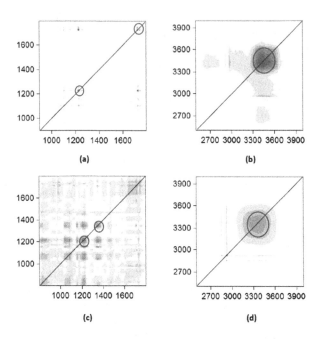

Fig. 2. Synchronous 2D correlation maps using the FTIR spectra of DMPC and DMPC+EGCG liposomes in the wavenumber region of 900–1800 cm^{-1} (a, c) and 2700–3900 cm^{-1} (b, d), respectively. Blue and red dots means negative and positive correlation, respectively. Red circles indicate the autopeaks. (Color figure online)

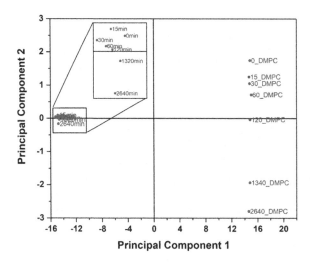

Fig. 3. PCA of the IR spectra of the DMPC (dots on the right side) and DMPC+EGCG (grouped dots on the left side) liposomes over irradiation time. The biplot of the first two principal components showed 99.79% of the cumulative variance.

5 Conclusions

This work demonstrated the feasibility of use EGCG as a natural agent to protect lipid membranes from oxidative stress induced by UV radiation exposure. The IR and the 2D correlation spectroscopies revealed that the polar moiety of the DMPC phospholipid is the most vulnerable and sensitive structural target of UV radiation. A prolonged exposure to UV radiation alters the morphology of liposome and the structure of the lipid, causing the rupture of liposomes, fact evidenced by the loss of water molecules. On the basis of the PCA calculations, the EGCG was seen to slows down the cascade of the oxidant-events responsible for damaging the lipid. This work also showed that, after a certain irradiation time, the shield of EGCG molecules that protects lipid membranes, was disrupted by the UV light, since this external agent attacks the gallate group of the EGCG (ring responsible to bind EGCG to the polar moiety of the DMPC lipid). In future work, chromatography techniques will be used to reveal the photoproducts found in the photodegradation of DMPC, in the absence and presence of EGCG, in order to better understand the antioxidant role of EGCG in lipid membranes. Furthermore, data can also be analysed by non-linear PCA analysis and by advanced signal processing methods such as compressed sensing.

Acknowledgments. The authors acknowledge the financial support from FEDER, through Programa Operacional Factores de Competitividade COMPETE and Fundação para a Ciência e a Tecnologia through research project grants PEst-OE/FIS/UI0068/2011, UID/FIS/00068/2013 and PTDC/FIS-NAN/0909/2014. FP acknowledges the fellowship grant PD/BD/106036/2015 from RABBIT Doctoral Programme (Portugal).

References

1. Lushchak, V.I.: Free radicals, reactive oxygen species, oxidative stress and its classification. Chem. Biol. Interact. **224**, 164–1750 (2014)
2. Grzesik, M., Naparło, K., Bartosz, G., Sadowska-Bartosz, I.: Antioxidant properties of catechins: comparison with other antioxidants. Food Chem. **241**, 480–492 (2018)
3. Lu, Y.-P., et al.: Topical applications of caffeine or (-)-epigallocatechin gallate (EGCG) inhibit carcinogenesis and selectively increase apoptosis in UVB-induced skin tumors in mice. Proc. Nat. Acad. Sci. **99**(19), 12455–12460 (2002)
4. Ikehata, H., Ono, T.: The mechanisms of UV mutagenesis. J. Radiat. Res. **52**(2), 115–125 (2011)
5. Katiyar, S.K., Perez, A., Mukhtar, H.: Green tea polyphenol treatment to human skin prevents formation of ultraviolet light B-induced pyrimidine dimers in DNA. Clin. Cancer Res. **6**(10), 3864–3869 (2000)
6. Meeran, S.M., Akhtar, S., Katiyar, S.K.: Inhibition of UVB-induced skin tumor development by drinking green tea polyphenols is mediated through DNA repair and subsequent inhibition of inflammation. J. Investig. Dermatol. **129**(5), 1258–1270 (2009)
7. Abram, V., Berlec, B., Ota, A., Šentjurc, M., Blatnik, P., Ulrih, N.P.: Effect of flavonoid structure on the fluidity of model lipid membranes. Food Chem. **139**(1–4), 804–813 (2013)

8. Raposo, M., et al.: DNA damage induced by carbon ions (C^3+) beam accessed by independent component analysis of infrared spectra. Int. J. Radiat. Biol. **90**(5), 344–350 (2014)
9. Jackson, J.E.: A user's Guide to Principal Components, vol. 587. Wiley, Hoboken (1999)
10. Pires, F., Magalhães-Mota, G., Ribeiro, P.A., Raposo, M.: Effect of UV radiation on DPPG and DMPC liposomes in presence of catechin molecules. In: Bracciali, A., Caravagna, G., Gilbert, D., Tagliaferri, R. (eds.) CIBB 2016. LNCS, vol. 10477, pp. 172–183. Springer, Cham (2017). https://doi.org/10.1007/978-3-319-67834-4_14

Effect of EGCG on the DNA in Presence of UV Radiation

Thais P. Pivetta$^{(\boxtimes)}$ (ID), Filipa Pires (ID), and Maria Raposo (ID)

CEFITEC, Departamento de Física,
Faculdade de Ciências e Tecnologia da Universidade Nova de Lisboa,
2829-516 Caparica, Portugal
{t.pivetta,af.pires}@campus.fct.unl.pt, mfr@fct.unl.pt

Abstract. The exposure to ultraviolet (UV) radiation is clearly a current concern since it damages the deoxyribonucleic acid (DNA) and increases the likelihood of developing skin cancer. On the other hand, green tea compounds such as (-)epigallocatechin-3-gallate (EGCG) present several biological properties and, are well-known for its antioxidant activity. The aim of this work is evaluate the effect of the UV radiation on DNA in presence of EGCG molecules. Results of the evolution of the UV-visible spectra with the UV irradiation suggest that EGCG act like an intercalant molecule and a micromolar concentration of EGCG is effective to induce a strong degradation on the DNA pyrimidines bases under UV radiation. This achievement can lead to a novel class of non-binding safe molecules capable of affinity interaction with the DNA as intercalant molecule which can be used as anti-tumor drugs.

Keywords: DNA · EGCG · UV radiation

1 Introduction

The study of the DNA damage is an important topic nowadays because of its role in cancer development and also as target for the development of new medicines. Due to several side effects generated by conventional chemotherapy, many natural molecules have been explored in order to obtain new alternative drugs for the treatment of cancer. EGCG is a polyphenol present in the green tea and compared to other compounds from the catechin family, EGCG is the largest one. Due to the structure, EGCG is able to create a stable conjugation with the DNA, property that could be useful to enhance DNA damage [1]. Catechins such as EGCG usually are associated with antioxidant effect which could provide a protection of the DNA however there are studies showing that these compounds can present pro-oxidant effect depending mainly of the pH and concentration [2]. In our study, the focus was the investigation of the effect of EGCG on the DNA damage when exposed to UV radiation and evaluate if there was a protection or damage in the DNA molecule. Results of the irradiation showed that the characteristic absorbance band of the DNA was more degraded in the presence of EGCG, showing a possible intercation with the DNA helix that could lead to the DNA damage.

M. Raposo et al. (Eds.): CIBB 2018, LNBI 11925, pp. 303–308, 2020.
https://doi.org/10.1007/978-3-030-34585-3_27

2   Scientific Background

Several environmental factors can generate damages in the DNA and among them, one of the most aggressive is the ultraviolet (UV) radiation. UV is part of the electromagnetic radiation spectrum, having smaller wavelengths than UV visible (400–700 nm), being divided in three ranges of radiation that is emitted by the sun: UVA (315–400 nm), UVB (280–315 nm) and UVC (200–280 nm) [3]. Part of the UV radiation emitted by the sun is well absorbed by the ozone present in our atmosphere however, due to ozone depletion, UV radiation level has been increasing and, consequently, its harmful effects on biological systems [4,5]. UV radiation can generate severe damages in the DNA, therefore excessive exposure to sunlight is associated with skin cancer development [3]. There was an increase in the occurrence of skin cancer over the past decades. World Health Organization estimates that each year occurs 2–3 million new cases of non-melanoma skin cancers and 132,000 of melanoma cancers. Also, in every three new diagnosis of cancer one is skin cancer and, for this reason, it is the most common type of cancer [6]. The radiation damage on the DNA molecules occurs through several mechanisms such as the DNA double strand breaks and formation of the reactive oxygen species [4]. Many researchers are interested in studying antioxidants molecules like (-)-epigallocatechin-3-gallate (EGCG) and its possible photo-protective effect. EGCG is a polyphenol and the main component of green tea. This beverage is widely consumed in the world and is usually associated with some beneficial biological properties as anti-inflammatory, antimicrobial, antioxidant properties, being also known for its anti-carcinogenic role in different types of cancer [7–9]. Recently, *in vitro* and *in vivo* studies showed that EGCG is able to prevent the DNA damage generated by UV radiation and carcinogenesis process, revealing the protective nature of catechins as well its ability to interact with biological molecules [10,11]. In this context, the main goal of this work was to assess the effect of UV radiation on DNA in presence of EGCG molecules, in order to discern about the photoprotection or photosensitization role of EGCG in the DNA.

3   Materials and Methods

Calf thymus DNA and EGCG were purchased from Sigma-Aldrich. An aqueous stock solution of calf thymus DNA with a concentration of 500 µg/mL was prepared and stirred during 48 h at room temperature in dark conditions, to solubilize the fibers. EGCG (450 µg/mL) were dissolved in ultrapure water. These stock solutions were used to prepare the mixture of DNA+EGCG with a final concentration of 100 µg/mL and 40 µg/mL, respectively. For the irradiation experiments, the solutions were placed into the quartz closed cuvettes and they were exposed to an UVC light of 254 nm (PHILIPS TUV PL-S 5W/2P Hg lamp). The samples were placed at 18.5 cm under UV source light and the irradiance, measured in Delta OHM Photo/Radiometer (HD 2102.2), take an average value of 3.09 W/m². Absorbance spectra of these samples were measured in a

Shimadzu spectrophotometer (UV-2101PC), scanning from 900 nm to 190 nm, using Milli-Q water as reference (Fig. 1).

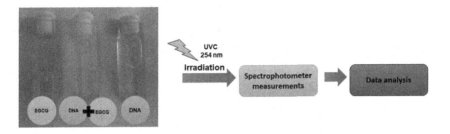

Fig. 1. Graphical ilustration of the experimental work.

4   Experimental Results

To discern about the effect of EGCG on DNA molecules, the absorbance spectra of DNA and DNA+EGCG aqueous solutions were measured as a function of the UV irradiation period of time. The obtained spectra are displayed in Fig. 2(A) and (B), respectively, for both type of samples. Analyzing the absorbance spectra of DNA sample after UV exposure, shown in Fig. 2(A), is clearly notice the vulnerability and the significantly damage on DNA bases after 6.5 h of UV irradiation. Moreover, the absorbance spectrum of DNA in aqueous solution has a broad band in the wavelength region between 230 and 300 nm, which was deconvoluted in two components, one at 270 nm and other at 255 nm, associated to pyrimidines and to all DNA bases, respectively. These bands are related with π-π^* transitions that is characteristic of heterocyclic compounds.

The addition of EGCG into DNA sample, Fig. 2(B), slightly alters the band shape at 270 nm. In fact, in the case of DNA+EGCG samples, the spectra in the region of 270 nm is due to the DNA bands (270 nm) but also is associated to the n-π^* transition involving oxygen electrons from the gallic acid group of EGCG (274 nm) [12]. Considering the strong antioxidant and radical scavenger activity of EGCG it will be expected that the UV induced damage on DNA bases was much lower than in pure DNA. Curiously, in the presence of EGCG, we observed a considered degradation of the band at 270 nm after approximately 3 h of irradiation. Looking to the Fig. 2(B) is also possible to observe a shifting of the band at 270 nm to lower wavelengths (265 nm) with increasing of irradiation time.

Considering the deconvolution of the UV-Vis spectra using Gaussian curves, it was possible to define peaks and calculate the area associated to each band in the region between 230 and 300 nm. Figure 3(A) and (B) show the calculated peak areas values as a function of irradiation time for DNA and DNA+EGCG samples, respectively. From these figures we can see that the species decrease

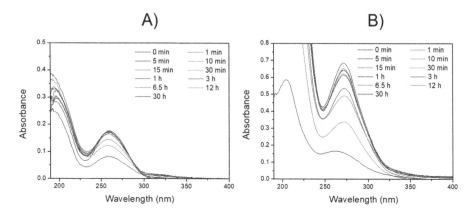

Fig. 2. Absorbance spectra of (A) DNA 100 μg/mL and (B) mixture of EGCG 40 μg/mL with DNA 100 μg/mL along different times of irradiation.

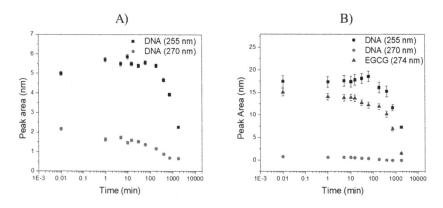

Fig. 3. Peak area behavior with different times of irradiation on the samples of (A) DNA 100 μg/mL and (B) mixture of EGCG 40 μg/mL with DNA 100 μg/mL.

with the irradiation time exhibiting different profiles of degradation. To conclude about which specie is more effectively damaged by radiation, the peak areas were normalized and plotted together in Fig. 4. Here we can see that the behavior of the DNA band at 255 nm is practically independent of EGCG while the area of the band at 270 nm is strongly affected by the presence of EGCG. The supposed attack on the DNA molecule may be on the pyrimidines since they correspond to the main target of the UV damage. About the role of the EGCG is possible to presume that at the concentration of 40 μg/mL the catechin enhances the damage. The behavior of the EGCG as antioxidant or pro-oxidant is under debate since there are some experimental evidence for and against each statement. In fact, this behavior is strongly influenced by external factors such as pH, temperature, concentration and presence of other molecules (metals, antioxidants) in medium, since they trigger rearrangements and influence the stereochemistry of

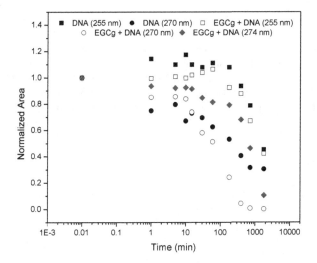

Fig. 4. Relation of the peak area ratio, calculated from the deconvolution of absorbance data measured in DNA and mixture of EGCG+DNA, with the irradiation period of time.

the EGCG and, consequently, influence the interaction with the biomolecules. Our results show that a micromolar concentration of EGCG is effective to damage DNA molecules, opening the way to new approach to treat cancer.

5   Conclusions

In this paper, we demonstrate that the presence of EGCG can induce strong degradation on the pyrimidines bases of DNA molecules. Possibly, we have found a novel class of non-binding molecules capable of affinity interaction with the DNA as intercalant molecule, allowing the DNA damage and having the same function of chemotherapeutic anti-tumor drugs which target the cellular DNA of tumor cells, creating bulky lesions that hindering the DNA repair, replication and transcription to induce the apoptosis of cancer cells. Most of these drugs are synthetic bringing very toxic and severe side-effects to patient. We expects that this work can drive new ways to treat cancer, since EGCG is a natural molecule that seems to be an intercalant molecule able to insert in the double helix environment of DNA, without forming covalent bonds. Since pH of the tumor environment differ from the healthy cell pH, in the future, we will study a range of concentrations of EGCG in DNA solutions at different pH in order to reveal the ideal concentration of EGCG to damage harmful DNA of cells. In addition, toxicity studies will be performed in order to reveal the ideal dose of EGCG that damage DNA, but do not induce toxicity in organs like kidney, intestine and liver.

Acknowledgments. The authors acknowledge the financial support from FEDER, through Programa Operacional Factores de Competitividade COMPETE. This work was also supported by Fundação para a Ciência e a Tecnologia (FCT-MCTES), Radiation Biology and Biophysics Doctoral Training Programme (RaB-BiT, PD/00193/2012); UID/Multi/04378/2013 (UCIBIO); UID/FIS/00068/2013 (CEFITEC); PEst-OE/FIS/UI0068/2011; PTDC/FIS-NAN/0909/2014. TP and FP acknowledge the scholarship grants PD/BD/142829/2018 and PD/BD/106036/2015 from RABBIT Doctoral Programme (Portugal).

References

1. Chanphai, P., Tajmir-Riahi, H.A.: Structural dynamics of DNA binding to tea catechins. Int. J. Biol. Macromol. **125**, 238–243 (2019)
2. Zhou, L., Elias, R.J.: Antioxidant and pro-oxidant activity of (-)-epigallocatechin-3-gallate in food emulsions: influence of pH and phenolic concentration. Food Chem. **138**, 1503–1509 (2013)
3. Yu, S.-L., Lee, S.-K.: Ultraviolet radiation: DNA damage, repair, and human disorders. Mol. Cell Toxicol. **13**, 21–28 (2017)
4. Greinert, R., et al.: UVA-induced DNA double-strand breaks result from the repair of clustered oxidative DNA damages. Nucleic Acids Res. **40**, 10263–10273 (2012)
5. McMichael, A.J., Lucas, R., Ponsonby, A.-L., Edwards, S.J.: Stratospheric ozone depletion, ultraviolet radiation and health. In: McMichael, A.J., Campbell-Lendrum, D.H., Corvalán, C.F., Ebi, K.L., Githeko, A., Scheraga, J.D., Woodward, A. (eds.) Climate Change and Human Health: Risks and Responses, pp. 159–180. World Health Organization, Geneva (2003)
6. WHO: World Health Organization. Ultraviolet radiation (UV). www.who.int/uv/faq/skincancer/en/index1.html. Accessed 19 June 2018
7. Steinmann, J., Buer, J., Pietschmann, T., Steinmann, E.: Anti-infective properties of epigallocatechin-3-gallate (EGCG), a component of green tea. Br. J. Pharmacol. **168**, 1059–1073 (2013)
8. Oz, H.S.: Chronic inflammatory diseases and green tea polyphenols. Nutrients **9**, 1–14 (2017)
9. Zhang, J., et al.: Epigallocatechin-3-gallate(EGCG) suppresses melanoma cell growth and metastasis by targeting TRAF6 activity. Oncotarget **7**, 79557–79571 (2016)
10. Morley, N., Clifford, T., Salter, L., Campbell, S., Gould, D., Curnow, A.: The green tea polyphenol (-)-epigallocatechin gallate and green tea can protect human cellular DNA from ultraviolet and visible radiation-induced damage. Photodermatol. Photoimmunol. Photomed. **21**, 15–22 (2005)
11. Meeran, S.M., Mantena, S.K., Katiyar, S.K.: Prevention of ultraviolet radiation - induced immunosuppression by (-)-epigallocatechin-3-gallate in mice is mediated through interleukin 12-dependent DNA repair. Clin. Cancer Res. **12**, 2272–2280 (2006)
12. Pires, F., Geraldo, V.P.N., Antunes, A., Marletta, A., Oliveira Jr., O.N., Raposo, M.: On the role of epigallocatechin-3-gallate in protecting phospholipid molecules against UV irradiation. Colloids Surf. B: Biointerfaces **173**, 312–319 (2018)

Non-thermal Atmospheric Pressure Plasmas: Generation, Sources and Applications

Sara Pereira⊙, Érica Pinto, Paulo António Ribeiro⊙, and Susana Sério^(✉)⊙

CEFITEC, Departamento de Física, Faculdade de Ciências e Tecnologia,
Universidade NOVA de Lisboa, 2829-516 Caparica, Portugal
`susana.serio@fct.unl.pt`

Abstract. Non-thermal atmospheric pressure plasmas, also known as Cold Atmospheric Plasmas (CAPs) are emerging as a potential alternative for cancer treatment since they can be generated at atmospheric conditions and their low temperatures allow the interaction with living tissues without thermal damage. This article focus on the study of the interaction between plasma and a non-cancerous cell line, particularly, VERO cells. Some in-vitro experiments were performed with a custom-made device in order to better understand how CAPs affect non-cancerous cells. It was also studied the influence of several factors such as the distance from the device (gap), the duration and the type of treatment, direct or indirect, on the cell viability after exposure to CAPs treatments. The obtained results revealed the importance of the determination of the optimal relation between gap and treatment time, since small variations in each one of them can lead to different results in the cell viability.

Keywords: Cold plasma · VERO cells · Plasma jet device · Plasma medicine

1 Introduction

In physics plasma, also known as the fourth state of matter, is defined as a partially ionized gas containing neutral particles and an equivalent number of electrons and positive ions [1–3]. Plasmas can be classified into two major categories: thermal and nonthermal ones [4]. Thermal plasmas are those in which ions and electrons are in thermodynamic equilibrium, which means that all the present particles have the same temperature. Contrariwise, non-thermal ones have electrons at a higher temperature than the ions and neutrals, and for this reason they are known as cold plasmas [5,6]. These cold plasmas have gained great interest in terms of biomedical applications because they can operate stably under open atmospheric conditions and have small working temperatures, which means that there is no risk of thermal destruction at the tissue-plasma contact zone [2,3].

© Springer Nature Switzerland AG 2020
M. Raposo et al. (Eds.): CIBB 2018, LNBI 11925, pp. 309–318, 2020.
https://doi.org/10.1007/978-3-030-34585-3_28

Nowadays, non-thermal plasmas are being widely used in wound healing, genetic transfection, dentistry, treatment of implants to improve their biocompatibility and also in dermatology. Despite the wide range of plasma applications in biomedical fields, prior studies have focused mainly on its bactericidal and sterilizing effects. However in the last years a number of studies have proposed its application for cancer treatment, showing selective tumor eradication capabilities, apoptotic signaling pathway deregulation and DNA damage through formation of intracellular ROS [2]. Cellular necrosis and senescence have also been proposed to explain the mechanism of CAPs treatment on cancer cells, although the specific effects resulting from the interaction between CAPs and cells remains not well understood.

2   Scientific Background

2.1   Generation of Cold Atmospheric Plasmas

In direct current discharge systems, gas breakdown starts in a stationary electric field created between two different electrodes. In the absence of an electrical field the total charge of electrons and protons, atoms or molecules in gases are equal. However, when an external field is applied, a difference of potential between the cathode and the anode is created, and as consequence a very weak current is produced and the voltage starts to increase [7,8]. As a result, energy is imparted to the electrons, which can gain enough kinetic energy to produce more electrons (secondary electrons), through collisions with the electrodes and the neutral gas atoms, in a process known as the avalanche breakdown or *Townsend* mechanism. In the particular case of low pressures, the electron mean free path is high and the avalanche process is observed until the plasma extends across the entire region between the electrodes [8,9].

To facilitate the understanding of *Townsend* mechanism, a direct current system of two parallel electrodes [10], separated by a distance d to which a V voltage is applied, providing an electric field E, is considered (Eq. 1) [11]:

$$E = \frac{V}{d} \tag{1}$$

According to *Townsend* discharge model, each primary electron produced near the cathode drifts to the anode generating positive ions by collision with the species supplied to the system, that are moving in the opposite direction (Fig. 1). These in turn will contribute to the extraction of the electrons from the cathode due to the emission of secondary electrons.

Although, *Townsend* was the first researcher to develop a model to explain the breakdown of gas molecules in the presence of an electrical field, it was F. Paschen who first studied the breakdown conditions at different gas pressures. The main outcome of Paschen studies is the well-known experimental Paschen curve relating the applied voltage V_B with the product of gas pressure and electrode separation (pd) [8,12,13]. The experimental Paschen curve for different gases is represented in Fig. 2, where it can be seen that all curves present a minimum voltage point, corresponding to the most favorable breakdown condition.

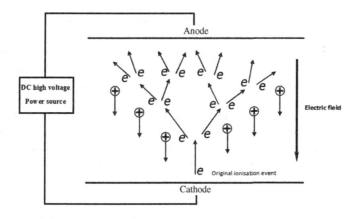

Fig. 1. Representation of the electron avalanche adapted from [9].

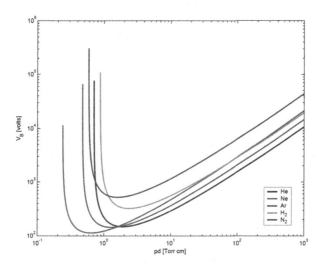

Fig. 2. Paschen's curve from [14].

2.2 Cold Atmospheric Plasma Sources

Since the first Dielectric Barrier Discharge (DBD) reactor was designed in the 19th century by W. von Siemens, a large number of different sources for CAPs production have been developed. Nowadays, in the field of plasma medicine research, there are two different configurations that are widely used for CAPs production: DBDs and plasma jet devices [15]. Plasma jets are generally composed by two different electrodes and a nozzle through which the gas flows. The two electrodes, normally a sharp point and a hollow cylindrical, are usually mounted in a coaxial arrangement. The plasma, formed inside the tube, is transported to the outside through the gas flow [16]. When compared to DBDs, plasma jets have the advantage of allowing more localized treatments and present greater inherent stability since the gas flow can be controlled more readily [4]. DBDs, on the other hand, allow the treatment of larger areas and often do not require a gas source, once most of these devices can work using only atmospheric air. However, in this last situation, the stability of DBDs will depend not only on factors such as the distance between the electrodes, and the conductivity of the tissue to be treated, but also on the ambient air humidity [15,17,18].

In this paper, a brief discussion of the breakdown phenomena leading to the cold plasma formation will be undertaken, and a short presentation of one of the most used Cold Atmospheric Pressure plasma (CAPs) setups will be done. After that, a cold plasma jet device, developed by us, dedicated to in vitro treatments will be described and the obtained results will be presented. Cell viability after CAPs treatment (direct and indirect) was determined by the resazurin assay, in which metabolically active cells convert resazurin into resorufin. To study the susceptibility of VERO to CAPs two different types of treatments were performed in this study: direct treatments, in which cells were directly exposed to the plasma, and indirect treatments, in which cell culture media was treated for different times, and only after the treatment, the liquid was applied to the VERO cells, whose viability was investigated.

3   Materials and Methods

Plasma properties, such as the density of charged particles and their energies, will depend on the applied power and its type (e.g. AC, DC, RF, etc) and also on the feeding gas [6]. For this reason, a dedicated plasma jet device was developed to investigate the effects behind the interactions between cold plasmas and living cells.

3.1   Plasma Jet Device

The cold atmospheric plasma jet device developed consists of a hand-held unit composed by a borosilicate capillary with an outer diameter of 6.93 mm and an inner diameter of 4.94 mm, with an electrode in its center (2 mm diameter), and a second electrode around it. The last one, connected to a custom-made DC high voltage power supply (2.5 mA, 20 kV) (Fig. 3). The jet operates at an Argon (99% purity, Air Liquide) flow rate of 3 slm (standard liters per minute) controlled by a flowmeter (Dynamal Argon 0–15 L/min, Air Liquide), which ensures a vigorous mixing of the medium during the plasma treatments and drives the plasma stream from the top of the pin-type electrode to the surrounding air outside the borosilicate tube.

3.2   Cell Line and Cell Culture

The present study was carried out using VERO cells. This cell line was chosen since it can be replicated through many cell division cycles without become senescent [19]. Cells were cultured in 75 cm^2 flasks (Corning®, 4314640) with

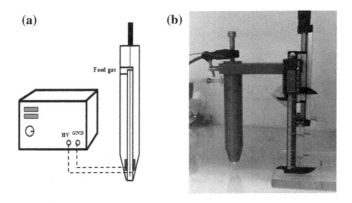

Fig. 3. (a) Schematic representation of the custom made Argon plasma jet device; (b) The customized jet device.

complete DMEM [Dulbecco's Modified Eagle's Medium (DMEM, Sigma-Aldrich, D5030), supplemented with 10% of Fetal Bovine Serum (FBS, Gibco, 10270106), 1.0 g/L D-glucose (Gibco, 15023-021), 3.7 g/L sodium bicarbonate (Sigma-Aldrich (S5761), 1% GlutaMAX^{TM} (L-alanyl-L-glutamine dipeptide, Life Technologies, 35050-038), 1% sodium pyruvate (Gibco, 11360039), penicillin (100 U/mL) and streptomycin (100 μg/mL)(Invitrogen, 15140122)]. After, reaching confluence, the cells had to be transferred into new flasks by an enzymatic method. VERO is an adherent cell line and it was maintained under standard conditions (37°C in a humidified atmosphere of 5% CO_2).

3.3   Plasma Treatments

Two different types of treatments were performed in this study: direct and indirect ones. In direct treatments, cells cultured in different wells of a 24-well microplate were directly exposed to the cold plasma. On the other hand, in indirect plasma treatments, the plasma jet device was used to vertically irradiate the medium in different wells of a 24-well microplate, and only then the treated medium was transferred to the cells previously cultured in different wells of a 96-well microplate (see Fig. 4). In both approaches, plasma treatments were performed for different times, namely 30, 60, 90 and 120 s for the case of direct treatments, and 30, 60, 90, 120 and 150 s for indirect plasma treatments. Moreover, two different distances between the borosilicate tube and the upper edge of the wells where the treatments were carried out were also tested. Untreated medium was used as positive control, both in direct and in indirect treatments.

3.4   Cell Viability Assays

Cell viability was assessed 48 h after plasma treatments through the resazurin assay, which is based on the ability of the dehydrogenase enzyme, present in

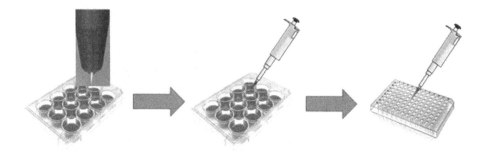

Fig. 4. Schematic representation of the indirect plasma treatment stages involved. First, CAPs vertically irradiated 2 mL of culture medium in different wells of a 12-well microplate. Then, 150 μL of the treated medium was transferred for different wells of a 96-well microplate, in which VERO cells were previously cultured.

metabolic active cells to reduce the resazurin (7-Hydroxy-3H-phenoxazin-3-one 10-oxide) blue dye into a pink colored resorufin (3H-phenoxazin-3-one) product [20]. The quantity of resorufin produced will be proportional to the number of viable cells, which can be easily quantified using a microplate reader. To assess cell viability, the medium in the wells was discarded and 150 μL of resazurin were added to all the wells under test. After incubation in the dark for four hours, at 37 °C in an humidified atmosphere of 5% CO_2, the absorbance was measured in a microplate reader (BioTec, ELX800), using a wavelength of 570 nm and a reference of 600 nm.

3.5   Statistical Analysis

All data are expressed as mean $\pm$ standard deviation of at least three independent experiments. The statistical significance of the differences was evaluated using the Student's t-test and statistical significance was recognized as * for $p < 0.05$, ** for $p < 0.01$ and *** for $p < 0.005$.

4   Experimental Results

As previously referred with this study, it was pretended to evaluate the susceptibility of VERO cells to two different kinds of plasma treatments, direct and indirect, and also the influence of using different working parameters during the treatments, such as different gaps and treatment times. The obtained results will be presented in this section.

4.1   In Vitro CAPS Direct Treatments

Gap Influence. To try to understand if the distance between the end of the borosilicate tube and the bottom of the wells (gap) where cells are cultured have a significant influence on plasma treatments, two different gaps, having a difference of 7 mm between them were used in direct treatments: 1.5 and 2.1 cm. For this part of the work, cells added to the wells of a 24-well microplate surrounded by a few microliters of culture medium, were directly irradiated by the cold plasma. As a result, higher viabilities were observed for tests performed using the highest of the two studied gaps. Moreover, according to the obtained results, it can be seen that for a direct plasma treatment of VERO cells with a duration of only 30 s the gap used for the treatment has a significant influence on the obtained results since a p-value below 0.005 was found between them. However, the differences in cell viability obtained for the direct plasma treatments performed using the two referred gaps seemed to become less significant for longer treatment times. For example, as it can be seen in Fig. 5 after 150 s of treatment, no significant differences in cell viability were found between the two studied gaps. Thus, it can indicate that the gap used to carried out the treatments must be carefully chosen, and a compromise between the used gap and the treatment time should be found.

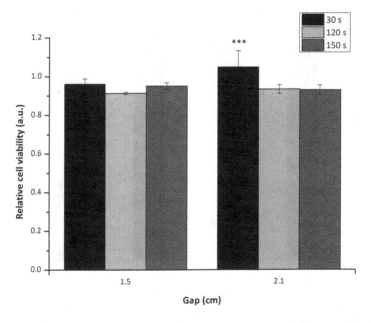

Fig. 5. Cell viability for different time periods as function of the gap, obtained for direct plasma treatments. Results are presented as the mean ± s.d. of three repeated experiments. Student's t-test was performed, and the significance compared with the first bar (gap of 1.5 cm) is indicated as * for p < 0.05, ** for p < 0.01 and *** for p < 0.005.

Influence of Treatment Time. To investigate the influence that direct application of cold plasmas might have in the cells, two different approaches were considered. First, plasma jet directly irradiated cells suspended in 1 mL of culture
medium, and second CAPs irradiated cells added in different wells of a 24-well
microplate. In this last approach, cells were surrounded by a few microliters of
culture medium to avoid that they start to die by stress. From the analysis of the
obtained results, a slight decrease in cell viability as the treatment time increases
can be seen (Fig. 6). This reduction in viability when compared to the control
group seems to be more relevant when plasma is applied to the cells added to the
microplate wells. For the direct treatments performed to cells in suspension no
significant differences in cell viability were found between the tested treatment
times and the control group. The same did not happen when treatments were
carried out to cells added to the plates. In this last case, significant differences
relative to the control group were found for all the tested treatment times. However, interestingly, no significant differences were detected between the different
tested treatment times.

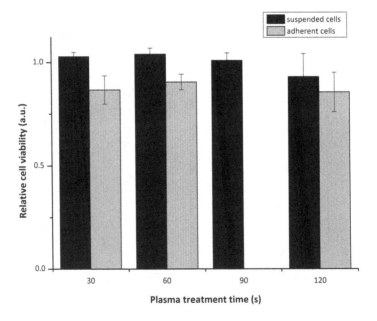

Fig. 6. Cell viability for different time periods for suspended and adherent cells,
obtained for direct plasma treatments.

4.2   Effectiveness of Indirect Treatments

Concerning to indirect CAPs application, until a treatment of 120 s, a slight
increase in cell viability could be observed as the plasma treatment time increases
(see Fig. 7). For all performed treatments, the final cell viability was higher than
the viability of the control group.

With respect to the influence of gap, no significant differences were found for any of the tested treatment times.

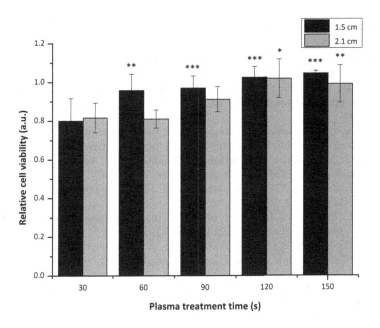

Fig. 7. Cell viability for different time periods as function of the gap, obtained for indirect plasma treatments. Results are presented as the mean $\pm$ s.d. of three repeated experiments. Student's t-test was performed, and the significance compared with the first bar (plasma treatment of 30 s) is indicated as * for $p < 0.05$, ** for $p < 0.01$ and *** for $p < 0.005$.

5 Conclusions

In summary, one can conclude that the plasma produced using the developed plasma jet does not critically affect the viability of non-cancerous cell lines, particularly VERO cells. It was observed, that the effects of cold plasma treatments are dose-dependent, and the distance from the plasma device to the zone to be treated must be taken into account when performing plasma treatments. Moreover, no significant differences could be found when comparing direct and indirect treatments. This last approach, has the advantage that treated medium can be injected into the body and therefore can reach tissues inaccessible through direct irradiation.

Acknowledgments. This work was supported by Fundação para a Ciência e a Tecnologia (FCT), within the Radiation Biology and Biophysics Doctoral Training Programme (RaBBiT, PD/00193/2012), through the scholarship grant number PD/BD/1144 44/2016 (S. Pereira), the project UID/Multi/04378/2013 (UCIBIO) and the project UID/FIS/00068/2013 (CEFITEC). The authors acknowledge Professor Jorge Carvalho Silva from Physics Department, FCT/UNL for the use of TELab - Tissue Engineering Laboratory facilities.

References

1. Burm, K.: Plasma: the fourth state of matter. Plasma Chem. Plasma Process. **32**(2), 401–407 (2012)
2. Stoffels, E., Sakiyama, Y., Graves, D.B.: Cold atmospheric plasma: charged species and their interactions with cells and tissues. IEEE Trans. Plasma Sci. **36**(4), 1441–1457 (2008)
3. Hoffmann, C., Berganza, C., Zhang, J.: Cold atmospheric plasma: methods of production and application in dentistry and oncology. Med. Gas Res. **3**(1), 3–21 (2013)
4. Fridman, G., Friedman, G., Gutsol, A., Shekhter, A.B., Vasilets, V.N., Fridman, A.: Applied plasma medicine. Plasma Process. Polym. **5**(6), 503–533 (2008)
5. Morfill, G.E., Kong, M.G., Zimmermann, J.L.: Focus on plasma medicine. New J. Phys. **11**(11), 115011 (2009)
6. Bárdos, L., Baránková, H.: Cold atmospheric plasma: sources, processes, and applications. Thin Solid Films **518**(23), 6705–6713 (2010)
7. Staack, D., Farouk, B., Gutsol, A., Fridman, A.: Characterization of a DC atmospheric pressure normal glow discharge. Plasma Sour. Sci. Technol. **14**(4), 700–711 (2005)
8. Fridman, A., Yang, Y., Cho, Y.I.: Plasma Discharge in Liquid: Water Treatment and Applications. CRC Press, Boca Raton (2012)
9. Fridman, A., Chirokov, A., Gutsol, A.: Non-thermal atmospheric pressure discharges. J. Phys. D Appl. Phys. **38**(2), R1–R24 (2005)
10. Townsend, J.S.: Electricity in gases (1915)
11. Cobine, J.D.: Gaseous Conductors: Theory and Engineering Applictaions. Dover, Downers Grove (1958)
12. Bazelyan, E.M., Raizer, Y.P.: Spark Discharge. CRC Press, Boca Raton (1997)
13. Lisovsky, V.A., Yakovin, S.D., Yegorenkov, V.D.: Low-pressure gas breakdown in uniform dc electric field. J. Phys. D Appl. Phys. **33**(21), 2722 (2000)
14. Lieberman, M.A., Lichtenberg, A.J.: Principles of plasma discharges and materials processing. MRS Bull. **30**, 899–901 (1994)
15. Laroussi, M.: Plasma medicine: a brief introduction. Plasma **1**(1), 47–60 (2018)
16. Daeschlein, G., et al.: In vitro susceptibility of important skin and wound pathogens against low temperature atmospheric pressure plasma jet (APPJ) and dielectric barrier discharge plasma (DBD). Plasma Process. Polym. **9**(4), 380–389 (2012)
17. Balzer, J., et al.: Non-thermal dielectric barrier discharge (DBD) effects on proliferation and differentiation of human fibroblasts are primary mediated by hydrogen peroxide. PLos One **10**(12), e0144968 (2015)
18. Kaushik, N., Kumar, N., Kim, C.H., Kaushik, N.K., Choi, E.H.: Dielectric barrier discharge plasma efficiently delivers an apoptotic response in human monocytic lymphoma. Plasma Process. Polym. **11**(12), 1175–1187 (2014)
19. Desmyter, J., Melnick, J.L., Rawls, W.E.: Defectiveness of interferon production and of rubella virus interference in a line of African green monkey kidney cells (Vero). J. Virol. **2**(10), 955–961 (1968)
20. Anoopkumar-Dukie, S., Carey, J.B., Conere, T., O'sullivan, E., Van Pelt, F.N., Allshire, A.: Resazurin assay of radiation response in cultured cells. Br. J. Radiol. **78**(934), 945–947 (2005)

Adsorption of Triclosan on Sensors Based on PAH/PAZO Thin-Films: The Effect of pH

Joao Pereira-da-Silva[✉][iD], Paulo M. Zagalo[iD], Goncalo Magalhães-Mota,
Paulo A. Ribeiro[iD], and Maria Raposo[iD]

CEFITEC, Departamento de Física, Faculdade de Ciências e Tecnologia,
Universidade Nova de Lisboa, 2829-516 Caparica, Portugal
{jvp.silva,p.zagalo,g.barreto}@campus.fct.unl.pt, {mfr,pfr}@fct.unl.pt

Abstract. Triclosan (TCS) is a broad-spectrum antimicrobial, preservative agent widely used in pharmaceuticals and personal care products, considered as a troubling contaminant from the environmental point of view because of its toxicity, bacterial resistance promotion, and estrogenic effects. Under this compliance, the pernicious presence of TCS in the environment is requiring the development of molecular dedicated sensors, which in turn leads to the need to find adequate molecular systems capable of giving rise to a transduction. In this work, in order to investigate the affinity of TCS to common polyelectrolytes in an aqueous environment the adsorption of TCS on thin layer-by-layer (LbL) films of poly(1-(4-(3-carboxy-4-hydroxyphenylazo) benzene sulfonamido) -1,2ethanediyl, sodium salt) (PAZO) and poly (allylamine hydrochloride) (PAH) polyelectrolytes at different values of pH of the solution and changing the outer layer, PAZO and PAH, was investigated. Results demonstrated that the PAH layer is the most indicated to better adsorb TCS molecules. These results are of great importance for the development of TCS sensors based on LbL films, since it indicates that the outer layers of LbL films should be positive electrically charged.

Keywords: Triclosan · Adsorption · pH · Thin-film · PAH

1 Introduction

Triclosan [5-chloro-2-(2,4-dichlorophenoxy)-phenol] (TCS) is a broad-spectrum antibacterial and antifungal agent frequently used in pharmaceuticals, polymer and textile manufacturing industry, and personal care products (PPCPs), such as toothpastes, detergents, shampoos, body washes, and deodorants [2, and references therein]. The presence of TCS in these industrial and day-to-day products results in its discharge into surface waters and also in its percolation through the soil and into groundwaters. Even at low concentrations it can have a serious impact on an environmental scale due to its high and acute levels of toxicity to fishes, plants and other water living organisms [2, 3, 7, 10, and references therein].

© Springer Nature Switzerland AG 2020
M. Raposo et al. (Eds.): CIBB 2018, LNBI 11925, pp. 319–325, 2020.
https://doi.org/10.1007/978-3-030-34585-3_29

This agent has been detected as a contaminant in samples of human breast milk, blood, and urine, and has been reported to impair important physiological processes such as thyroid hormone homeostasis and estrogen-dependent responses [3,10]. Furthermore, due to TCS's ability to block lipid biosynthesis, by inhibiting the enzyme enoyl-acyl carrier protein reductase, it may instigate a bacterial resistance development [2,3, and references therein].

2   Scientific Background

TCS is an ionizable organic chemical with a pKa of 7.9. The fraction of non-dissociated and dissociated species for organic acids can be estimated by the equations $f_A^N = (1 + 10^{pH - pK_a})^{-1}$ and $f_I^N = (1 + 10^{pK_a - pH})^{-1}$, respectively [10]. At a typical environmental pH ($pH = 8$), TCS exists both in its neutral and ionized forms and its solubility increases as pH becomes more alkaline [6,9]. Although a hydroxyl functional group is present, it is hydrophobic with octanol-water partitioning coefficient (logKow) of 4.76, and consequently TCS is adsorbed on suspended matter. For this reason, this method has been widely put into practice in the removal of this compound in wastewater treatment processes [6,9,10]. There is a direct dependence between the influence of pH on organic chemical adsorption and the attractive and repulsive forces as well as certain attractive interactions. In this context, it is of the utmost importance to develop and optimize technologies and techniques aiming not only to eliminate but also to detect TCS and its degradation products in aqueous environments [3,5,10].

Recently it was demonstrated that a set of layer-by-layer films can be used as an electronic tongue to detect TCS in water [5]. For example, LbL films prepared with the polyelectrolytes poly[1-[4-(3-carboxy-4-hydroxyphenylazo) benzene sulfonamido]-1,2ethanediyl, sodium salt] (PAZO) and poly (allylamine hydrochloride) (PAH), namely (PAH/PAZO)$_n$ LbL films, where n is the integer number of PAH/PAZO bilayers, were seen to be suitable for use in TCS detection in an electronic tongue based device. However, given that TCS is ionized at higher pHs, with those films presenting a last electrically negative charged layer, it is expected that the amount of adsorbed TCS will not be significant. On the other hand, in the case of ending the LbL films with a PAH positive surface, more TCS molecules can be adsorbed resulting in a much more sensitive sensor to TCS.

The goal of this work is to analyse the amount of TCS adsorbed molecules per unit of area on (PAH/PAZO)$_{10}$ and (PAH/PAZO)$_{10}$/PAH LbL films when these films are immersed in TCS aqueous solutions with the same concentration and different pHs.

3   Materials and Methods

TCS and both polyelectrolytes used were acquired from *Sigma-Aldrich*. A stock solution was attained by dissolving 2.895 mg of TCS into 10 mL of pure methanol and stored at 4 °C. The sample solutions for all adsorption tests were prepared by diluting the stock solution in Milli-Q ultra-pure water, to achieve a solution of TCS with 10^{-4} M of concentration. The solutions' pH was adjusted to 6.5, 7.9 and 10.1 adding drops of an aqueous solution of 0.10 M sodium hydroxide (NaOH) to the TCS solutions. The pH of the final solutions was measured using a Digital pH Meter ProLab 1000.

3.1 Layer-by-Layer (LbL) Thin Film Preparation

Thin films of PAH/PAZO adsorbed on quartz substrates were produced through (LbL) technique [5]. Solutions of PAH and PAZO were prepared with a concentration of 10^{-2} M through the dissolution of the compounds in Milli-Q ultra-pure water. The production of the thin films consisted on the alternated adsorption of PAZO and PAH layers from aqueous solutions on the surface of quartz substrates, with a washing stage with Milli-Q ultra-pure water between the immersion in the solutions and a drying stage with a gentle nitrogen flux at the end of each bilayer. The immersion time in each polyelectrolyte solution and also in the ultra-pure water was of 30 s. This method was repeated multiple times in order to obtain the desired number of bilayers in the films, namely $(PAH/PAZO)_{10}$ and $(PAZO/PAH)_{10}/PAH$ films.

3.2 Adsorption Experiments

The TCS adsorbed amount per unit of area on the LbL films was achieved by measuring, before and after 45 min of adsorption, the ultraviolet–visible (UV-Vis) absorbance spectra with a UV 2101 PC Scanning Spectrophotometer, from the difference between both spectra, calculating the TCS absorption coefficients and using the Beer Lambert law.

4 Results

Figure 1 presents the absorbance spectra of $(PAH/PAZO)_{10}$ and $(PAH/PAZO)_{10}$ /PAH LbL films before and after immersion in the TCS solution at pH 6.5, pH 7.9 and pH 10.1.

By subtracting the curves in each graph one can obtain the absorbance related with TCS molecules in $(PAH/PAZO)_{10}$ and $(PAH/PAZO)_{10}/PAHLbL$ films, the resultant spectra are displayed in Fig. 2(a) and (b), respectively. From these graphs it is possible to observe that the absorbance is higher for the case of triclosan adsorbed on $(PAH/PAZO)_{10}/PAH$ films with a pH of 6.5 and decreases as the pH rises. Although the TCS absorbance values appear to be smaller for the case of $(PAH/PAZO)_{10}$, the behaviour of the absorbance with the pH is similar to the films that present PAH as the outer layer. It is important to note that the spectra of TCS in aqueous solutions follows the behaviour of the curve presented in the inset of Fig. 2(a). This spectrum which corresponds to that of a TCS aqueous solution with a concentration of 10^{-4} M reveals two main bands; one at 230 nm and another one at 280 nm. The absorbance at 280 nm was not considered in the calculations because this band is strongly affected by pH.

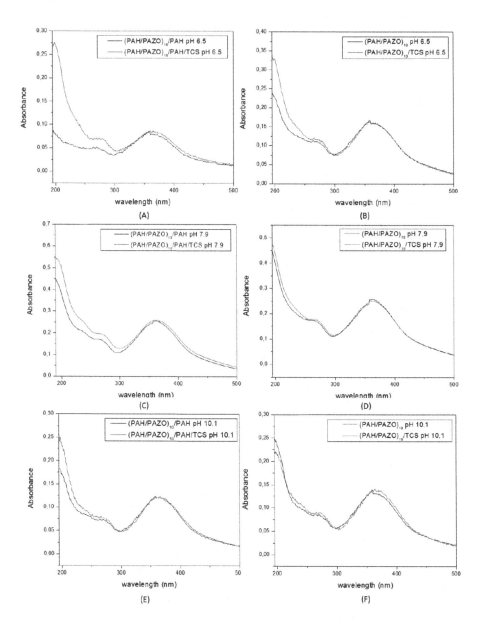

Fig. 1. UV-Vis spectra of (PAH/PAZO)$_{10}$/PAH film before and after immersion in TCS solution 10^{-4} M at: (A) pH = 6.5; (C) pH = 7.9 and (E) pH = 10.1. UV-Vis spectra of (PAH/PAZO)$_{10}$ film before and after immersion in TCS solution 10^{-4} M at: (B) pH = 6.5; (D) pH = 7.9 and (F) pH = 10.1.

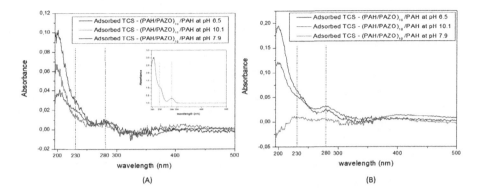

Fig. 2. (A) Absorbance of TCS on the $(PAH/PAZO)_{10}/PAH$ films at pHs 6.5, 7.9 and 10.1; (B) Absorbance of TCS on the $(PAH/PAZO)_{10}$ films at pHs 6.5, 7.9 and 10.1

From this spectrum the absorption coefficient at 230 nm can be calculated in order to estimate the adsorbed amount per unit of area of TCS on the LbL films through the Beer-Lambert law. Given that, TCS is adsorbed in two surfaces of the film, the absorbance was divided by two, i.e., during the measurement of the absorbance spectra the light passes per two LbL films since it is adsorbed on the both sides of quartz supports.

From the values of absorbance at 230 nm (on the shoulder) in the UV-Vis spectrum of the TCS solution with a concentration of 10^{-4} M (inset of Fig. 2(a)), the molar absorption coefficient ($\epsilon(\lambda_{230})$ was calculated. In this sense, adsorbed amount of TCS per unit of area in all LbL thin-films was calculated, revealing that it linearly decreases with pH both in the PAH and PAZO layers as demonstrated in Fig. 3.

Figure 2(a) and (b) show that the adsorption of TCS on $(PAH/PAZO)_{10}$ and $(PAH/PAZO)_{10}/PAH$ thin films is strongly influenced by pH of the TCS solution. It should be remarked here that pH is known to affect the dissociation of the TCS molecules [3] since TCS begins to ionize at pH above 6.14 and it becomes completely ionized at pH higher than 10.14. However, when $pH < pKa$, the neutral species are the dominant and when $pH > pKa$, the dissociated species (anions) are the ones that dominate [3]. At pH 6.5, the PAH presents a positive surface charge and TCS is nearly non-dissociated. Therefore, an improved adsorption capacity was observed due to the increase of attractive interactions, namely on the PAH layer (H-bond formation and hydrophobic interaction) [10]. Additionally, protonated TCS molecules are more hydrophobic than the deprotonated anions, consequently increased adsorption is likely to occur on the PAH and PAZO layers at lower pH [1, 8]. As the pH increases, the negatively charged TCS species and the positively charged surface of PAH probably tend to form electrostatic interactions, namely near pH 7.9. On the other hand, with the increase of pH, the ionized TCS could not provide hydroxyl hydrogen atoms from the phenolic groups to nitrogen atoms of PAH, hence the hydrogen bonding between PAH and TCS molecules weakens and disappears completely at pH

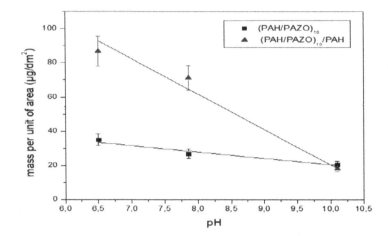

Fig. 3. The adsorbed mass of TCS per unit of area in the $(PAH/PAZO)_{10}$ $(PAH/PAZO)_{10}/PAH$ thin-films *vs* pH

higher than 10.1 [4]. Additionally, the PAH polyelectrolytes tend to lose the positive charge and the PAZO degree of ionization increases at pH higher than 8. Hence, the electrostatic repulsion counteracts the attractive forces and become one of the dominant mechanisms between the deprotonated TCS and the negatively charged PAZO layer [10]. For these reasons, the adsorption of TCS is lower at high pH.

5   Conclusions

It has been demonstrated that TCS adsorbed amount per unit of area is strongly dependent on the solution pH and also on the degree of ionization of the last polyelectrolyte of the LbL film. In order to optimize the detection of TCS, it was shown that the outer layer of the LbL film should be PAH. Additionally, in the range of pHs analysed, pH 6.5 is the most efficient to promote TCS adsorption. These results are to be taken into account in the designing of molecular sensor devices based on LbL films.

Acknowledgments. The authors acknowledge the financial support from FEDER, through Programa Operacional Factores de Competitividade–COMPETE and Fundação para a Ciência e a Tecnologia–FCT, by the project PTDC/FIS-NAN/0909/ 2014 and for the Portuguese research Grant No. PEst-OE/FIS/UI0068/2011 and UID/FIS/00068/2013. J. Pereira-da-Silva and P.M. Zagalo acknowledge the fellowships PD/BD/142768/2018 and PD/BD/142767/2018 from RABBIT Doctoral Programme, respectively.

References

1. Abegão, L.M.G., Ribeiro, J.H.F., Ribeiro, P.A., Raposo, M.: Nano-molar deltamethrin sensor based on electrical impedance of PAH/PAZO layer-by-layer sensing films. Sensors (Switzerland) **13**(8), 10167–10176 (2013). https://doi.org/10.3390/s130810167

2. Aranami, K., Readman, J.W.: Photolytic degradation of triclosan in freshwater and seawater. Chemosphere **66**(6), 1052–1056 (2007). https://doi.org/10.1016/j.chemosphere.2006.07.010

3. Behera, S.K., Oh, S.Y., Park, H.S.: Sorption of triclosan onto activated carbon, kaolinite and montmorillonite: effects of pH, ionic strength, and humic acid. J. Hazard. Mater. **179**(1–3), 684–691 (2010). https://doi.org/10.1016/j.jhazmat.2010.03.056

4. Lei, C., Hu, Y.Y., He, M.Z.: Adsorption characteristics of triclosan from aqueous solution onto cetylpyridinium bromide (CPB) modified zeolites. Chem. Eng. J. **219**, 361–370 (2013). https://doi.org/10.1016/j.cej.2012.12.099

5. Marques, I., Magalhães-Mota, G., Pires, F., Sério, S., Ribeiro, P., Raposo, M.: Detection of traces of triclosan in water **421** (2016)

6. Nghiem, L.D., Coleman, P.J.: NF/RO filtration of the hydrophobic ionogenic compound triclosan: transport mechanisms and the influence of membrane fouling. Sep. Purif. Technol. **62**(3), 709–716 (2008). https://doi.org/10.1016/j.seppur.2008.03.027

7. Postigo, C., Barceló, D.: Synthetic organic compounds and their transformation products in groundwater: occurrence, fate and mitigation. Sci. Total Environ. **503–504**, 32–47 (2015). https://doi.org/10.1016/j.scitotenv.2014.06.019

8. Tong, Y., Mayer, B.K., McNamara, P.J.: Triclosan adsorption using wastewater biosolids-derived biochar. Environ. Sci.: Water Res. Technol. **2**(4), 761–768 (2016). https://doi.org/10.1039/c6ew00127k

9. Wu, W., Hu, Y., Guo, Q., Yan, J., Chen, Y., Cheng, J.: Sorption/desorption behavior of triclosan in sediment-water-rhamnolipid systems: effects of pH, ionic strength, and DOM. J. Hazard. Mater. **297**, 59–65 (2015). https://doi.org/10.1016/j.jhazmat.2015.04.078

10. Zhou, S., Shao, Y., Gao, N., Deng, J., Tan, C.: Equilibrium, kinetic, and thermodynamic studies on the adsorption of triclosan onto multi-walled carbon nanotubes. CLEAN Soil Air Water **41**(6), 539–547 (2013). https://doi.org/10.1002/clen.201200082

Detection of Triclosan Dioxins After UV Irradiation – A Preliminar Study

Gonçalo Magalhães-Mota$^{(\boxtimes)}$ ⓘ, Filipa Pires ⓘ, Paulo A. Ribeiro ⓘ, and Maria Raposo ⓘ

CEFITEC, Department of Physics, FCT-UNL, Universidade Nova de Lisboa,
2829-516 Caparica, Portugal
`g.barreto@campus.fct.unl.pt, mfr@fct.unl.pt`

Abstract. Triclosan (TCS) by itself represents a major health and environmental problem. Also concerning are its photoproducts, various dioxins, which are even more dangerous, creating a need and opportunity to develop dedicated sensors to detect their presence in water. By treating featured data through principal component analysis (PCA), the footprint of the dangerous TCS products after irradiation can be clearly outlined. This result allow us to conclude that a TCS sensor device based on electronic tongue concept can be envisaged.

Keywords: Triclosan · Dioxins · Environmental · Impedance ·
Electronic tongue

1 Introduction

Triclosan $[5 - chloro - 2 - (2, 4 - dichlorophenoxy)phenol]$ (TCS), belonging to the class of Pharmaceuticals and Personal Care Products (PPCPs) [1], is a broad-spectrum antimicrobial agent and bactericide. Because of its antimicrobial efficacy, it is widely used in personal health and skin care products, such as soaps, detergents, hand cleansers, cosmetics, toothpastes, etc. However, it has been considered to disrupt the endocrine system, for instance, thyroid hormone homeostasis and possibly the reproductive system. For these reasons, the use of TCS in personal care products is under close scrutiny [2].

The widespread use of triclosan has raised concerns about the environmental impact of triclosan residues as well as food safety problems caused by its presence in foods. Triclosan has been detected in river water samples in both North America and Europe and is likely widely distributed wherever triclosan-containing products are used. Presumably because of its widespread usage, Kolpin found that triclosan is among the most commonly found chemicals in U.S. waterways [3]. Similar studies in Sweden also demonstrated significant contamination of waterways. Although significant amounts are removed in sewage plants, considerable quantities remain in the sewage effluent, initiating widespread environmental contamination. Triclosan undergoes bioconversion to methyl-triclosan,

© Springer Nature Switzerland AG 2020
M. Raposo et al. (Eds.): CIBB 2018, LNBI 11925, pp. 326–332, 2020.
https://doi.org/10.1007/978-3-030-34585-3_30

which has been demonstrated to bioaccumulate in fish. In addition, triclosan has been found in human urine samples with a mean of 127 ng/mL from 30 persons with no known industrial exposure and was found in significant amounts, three of five samples, of mother's milk, clearly demonstrating its presence in humans. The action of sunlight in river water has been reported to convert triclosan into dioxin derivatives and raises the possibility of pharmacological dangers not envisioned when the compound was originally utilized [4]. Furthermore, the standard procedure to treat common drinking water and wastewater is to use chlorine, unfortunately triclosan reacts to it, creating chlorinated triclosan derivatives. Both methyl-triclosan and the chlorinated triclosan derivatives when exposed to sunlight originate several different and dangerous dioxins [5].

2   Scientific Background

Recently, we demonstrated that is possible to detect TCS concentrations in water down to 10^{-12} M [6]. This paper will focus on the 2,8-dichlorodibenzo- p-dioxin (2,8-DCDD), which is a dioxin that comes from the direct photolysis of triclosan [7] and one of the most common dioxins of triclosan found on surface waters [8], to achieve this, two different spectroscopy techniques were used in conjunction with the statistical procedure called Principal Component Analysis (PCA).

The aim of this paper is to bring us one step closer to a fully functional TCS sensor device on the electronic tongue.

3   Materials and Methods

Triclosan was acquired from Sigma-Aldrich. Due to its poor solubility in water, TCS was first dissolved in methanol and then diluted with ultrapure water into the concentration range of 10^{-18} to 10^{-6} while maintaining the proportion of methanol to water at 1%. A solution of water containing 1% of metanol was used as control.

In order to produce the 2,8-DCDD dioxin, TCS solutions were exposed to a 254 nm UVC germicide lamp (Philips TUV PL-S 5W/2P 1CT) with a irradiance of $1.9\,W/m^2$. The irradiation times were 5, 15, 30, 60, 180, 360, 720, 1440 and 4320 min.

An ultraviolet–visible (UV–vis) spectrophotometer model Evolution 300 (Thermo Scientific, Waltham, MA, USA) was used for monitoring the conversion of TCS into 2,8-DCDD along the irradiation time.

The electrical characterisation of the irradiated solutions were measured in a Solartron 1260 impedance analyser equipped with a Solartron 1296 dielectric interface. Applying a voltage of 1 Volt, in the frequency range of 0.01 Hz to 3.2 MHz, to the gold-deposited interdigitated electrodes (IEs), which were acquired from DropSens (Asturias, Spain).

The impedance data was processed by Principal component analysis (PCA) method in order to emphasize not only the differences between the TCS concentrations but also the new products that are being produced during the different irradiation periods of time.

4   Experimental Results

In this section, the acquired experimental data of the performed techniques will be presented, on which our results are based on. Beginning with the ultraviolet-visible spectrophotometry, followed by the impedance spectroscopy, and ending with the Principal Component Analysis (PCA) applied to the data.

4.1   UV-Visible Spectra

The absorbance spectra of a 10^{-6} M triclosan solution irradiated for different periods of time is shown in Fig. 1. Similar behaviour is achieved for the other TCS solutions, and thus, those results will not be shown.

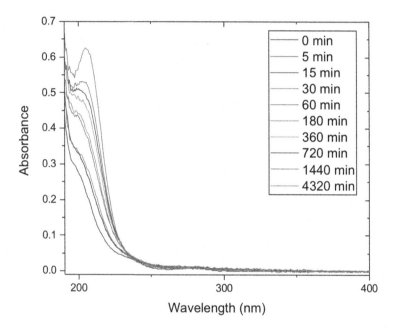

Fig. 1. UV-visible spectra of triclosan with a concentration of 10^{-6} M before and after being exposed to UV radiation.

The exposure to UV radiation creates a peak at roughly 205 nm, suggesting that this might be an absorbance peak characteristic of the dioxin. Figure 2 shows the evolution of this peak through irradiation time.

The data present in Fig. 2 seems to be fitted as an exponential curve with the following expression:

$$A = A_0 + A_D \left(1 - e^{-\sqrt{\frac{t}{\tau}}}\right) \tag{1}$$

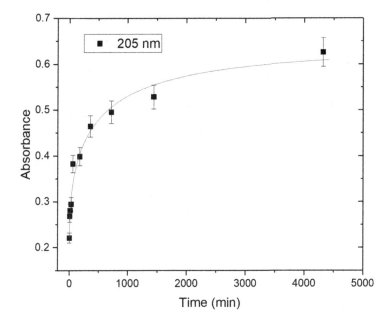

Fig. 2. Evolution of the absorbance at 205 nm of triclosan at a concentration of 10^{-6} M in function of the irradiation time.

Where A is the absorbance, A_0 is the initial absorbance, A_D is the increase absorbance value associated with the dioxin formation, t is the irradiation time and τ is the characteristic dioxin formation time, which in the present case takes a value of 500 min.

4.2 Impedance Spectroscopy

Impedance spectroscopy was used to characterise the triclosan solutions, by measuring the capacitance, loss tangent and impedance (both real and imaginary) in a range of frequency of 0.01 Hz to 3.2 MHz. Like the previous UV-vis spectras, the behaviours between the different concentrations of triclosan are very similiar and only the 10^{-6} M concentration (Fig. 3) will be shown.

Figure 3 proves that impedance spectroscopy can discriminate the different triclosan solution irradiation times, however, the sheer amount of data allied to the proximity and occasionally the overlap of graph lines, shows the need to resort to statistical methods, such as PCA.

4.3 Principal Component Analysis

Principal Component Analysis (PCA) was applied to analyse the large volume of raw data. PCA also reveals correlations between the raw data that would consume too much time or might be imperceptible. The PCA of the combine

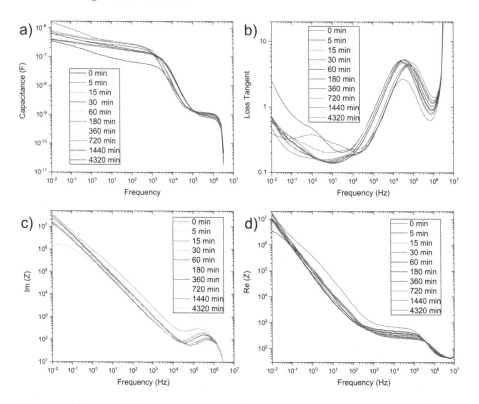

Fig. 3. Capacitance (A), Loss Tangent (B), Imaginary (C) and Real impedance (D) versus frequency for a concentration of 10^{-6} M of triclosan before and after being exposed to UV radiation.

raw data of capacitance, loss tangent and both real and imaginary impedance across the range in frequency is displayed in Fig. 4.

Figure 4 was marked with 5 different colours in order to highlight the different groups of data. Blue was used for the initial TCS solutions before any UV irradiation, yellow for the 5 min irradiated solutions, red for the 4320 min solutions, while purple was used for the water solutions, regardless of the irradiation time, and finally, green for the remaining irradiation times of the TCS solutions.

Although the blue, red and purple marks bound all their respective concentrations (or all the water solutions in the purple case). In the yellow mark however, there are two irradiation times missing, the two lowest concentrations of triclosan (10^{-15} and 10^{-18} M). One possible explanation might be, since it were the two lowest concentrations, that those 5 min of irradiation are sufficient to transform all or at least a large part of TCS into 2,8-DCDD, with the more concentrated TCS solutions requiring a bit more time. This would also explain why those two data points can be found in the green mark region. This is corroborated by Latch's study, in which he found that a TCS solution of 16.2×10^{-6} M

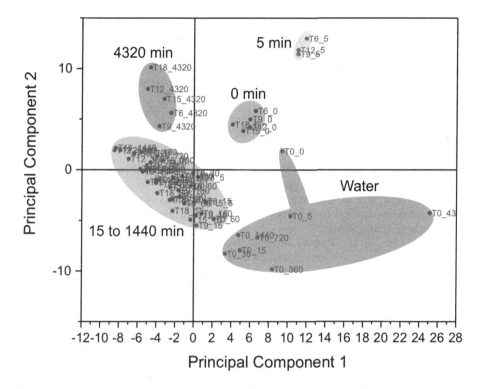

Fig. 4. Principal Component Analysis (PCA) of the triclosan solutions over the UV irradiation time (Color figure online)

was almost totally transform to 2,8-DCDD after only 360 seconds of irradiation time [7].

As for the green mark, further investigation is required in order to better discern the irradiation times embedded in it. Despite the fact that the lower irradiated times are located in the lower region of the green mark, and as they go up, the remaining irradiation times follow the trend.

5   Conclusion

In this study, TCS was successfully distinguish from the water control solution and from the 2,8-DCDD dioxin. Long UV irradiation exposure (4320 min) was also successfully distinguish. With the remaining times (although the 5 min irradiation time to a lesser extend) needing further investigation, despite leaving a very optimistic first impression. In the future, to further increase the affinity and sensibility of the system, thin films specifically design to 2,8-DCDD dioxin should be made, since it will greatly improve its detection, thus bringing further clarification.

Acknowledgments. The authors acknowledge the financial support from FEDER, through Programa Operacional Factores de Competitividade–COMPETE and Fundação para a Ciência e a Tecnologia–FCT, by the project PTDC/FIS-NAN/0909/ 2014 and for the Portuguese research Grant No. PEst-OE/FIS/UI0068/2011 and UID/FIS/00068/2013. Filipa Pires acknowledges the fellowship PD/BD/106036/2015 from RABBIT Doctoral Programme (Portugal).

References

1. Daughton, C.G., Ternes, T.A.: Pharmaceuticals and personal care products in the environment: agents of subtle change? (1999)
2. Ma, H., Wang, L., Liu, H., Luan, F., Gao, Y.: Application of a non-aqueous capillary electrophoresis method to the analysis of triclosan in personal care products. Anal. Methods **6**(4723) (2014)
3. Kolpin, D.W., et al.: Pharmaceuticals, hormones, and other organic wastewater contaminants in us streams, 1999–2000: a national reconnaissance. Environ. Sci. Technol. **36**(6), 1202–1211 (2002)
4. Shelver, W.L., Kamp, L.M., Church, J.L., Rubio, F.M.: Measurement of triclosan in water using a magnetic particle enzyme immunoassay. J. Agric. Food Chem. **55**(10), 3758–3763 (2007)
5. Yueh, M.-F., Tukey, R.H.: Triclosan: a widespread environmental toxicant with many biological effects. Annu. Rev. Pharmacol. Toxicol. **56**, 251–272 (2016)
6. Marques, I., Magalhães-Mota, G., Pires, F., Sério, S., Ribeiro, P.A., Raposo, M.: Detection of traces of triclosan in water. Appl. Surf. Sci. **421**, 142–147 (2017)
7. Latch, D.E., Packer, J.L., Arnold, W.A., McNeill, K.: Photochemical conversion of triclosan to 2, 8-dichlorodibenzo-p-dioxin in aqueous solution. J. Photochem. Photobiol., A **158**(1), 63–66 (2003)
8. Buth, J.M., et al.: Dioxin photoproducts of triclosan and its chlorinated derivatives in sediment cores. Environ. Sci. Technol. **44**(12), 4545–4551 (2010)

Correction to: Computational Intelligence Methods for Bioinformatics and Biostatistics

Maria Raposo⬤, Paulo Ribeiro⬤, Susana Sério⬤,
Antonino Staiano⬤, and Angelo Ciaramella⬤

Correction to:
M. Raposo et al. (Eds.): *Computational Intelligence Methods*
for Bioinformatics and Biostatistics, **LNBI 11925,**
https://doi.org/10.1007/978-3-030-34585-3

In the original version of the book, the affiliations of Antonino Staiano and Angelo Ciaramella were wrong. Both affiliations have been corrected to:
Università degli Studi di Napoli Parthenope.

In the original version of the chapter "Efficient and Settings-Free Calibration of Detailed Kinetic Metabolic Models with Enzyme Isoforms Characterization" the name of the author "Andrea Tangherloni" was incorrect. This has now been corrected.

The updated version of the book can be found at
https://doi.org/10.1007/978-3-030-34585-3
https://doi.org/10.1007/978-3-030-34585-3_17

Author Index

Printed in the United States
by Baker & Taylor Publisher Services